Springer Finance

T0230018

# Springer Finance

*Springer Finance* is a programme of books addressing students, academics and practitioners working on increasingly technical approaches to the analysis of financial markets. It aims to cover a variety of topics, not only mathematical finance but foreign exchanges, term structure, risk management, portfolio theory, equity derivatives, and financial economics.

*Ammann M.*, Credit Risk Valuation: Methods, Models, and Application (2001)
*Back K.*, A Course in Derivative Securities: Introduction to Theory and Computation (2005)
*Barucci E.*, Financial Markets Theory. Equilibrium, Efficiency and Information (2003)
*Bielecki T.R. and Rutkowski M.*, Credit Risk: Modeling, Valuation and Hedging (2002)
*Bingham N.H. and Kiesel R.*, Risk-Neutral Valuation: Pricing and Hedging of Financial Derivatives (1998, 2nd ed. 2004)
*Brigo D. and Mercurio F.*, Interest Rate Models: Theory and Practice (2001, 2nd ed. 2006)
*Buff R.*, Uncertain Volatility Models – Theory and Application (2002)
*Carmona R.A. and Tehranchi M.R.*, Interest Rate Models: An Infinite Dimensional Stochastic Analysis Perspective (2006)
*Cesari G., Aquilina J., Charpillon N., Filipović Z., Lee G., Manda I.*, Modelling, Pricing, and Hedging Counterparty Credit Exposure (2009)
*Dana R.-A. and Jeanblanc M.*, Financial Markets in Continuous Time (2003)
*Deboeck G. and Kohonen T. (Editors)*, Visual Explorations in Finance with Self-Organizing Maps (1998)
*Delbaen F. and Schachermayer W.*, The Mathematics of Arbitrage (2005)
*Elliott R.J. and Kopp P.E.*, Mathematics of Financial Markets (1999, 2nd ed. 2005)
*Fengler M.R.*, Semiparametric Modeling of Implied Volatility (2005)
*Filipović D.*, Term-Structure Models (2009)
*Fusai G. and Roncoroni A.*, Implementing Models in Quantitative Finance (2008)
*Geman H., Madan D., Pliska S.R. and Vorst T. (Editors)*, Mathematical Finance – Bachelier Congress 2000 (2001)
*Gundlach M. and Lehrbass F. (Editors)*, CreditRisk$^+$ in the Banking Industry (2004)
*Jeanblanc M., Yor M., Chesney M.*, Mathematical Methods for Financial Markets (2009)
*Jondeau E.*, Financial Modeling Under Non-Gaussian Distributions (2007)
*Kabanov Y.A. and Safarian M.*, Markets with Transaction Costs (2009)
*Kellerhals B.P.*, Asset Pricing (2004)
*Külpmann M.*, Irrational Exuberance Reconsidered (2004)
*Kwok Y.-K.*, Mathematical Models of Financial Derivatives (1998, 2nd ed. 2008)
*Malliavin P. and Thalmaier A.*, Stochastic Calculus of Variations in Mathematical Finance (2005)
*Meucci A.*, Risk and Asset Allocation (2005, corr. 2nd printing 2007, Softcover 2009)
*Pelsser A.*, Efficient Methods for Valuing Interest Rate Derivatives (2000)
*Platen E. and Heath D.*, A Benchmark Approach to Quantitative Finance (2006, corr. printing 2010)
*Prigent J.-L.*, Weak Convergence of Financial Markets (2003)
*Profeta C., Roynette B., Yor M.*, Option Prices as Probabilities (2010)
*Schmid B.*, Credit Risk Pricing Models (2004)
*Shreve S.E.*, Stochastic Calculus for Finance I (2004)
*Shreve S.E.*, Stochastic Calculus for Finance II (2004)
*Yor M.*, Exponential Functionals of Brownian Motion and Related Processes (2001)
*Zagst R.*, Interest-Rate Management (2002)
*Zhu Y.-L., Wu X., Chern I.-L.*, Derivative Securities and Difference Methods (2004)
*Ziegler A.*, Incomplete Information and Heterogeneous Beliefs in Continuous-time Finance (2003)
*Ziegler A.*, A Game Theory Analysis of Options (2004)

Christophe Profeta · Bernard Roynette · Marc Yor

# Option Prices as Probabilities

## A New Look at Generalized Black-Scholes Formulae

 Springer

Christophe Profeta
Université Nancy I
Inst. Élie Cartan (IECN)
Faculté des Sciences
54506 Vandoeuvre-les-Nancy CX
France
christophe.profeta@iecn.u-nancy.fr

Bernard Roynette
Université Nancy I
Inst. Élie Cartan (IECN)
Faculté des Sciences
54506 Vandoeuvre-les-Nancy CX
France
bernard.roynette@iecn.u-nancy.fr

Marc Yor
Université Paris VI
Labo. Probabilités et Modèles
Aléatoires
175 rue du Chevaleret
75013 Paris
France

ISBN 978-3-642-10394-0          e-ISBN 978-3-642-10395-7
DOI 10.1007/978-3-642-10395-7
Springer Heidelberg Dordrecht London New York

Library of Congress Control Number: 2010920154

Mathematics Subject Classification (2000): 60G44, 60J65, 60H99, 60J60

*Cover design*: VTEX, Vilnius

Printed on acid-free paper

Springer is part of Springer Science+Business Media (www.springer.com)

# Preface

Discovered in the seventies, Black-Scholes formula continues to play a central role in Mathematical Finance. We recall this formula. Let $(B_t, t \geq 0;\ \mathscr{F}_t, t \geq 0,\ \mathbb{P})$ denote a standard Brownian motion with $B_0 = 0$, $(\mathscr{F}_t, t \geq 0)$ being its natural filtration. Let $\left(\mathscr{E}_t := \exp\left(B_t - \frac{t}{2}\right), t \geq 0\right)$ denote the exponential martingale associated to $(B_t, t \geq 0)$. This martingale, also called geometric Brownian motion, is a model to describe the evolution of prices of a risky asset. Let, for every $K \geq 0$:

$$\Pi_K(t) := \mathbb{E}\left[(K - \mathscr{E}_t)^+\right] \tag{0.1}$$

and

$$C_K(t) := \mathbb{E}\left[(\mathscr{E}_t - K)^+\right] \tag{0.2}$$

denote respectively the price of a European put, resp. of a European call, associated with this martingale. Let $\mathscr{N}$ be the cumulative distribution function of a reduced Gaussian variable:

$$\mathscr{N}(x) := \frac{1}{\sqrt{2\pi}} \int_{-\infty}^{x} e^{-\frac{y^2}{2}} dy. \tag{0.3}$$

The celebrated Black-Scholes formula gives an explicit expression of $\Pi_K(t)$ and $C_K(t)$ in terms of $\mathscr{N}$:

$$\Pi_K(t) = K\mathscr{N}\left(\frac{\log(K)}{\sqrt{t}} + \frac{\sqrt{t}}{2}\right) - \mathscr{N}\left(\frac{\log(K)}{\sqrt{t}} - \frac{\sqrt{t}}{2}\right) \tag{0.4}$$

and

$$C_K(t) = \mathscr{N}\left(\frac{-\log(K)}{\sqrt{t}} + \frac{\sqrt{t}}{2}\right) - K\mathscr{N}\left(-\frac{\log(K)}{\sqrt{t}} - \frac{\sqrt{t}}{2}\right). \tag{0.5}$$

Comparing the expressions (0.4) and (0.5), we note the remarkable identity:

$$C_K(t) = K\Pi_{1/K}(t). \tag{0.6}$$

These formulae have been the starting point of the present monograph, which consists of a set of eight Chapters, followed by three Complements and three Appendices. We now summarize the contents of these different items.

## About Chapter 1

The processes $((K - \mathcal{E}_t)^+, t \geq 0)$ and $((\mathcal{E}_t - K)^+, t \geq 0)$ are submartingales; hence, $\Pi_K$ and $C_K$ are increasing functions of $t$. Furthermore, it is easily shown that:

$$\Pi_K(t) \xrightarrow[t \to \infty]{} K \qquad \text{and} \qquad C_K(t) \xrightarrow[t \to \infty]{} 1. \tag{1.1}$$

This motivates our question: can we exhibit, on our probability space, a positive r.v. $X^{(\Pi)}$ $\left(\text{resp. } X^{(C)}\right)$ such that $\left(\frac{1}{K}\Pi_K(t), t \geq 0\right)$, $\left(\text{resp. } (C_K(t), t \geq 0)\right)$ is the cumulative distribution function of $X^{(\Pi)}$ $\left(\text{resp. } X^{(C)}\right)$ ? We answer this question in Chapter 1 of the present monograph. Precisely, let:

$$\mathcal{G}_K^{(\mathcal{E})} := \sup\{s \geq 0;\ \mathcal{E}_s = K\} \tag{1.2}$$
$$(= 0 \text{ if this set } \{s \geq 0;\ \mathcal{E}_s = K\} \text{ is empty}).$$

Then:

$$\Pi_K(t) := K\mathbb{P}\left(\mathcal{G}_K^{(\mathcal{E})} \leq t\right) \tag{1.3}$$

and

$$C_K(t) := \mathbb{P}^{(\mathcal{E})}\left(\mathcal{G}_K^{(\mathcal{E})} \leq t\right) \tag{1.4}$$

where, in the latter formula, $\mathbb{P}^{(\mathcal{E})}$ is the probability obtained from $\mathbb{P}$ by "change of numéraire":

$$\mathbb{P}^{(\mathcal{E})}_{|\mathscr{F}_t} := \mathcal{E}_t \cdot \mathbb{P}_{|\mathscr{F}_t}. \tag{1.5}$$

## About Chapter 2

Formulae (1.3) and (1.4) may be extended when the martingale $(\mathcal{E}_t, t \geq 0)$ is replaced by a positive, continuous, local martingale such that:

$$\lim_{t \to \infty} M_t = 0 \quad \text{a.s.} \tag{2.1}$$

Indeed, in Chapter 2, we show that, with:

$$\mathcal{G}_K^{(M)} := \sup\{t \geq 0;\ M_t = K\}, \tag{2.2}$$

one has:

$$\mathbb{E}\left[(K-M_t)^+\right] = K\mathbb{P}\left(\mathscr{G}_K^{(M)} \leq t\right). \tag{2.3}$$

If, furthermore, $(M_t, t \geq 0)$ is a "true" martingale, (i.e. $\mathbb{E}[M_t] = 1$ for every $t \geq 0$), then:

$$\mathbb{E}\left[(M_t-K)^+\right] = \mathbb{P}^{(M)}\left(\mathscr{G}_K^{(M)} \leq t\right) \tag{2.4}$$

where the probability $\mathbb{P}^{(M)}$ is given by:

$$\mathbb{P}_{|\mathscr{F}_t}^{(M)} := M_t \cdot \mathbb{P}_{|\mathscr{F}_t}. \tag{2.5}$$

Of course, a formula such as (2.3) or (2.4) has practical interest only if we know how to compute the law of the r.v. $\mathscr{G}_K^{(M)}$. This is why, in Chapter 2, we have developed:

- general results (cf. Sections 2.3, 2.4 and 2.5), which allow the computation of the law of $\mathscr{G}_K^{(M)}$,
- explicit examples of computations of these laws.

Moreover, Chapter 2 also contains an extension of formulae (2.3) and (2.4) to the case where the martingale $(M_t, t \geq 0)$ is replaced by several orthogonal martingales $(M_t^{(i)}, t \geq 0)_{i=1,\dots,n}$.

## About Chapter 3

Formulae (2.3) and (2.4) show the central importance of the family of r.v. $\mathscr{G}_K^{(M)}$ defined by (2.2). However, these r.v's are not stopping times (e.g: with respect to the natural filtration of the martingale $(M_t, t \geq 0)$). Nonetheless, we wish to study the process $(M_t, t \geq 0)$ before and after $\mathscr{G}_K^{(M)}$. This study hinges naturally upon the progressive enlargement of filtration technique, i.e. upon the introduction of the filtration $(\mathscr{F}_t^K, t \geq 0)$, which is the smallest filtration containing $(\mathscr{F}_t, t \geq 0)$ and making $\mathscr{G}_K^{(M)}$ a $(\mathscr{F}_t^K, t \geq 0)$ stopping time, and upon the computation of Azéma's supermartingale $\left(\mathbb{P}\left(\mathscr{G}_K^{(M)} \geq t | \mathscr{F}_t\right), t \geq 0\right)$. These computations and the corresponding study of the process $(M_t, t \geq 0)$ before and after $\mathscr{G}_K^{(M)}$ are dealt with in Chapter 3.

The preceding discussion leads us to consider a slightly more general set-up, that of Skorokhod submartingales, i.e. continuous submartingales $(X_t, t \geq 0)$ such that:

$$X_t = -\mathscr{M}_t + L_t \qquad (t \geq 0) \tag{3.1}$$

with:

- $X_t \geq 0, \quad X_0 = 0,$      (3.2)
- $(L_t, t \geq 0)$ increases and $dL_t$ is supported by $\{t \geq 0; X_t = 0\}$,      (3.3)
- $(\mathscr{M}_t, t \geq 0)$ is a local martingale.      (3.4)

A sketch of study of the general set-up is found at the end of Chapter 3.

## About Chapter 4

Chapter 4 may be read independently from the other ones.

Let $\left(\mathscr{E}_t := \exp\left(B_t - \frac{t}{2}\right), t \geq 0\right)$ denote the geometric Brownian motion. We define, for every $x \geq 0$ and $K \geq 0$, the Black-Scholes type perpetuity:

$$\Sigma_K^{(x)} := \int_0^\infty (x\mathscr{E}_t - K)^+ \, dt. \tag{4.1}$$

Other perpetuities (see for instance Dufresne [21], Salminen-Yor [80], [81] and [82] or Yor [98]) have been defined and studied in the context of Mathematical Finance. In particular, it goes back to D. Dufresne that, for, $a \neq 0$ and $\nu > 0$:

$$\int_0^\infty \exp\left(aB_t - \nu t\right) dt \overset{(law)}{=} \frac{2}{a^2 \gamma_{\frac{2\nu}{a^2}}} \tag{4.2}$$

where $\gamma_b$ is a gamma variable with parameter $b$. In Chapter 4 of this monograph, we study the perpetuity $\Sigma_K^{(x)}$ in detail: its moments, its Laplace transform, asymptotic properties, and so on.

## About Chapter 5

We come back to the Brownian set-up. Let $(B_t, t \geq 0)$ denote a Brownian motion starting from 0 and $(\mathscr{F}_t, t \geq 0)$ its natural filtration. Let, for $\nu$ real, $\left(\mathscr{E}_t^{(\nu)}, t \geq 0\right)$ denote the exponential martingale defined by:

$$\left(\mathscr{E}_t^{(\nu)}, t \geq 0\right) := \left(\exp\left(\nu B_t - \frac{\nu^2 t}{2}\right), t \geq 0\right). \tag{5.1}$$

Formulae (1.3) and (1.4), or rather their generalization for a general index $\nu$,

$$\mathbb{E}\left[\left(K - \mathscr{E}_t^{(\nu)}\right)^+\right] := K\mathbb{P}\left(\mathscr{G}_K^{(\mathscr{E}^{(\nu)})} \leq t\right) \tag{5.2}$$

and

$$\mathbb{E}\left[\left(\mathscr{E}_t^{(\nu)} - K\right)^+\right] := \mathbb{P}^{(\mathscr{E}^{(\nu)})}\left(\mathscr{G}_K^{(\mathscr{E}^{(\nu)})} \leq t\right) \tag{5.3}$$

where: $\mathbb{P}^{(\mathscr{E}^{(\nu)})}_{|\mathscr{F}_s} = \mathscr{E}_s^{(\nu)} \cdot \mathbb{P}_{|\mathscr{F}_s}$ have a drawback: it is not possible to estimate $\mathbb{P}\left(\mathscr{G}_K^{(\mathscr{E}^{(\nu)})} \leq t\right)$ only from the observation of the Brownian trajectory up to time $t$ since

$$\mathscr{G}_K^{(\mathscr{E}^{(\nu)})} := \sup\left\{s \geq 0; \, \mathscr{E}_s^{(\nu)} = K\right\} \tag{5.4}$$

is not a stopping time. To counter this drawback, we introduce:

$$\mathscr{G}_K^{(\mathscr{E}^{(\nu)})}(t) := \sup\left\{s \le t; \ \mathscr{E}_s^{(\nu)} = K\right\}, \tag{5.5}$$

and, now, $\mathscr{G}_K^{(\mathscr{E}^{(\nu)})}(t)$ is $\mathscr{F}_t$-measurable (although, it is still not a stopping-time). The first part of Chapter 5 is then devoted (see Theorem 5.1) to write some analogues of formulae (5.2) and (5.3) where we replace $\mathscr{G}_K^{(\mathscr{E}^{(\nu)})}$ by $\mathscr{G}_K^{(\mathscr{E}^{(\nu)})}(t)$. This rewriting of formulae (5.2) and (5.3) leads to the interesting notion of past-future martingale. More precisely, let the past-future filtration $(\mathscr{F}_{s,t}, \ 0 \le s \le t)$ be defined by:

$$\mathscr{F}_{s,t} = \sigma\left(B_u, u \le s; \ B_h, h \ge t\right) \tag{5.6}$$

and note that, if $[s,t] \subset [s',t']$:

$$\mathscr{F}_{s,t} \supset \mathscr{F}_{s',t'}. \tag{5.7}$$

We then say that a process $(\Delta_{s,t}, \ s \le t)$ which is $\mathscr{F}_{s,t}$-adapted and takes values in $\mathbb{R}$, is a past-future martingale if, for every $[s,t] \subset [s',t']$:

$$\mathbb{E}\left[\Delta_{s,t}|\mathscr{F}_{s',t'}\right] = \Delta_{s',t'}. \tag{5.8}$$

The second part of Chapter 5 is devoted to the study of past-future martingales and to the description of the set of these two parameter "martingales".

## About Chapter 6

We come back to the price of a European put associated to a martingale $(M_t, t \ge 0)$ such that $M_0 = 1$ a.s.:

$$\Pi_M(K,t) := \mathbb{E}\left[(K - M_t)^+\right] \qquad (0 \le K \le 1, \ t \ge 0). \tag{6.1}$$

We saw (see Chapter 2) that, for $K = 1$, the application $t \mapsto \Pi_M(1,t)$ is the cumulative distribution function of the r.v. $\mathscr{G}_1^{(M)}$. On the other hand, we have:

$$\Pi_M(K,0) = 0 \qquad (0 \le K \le 1) \qquad ; \qquad \Pi_M(0,t) = 0 \qquad (t \ge 0), \tag{6.2}$$

$$\Pi_M(K,t) = K\mathbb{P}\left(\mathscr{G}_K^{(M)} \le t\right) \xrightarrow[t \to \infty]{} K. \tag{6.3}$$

Then, the following question arises naturally: is the function of two variables $(\Pi_M(K,t), K \in [0,1], t \ge 0)$ the cumulative distribution function of a 2-dimensional r.v. taking values in $[0,1] \times [0,+\infty[$ ? In other terms, does there exist a probability $\gamma_M$ on $[0,1] \times [0,+\infty[$ such that, for every $K \in [0,1]$ and $t \ge 0$:

$$\mathbb{E}\left[(K - M_t)^+\right] = \gamma_M\left([0,K] \times [0,t]\right) \quad ? \tag{6.4}$$

Since, from Fubini, one has:

$$\mathbb{E}\left[(K - M_t)^+\right] = \int_0^K \mathbb{P}(M_t \leq x)dx \qquad (6.5)$$

the existence of $\gamma_M$ is equivalent to the assertion:

"for every $x < 1$, the function $t \mapsto \mathbb{P}(M_t \leq x)$ is increasing." $\qquad (6.6)$

Of course, (6.6) is also equivalent to the existence, for every $x < 1$, of a r.v. $Y_x \geq 0$, such that:

$$\mathbb{P}(M_t \leq x) = \mathbb{P}(Y_x \leq t) \qquad (x < 1, t \geq 0) \qquad (6.7)$$

We call the family of r.v. $(Y_x, x \in [0,1[)$ a decreasing pseudo-inverse of the process $(M_t, t \geq 0)$. In Chapter 6, we show, for $\left(M_t = \mathcal{E}_t := \exp\left(B_t - \frac{t}{2}\right), t \geq 0\right)$ the existence of a decreasing pseudo-inverse of this martingale. This implies the existence of a probability $\gamma_{\mathcal{E}}$ on $[0,1] \times [0,+\infty[$ such that:

$$\mathbb{E}\left[(K - \mathcal{E}_t)^+\right] = \gamma_{\mathcal{E}}\left([0,K] \times [0,t]\right) \qquad (K \in [0,1], t \geq 0). \qquad (6.8)$$

We then describe, in several ways, the r.v. (taking values in $[0,1] \times [0,+\infty[$) which admits $\gamma_{\mathcal{E}}$ as cumulative distribution function.

## About Chapters 7 and 8

We say that a process $(X_t, t \geq 0)$, taking values in $\mathbb{R}^+$ and starting from $x \geq 0$, admits an increasing pseudo-inverse (resp. a decreasing pseudo-inverse) if:

i) for every $y > x$, $\lim_{t \to \infty} \mathbb{P}_x(X_t \geq y) = 1$, $\qquad (7.1)$

ii) for every $y > x$, the function from $\mathbb{R}^+$ into $[0,1] : t \to \mathbb{P}_x(X_t \geq y)$ is increasing, $\qquad (7.2)$

    resp. if:

i') for every $y < x$, $\lim_{t \to \infty} \mathbb{P}_x(X_t \leq y) = 1$, $\qquad (7.1')$

ii') for every $y < x$, the function from $\mathbb{R}^+$ into $[0,1] : t \to \mathbb{P}_x(X_t \leq y)$ is increasing. $\qquad (7.2')$

Conditions (i) and (ii) (resp. (i') and (ii')) are equivalent to the existence of a family of positive r.v. $(Y_{x,y}, y > x)$ (resp. $(Y_{x,y}, y < x)$) such that:

$$\mathbb{P}_x(X_t \geq y) = \mathbb{P}(Y_{x,y} \leq t) \qquad (y > x, t \geq 0), \qquad (7.3)$$

resp.

$$\mathbb{P}_x(X_t \leq y) = \mathbb{P}(Y_{x,y} \leq t) \qquad (y < x, t \geq 0). \qquad (7.3')$$

We call such a family of r.v. $(Y_{x,y}, y > x)$ (resp. $(Y_{x,y}, y < x)$) an increasing (resp. decreasing) pseudo-inverse of the process $(X_t, t \geq 0)$.

In the preceding Chapter 6, we noticed the importance of the notion of pseudo-inverse to be able to consider the price of a European put $\Pi_M(K,t)$ as the cumulative distribution function of a couple of r.v.'s taking values in $[0,1] \times [0,+\infty[$. This is the reason why, in Chapters 7 and 8, we try to study systematically this notion:

- in Chapter 7, we show that the Bessel processes with index $\nu \geq -\frac{1}{2}$ admit an increasing pseudo-inverse; we then extend this result to several processes in the neighborhood of Bessel processes. Finally, if $(Y_{x,y}, y > x)$ is the pseudo-inverse of a Bessel process, we show that the laws of the r.v.'s $(Y_{x,y}, y > x)$ enjoy remarkable properties, which we describe.
- In Chapter 8, we study the existence of pseudo-inverses for a class of real-valued diffusions. There again, we show that these pseudo-inverses have remarkable distributions, related with non-Markovian extensions of the celebrated theorem of J. Pitman about "$2S - X$". (See [73] for these extensions of Pitman's theorem).

Each of these eight chapters ends with Notes and Comments which make precise the sources – mainly recent preprints – of the contents of that chapter. Some of them contain Exercises and Problems, from which the reader may "feel" better some springs of our arguments. Typically, Problems develop thoroughly a given topic.

Finally, this monograph ends with two appendices A and B which consist respectively of three Complements and three Notes on Bessel items.

In the first Complement A.1, we study an example of a European call price associated to a strict local martingale (which therefore does not satisfy the hypotheses of Section 2.2). Then, in Complement A.2, we give some criterion to measure how much a random time (such as the last passage times we introduced previously) differs from a stopping time. The last Complement A.3 is dedicated to an extension of Dupire's formula to the general framework of positive, continuous martingales converging towards 0.

As for the Notes on Bessel items, B.1 recalls the definition (and some useful formulae) for the modified Bessel functions and the McDonald functions, while the two others B.2 and B.3 summarize some well-known results about Bessel and Squared Bessel processes.

As a conclusion this monograph provides some new looks at the generalized Black-Scholes formula, in the following directions:

- In Chapters 1 and 2, the prices of a European put and call are expressed in terms of last passage times.
- In Chapter 5, a further extension of the Black-Scholes formula in terms of last passage times is given when working under a finite horizon. This leads to the definition of the notion (and the detailed study) of past-future (Brownian) martingales.
- In the set-up of geometric Brownian motion, we associate to the set of prices of European puts, indexed by strikes $K$ and maturities $t$, a probability $\gamma$ on $[0,1] \times$

$[0, +\infty[$. The knowledge of the probability $\gamma$ is equivalent to that of the set of the prices of European puts. $\gamma$ is described completely in Chapter 6.
- The construction of $\gamma$ hinges upon the notion of pseudo-inverse of a process; this notion is defined and studied in detail in the general set-up of Bessel processes (Chapter 7) and linear diffusions (Chapter 8).

This kind of "new look" at European options may also be developed for exotic options, e.g. Asian options, a study we have engaged in, but which lies outside the scope of this monograph.

Here are a few indications about the genesis of this monograph: it really started in August 2007, with the question from M. Qian [69] to give a simple formula for $\int_0^\infty e^{-\lambda t} \mathbb{E}\left[(\mathscr{E}_t - 1)^+\right] dt$. This question, and its solution as developed in Chapter 1, then motivated the search for the various developments which we just presented in this Introduction. We thank M. Qian most sincerely for providing this starting point. Thanks are also due to J. Akahori (Ritsumeikan Univ.) who suggested to consider last hitting times of a martingale up to finite maturity.

To conclude, it turns out finally that a number of topics related to last passage times seem to find a natural niche in our discussions of generalized Black-Scholes formulae. We have not, intentionally, discussed about the importance of the Black-Scholes formula as a pillar of mathematical finance so far, but we hope that our last passage times interpretation shall help develop other up to now hidden aspects of this topic.

Nancy, Paris,                                                          *Christophe Profeta*
October 2009                                                          *Bernard Roynette*
                                                                          *Marc Yor*

# Contents

1   **Reading the Black-Scholes Formula in Terms of First and Last Passage Times** ...................................................... 1
   1.1   Introduction and Notation.................................... 1
      1.1.1   Basic Notation ....................................... 1
      1.1.2   Exponential Martingales and the Cameron-Martin Formula . 2
      1.1.3   First and Last Passage Times ......................... 2
      1.1.4   The Classical Black-Scholes Formula .................. 3
   1.2   The Black-Scholes Formula in Terms of First and Last Passage Times ............................................................. 5
      1.2.1   A New Expression for the Black-Scholes Formula......... 5
      1.2.2   Comments ........................................... 6
      1.2.3   Proof of Theorem 1.2 ................................ 7
      1.2.4   On the Agreement Between the Classical Black-Scholes Formula (Theorem 1.1) and our Result (Theorem 1.2) ..... 10
      1.2.5   A Remark on Theorem 1.2 and Time Inversion .......... 11
   1.3   Extension of Theorem 1.2 to an Arbitrary Index $\nu$ .............. 13
      1.3.1   Statement of the Main Result ......................... 13
      1.3.2   Some Comments on Theorem 1.3...................... 14
      1.3.3   A Short Proof of Theorem 1.3......................... 15
   1.4   Another Formulation of the Black-Scholes Formula .............. 16
      1.4.1   Statement of the Result .............................. 16
      1.4.2   First Proof of Theorem 1.4 ........................... 16
      1.4.3   A Second Proof of Theorem 1.4 ....................... 17
   1.5   Notes and Comments ....................................... 19

2   **Generalized Black-Scholes Formulae for Martingales, in Terms of Last Passage Times** ........................................... 21
   2.1   Expression of the European Put Price in Terms of Last Passage Times ..................................................... 21
      2.1.1   Hypotheses and Notation............................. 21

2.1.2   Expression of $\Pi(K,t)$ in Terms of $\mathscr{G}_K^{(M)}$ ................ 22

2.1.3   Proof of Theorem 2.1 ................................. 22

2.2   Expression of the European Call Price in Terms of Last Passage
Times ...................................................... 24

2.2.1   Hypotheses .......................................... 24

2.2.2   Price of a European Call in Terms of Last Passage Times ... 25

2.2.3   Proof of Theorem 2.2 ................................. 26

2.3   Some Examples of Computations of the Law of $\mathscr{G}_K^{(M)}$ ........... 27

2.4   A More General Formula for the Computation of the Law of $\mathscr{G}_K^{(M)}$ . 32

2.4.1   Hypotheses .......................................... 32

2.4.2   Description of the Law of $\mathscr{G}_K^{(M)}$ ....................... 33

2.4.3   Some Examples of Applications of Theorem 2.3 ......... 34

2.5   Computation of the Law of $\mathscr{G}_K$ in the Framework of Transient
Diffusions ................................................... 37

2.5.1   General Framework .................................. 37

2.5.2   A General Formula for the Law of $\mathscr{G}_K^{(X)}$ ................ 38

2.5.3   Case Where the Infinitesimal Generator is Given by its
Diffusion Coefficient and its Drift .................... 39

2.6   Computation of the Put Associated to a Càdlàg Martingale
Without Positive Jumps .................................... 41

2.6.1   Notation ............................................ 41

2.6.2   Computation of the Put Associated to the Martingale
$\left(M_a^{(\nu)}, a \geq 0\right)$ ......................................... 42

2.6.3   Computation of the Law of $\mathscr{G}_K^{(M^{(\nu)})}$ .................... 44

2.6.4   A More Probabilistic Approach of Proposition 2.2 ........ 45

2.6.5   An Application of Proposition 2.1 to the Local Times of
the Martingale $(\mathscr{E}_t, t \geq 0)$ ............................ 49

2.7   The case $M_\infty \neq 0$ ......................................... 50

2.7.1   Hypotheses .......................................... 50

2.7.2   A Generalization of Theorem 2.1 ...................... 51

2.7.3   First Proof of Theorem 2.5 ........................... 51

2.7.4   A Second Proof of Theorem 2.5 ....................... 52

2.7.5   On the Law of $S_\infty := \sup_{t \geq 0} M_t$ ............................. 53

2.8   Extension of Theorem 2.1 to the Case of Orthogonal Local
Martingales ................................................ 55

2.8.1   Statement of the Main Result ......................... 55

2.8.2   First Proof of Theorem 2.6, via Enlargement Theory ....... 56

2.8.3   Second Proof of Theorem 2.6, via Knight's Representation
of Orthogonal Continuous Martingales ................. 57

2.8.4   On the Law of $\bigvee_{i=1,\cdots,n} \mathscr{G}_{K_i}^{(i)}$ ................................ 59

2.9   Notes and Comments ....................................... 63

**3    Representation of some particular Azéma supermartingales** ........ 65
    3.1   A General Representation Theorem .......................... 65
          3.1.1   Introduction........................................ 65
          3.1.2   General Framework .................................. 66
          3.1.3   Statement of the Representation Theorem ............... 66
          3.1.4   Application of the Representation Theorem 3.1 to the
                  Supermartingale $\left(\mathbb{P}\left(\mathscr{G}_K > t \mid \mathscr{F}_t\right), t \geq 0\right)$, when $M_\infty = 0$ ..... 67
          3.1.5   A Remark on Theorem 3.2 ........................... 69
    3.2   Study of the Pre $\mathscr{G}_K$- and Post $\mathscr{G}_K$-processes, when $M_\infty = 0$ ........ 70
          3.2.1   Enlargement of Filtration Formulae .................... 70
          3.2.2   Study of the Post $\mathscr{G}_K$-Process ......................... 71
          3.2.3   Study of the Pre $\mathscr{G}_K$-Process .......................... 72
          3.2.4   Some Predictable Compensators....................... 73
          3.2.5   Expression of the Azéma supermartingale
                  $\left(\mathbb{P}\left(\mathscr{G}_K > t \mid \mathscr{F}_t\right), t \geq 0\right)$ when $M_\infty \neq 0$ ................... 76
          3.2.6   Computation of the Azéma Supermartingale.............. 77
    3.3   A Wider Framework: the Skorokhod Submartingales ............ 78
          3.3.1   Introduction........................................ 78
          3.3.2   Skorokhod Submartingales .......................... 79
          3.3.3   A Comparative Analysis of the Three Cases ............. 81
          3.3.4   Two Situations Where the Measure $\mathbb{Q}$ Exists ............ 82
    3.4   Notes and Comments ...................................... 87

**4    An Interesting Family of Black-Scholes Perpetuities** ............... 89
    4.1   Introduction ............................................. 89
          4.1.1   A First Example ................................... 89
          4.1.2   Other Perpetuities.................................. 90
          4.1.3   A Family of Perpetuities Associated to the Black-Scholes
                  Formula .......................................... 90
          4.1.4   Notation.......................................... 91
          4.1.5   Reduction of the Study ............................. 92
          4.1.6   Scaling Properties.................................. 92
          4.1.7   General Case of the Brownian Exponential Martingale of
                  Index $\nu \neq 0$ ...................................... 93
          4.1.8   Statement of the Main Results ....................... 93
    4.2   Proofs of Theorems 4.1, 4.2, 4.3 and 4.4 ................... 95
          4.2.1   A First Proof of Theorem 4.1 ........................ 95
          4.2.2   Second Proof of Theorem 4.1 ........................ 96
          4.2.3   Proof of Theorem 4.2 ............................... 98
          4.2.4   Proof of Theorem 4.3 ............................... 99
          4.2.5   Proof of Theorem 4.4 ............................... 102
    4.3   Asymptotic Behavior of $\mathbb{E}_1\left[\exp\left(-\frac{\theta}{2}\Sigma_1\right)\right]$ as $\theta \to \infty$ ............ 103
    4.4   Extending the Preceding Results to the Variables $\Sigma_k^{(\rho,x)}$ .......... 106
    4.5   Notes and Comments ...................................... 113

**5**    **Study of Last Passage Times up to a Finite Horizon** ............... 115
        5.1    Study of Last Passage Times up to a Finite Horizon for the
               Brownian Motion with Drift .................................. 115
               5.1.1    Introduction and Notation ........................... 115
               5.1.2    Statement of our Main Result ........................ 116
               5.1.3    An Explicit Expression for the Law of $G_x^{(\nu)}(t)$ ........... 123
        5.2    Past-Future (Sub)-Martingales ................................ 127
               5.2.1    Definitions ........................................ 127
               5.2.2    Properties and Characterization of PFH-Functions ........ 128
               5.2.3    Two Classes of PFH-Functions ....................... 131
               5.2.4    Another Characterization of PFH-Functions ............ 131
               5.2.5    Description of Extremal PFH-Functions ............... 133
        5.3    Notes and Comments ........................................ 141

**6**    **Put Option as Joint Distribution Function in Strike and Maturity** ... 143
        6.1    Put Option as a Joint Distribution Function and Existence of
               Pseudo-Inverses ............................................ 143
               6.1.1    Introduction ....................................... 143
               6.1.2    Seeing $\Pi_M(K,t)$ as a Function of 2 Variables ............ 144
               6.1.3    General Pattern of the Proof ........................ 144
               6.1.4    A Useful Criterion ................................. 145
               6.1.5    Outline of the Following Sections ................... 145
        6.2    The Black-Scholes Paradigm .................................. 146
               6.2.1    Statement of the Main Result ....................... 146
               6.2.2    Descriptions of the Probability $\gamma$ ..................... 149
               6.2.3    An Extension of Theorem 6.1 ........................ 155
               6.2.4    $\gamma$ as a Signed Measure on $\mathbb{R}^+ \times \mathbb{R}^+$ ................... 157
        6.3    Notes and Comments ........................................ 159

**7**    **Existence and Properties of Pseudo-Inverses for Bessel and Related**
        **Processes** ................................................... 161
        7.1    Introduction and Definition of a Pseudo-Inverse ............... 161
               7.1.1    Motivations ....................................... 161
               7.1.2    Definitions and Examples .......................... 162
               7.1.3    Aim of this Chapter ............................... 163
        7.2    Existence of Pseudo-inverses for Bessel Processes ............. 166
               7.2.1    Statement of our Main Result ....................... 166
               7.2.2    A Summary of some Results About Bessel Processes ...... 167
               7.2.3    Proof of Theorem 7.1 .............................. 172
               7.2.4    Interpretation in Terms of the Local Martingales
                        $(R_t^{-2\nu}, t \geq 0)$ ................................. 178
        7.3    Some Properties of the r.v.'s $(Y_{x,y}^{(\nu)}, y > x)$ $(\nu \geq -\frac{1}{2})$ ............. 179
               7.3.1    The Main Theorem .................................. 179
               7.3.2    Some Further Relations ............................ 185

7.4 Two Extensions of Bessel Processes with Increasing
Pseudo-Inverses . . . . . . . . . . . . . . . . . . . . . . . . . . . . . . . . . . . . . . . . . 192
7.4.1 Bessel Processes with Index $\nu \geq -\frac{1}{2}$ and Drift $a > 0$ . . . . . . 192
7.4.2 Squares of Generalized Ornstein-Uhlenbeck Processes,
also Called CIR Processes in Mathematical Finance . . . . . . . 193
7.4.3 A Third Example . . . . . . . . . . . . . . . . . . . . . . . . . . . . . . . . . . . . 195
7.5 The More General Family $\left(Y_{x,y}^{(\nu,\alpha)}; x < y, \nu \geq 0, \alpha \in [0,1]\right)$ . . . . . . . . 196
7.5.1 Some Useful Formulae . . . . . . . . . . . . . . . . . . . . . . . . . . . . . . . 196
7.5.2 Definition of $(G_y^{(\nu+\theta,\nu)}, y > 0, \nu, \theta \geq 0)$ and
$(T_y^{(\nu+\theta,\nu)}, y > 0, \nu, \theta \geq 0)$ . . . . . . . . . . . . . . . . . . . . . . . . . . 197
7.5.3 Existence and Properties of $\left(Y_{x,y}^{(\nu,\alpha)}; x < y, \nu \geq 0, \alpha \in [0,1]\right)$ 199
7.6 Notes and Comments . . . . . . . . . . . . . . . . . . . . . . . . . . . . . . . . . . . . . 201

8 Existence of Pseudo-Inverses for Diffusions . . . . . . . . . . . . . . . . . . . . . 203
8.1 Introduction . . . . . . . . . . . . . . . . . . . . . . . . . . . . . . . . . . . . . . . . . . . . 203
8.2 Pseudo-Inverse for a Brownian Motion with a Convex,
Decreasing, Positive Drift. . . . . . . . . . . . . . . . . . . . . . . . . . . . . . . . . . 205
8.3 Study of a Family of $\mathbb{R}^+$-Valued Diffusions . . . . . . . . . . . . . . . . . . . . 210
8.3.1 Definition of the Operator $T$ . . . . . . . . . . . . . . . . . . . . . . . . . 210
8.3.2 Study of the Family $(X^{(\alpha)})_{\alpha \geq 0}$ . . . . . . . . . . . . . . . . . . . . . 212
8.3.3 Existence of a Pseudo-Inverse when $\alpha = 0$ . . . . . . . . . . . . . . 217
8.4 Existence of Pseudo-Inverses for a $\mathbb{R}^+$-Valued Diffusion Started
at 0 . . . . . . . . . . . . . . . . . . . . . . . . . . . . . . . . . . . . . . . . . . . . . . . . . . . 220
8.4.1 Notations . . . . . . . . . . . . . . . . . . . . . . . . . . . . . . . . . . . . . . . . . 220
8.4.2 Biane's Transformation . . . . . . . . . . . . . . . . . . . . . . . . . . . . . . 222
8.4.3 Existence of Pseudo-Inverses . . . . . . . . . . . . . . . . . . . . . . . . . 225
8.4.4 A Second Proof of Theorem 8.3 . . . . . . . . . . . . . . . . . . . . . . . 228
8.5 Some Consequences of the Existence of Pseudo-Inverses . . . . . . . . . 232
8.5.1 Another Relation Between the Processes $X$ and $\overline{X}$ Started
from 0 . . . . . . . . . . . . . . . . . . . . . . . . . . . . . . . . . . . . . . . . . . . . 232
8.5.2 A Time Reversal Relationship . . . . . . . . . . . . . . . . . . . . . . . . . 233
8.5.3 Back to the Family $(X^{(\alpha)})_{\alpha \geq 0}$ . . . . . . . . . . . . . . . . . . . . . . . 235
8.6 Notes and Comments . . . . . . . . . . . . . . . . . . . . . . . . . . . . . . . . . . . . . 237

A Complements . . . . . . . . . . . . . . . . . . . . . . . . . . . . . . . . . . . . . . . . . . . . . . 239
A.1 Study of the Call Associated to a Strict Local Martingale (see
Yen-Yor [93]) . . . . . . . . . . . . . . . . . . . . . . . . . . . . . . . . . . . . . . . . . . . 239
A.1.1 Introduction . . . . . . . . . . . . . . . . . . . . . . . . . . . . . . . . . . . . . . . 239
A.1.2 Main Results . . . . . . . . . . . . . . . . . . . . . . . . . . . . . . . . . . . . . . . 239
A.1.3 An Extension . . . . . . . . . . . . . . . . . . . . . . . . . . . . . . . . . . . . . . 241
A.2 Measuring the "Non-Stopping Timeness" of Ends of Previsible
Sets (see Yen-Yor, [92]) . . . . . . . . . . . . . . . . . . . . . . . . . . . . . . . . . . . 242
A.2.1 About Ends of Previsible Sets . . . . . . . . . . . . . . . . . . . . . . . . . 242
A.2.2 Some Criterions to Measure the NST . . . . . . . . . . . . . . . . . . . 242

          A.2.3  Computations of Several Examples of Functions $m_L(t)$ ..... 244
     A.3  Some Connexions with Dupire's Formula ...................... 246
          A.3.1  Dupire's Formula (see [20, F]) ........................ 246
          A.3.2  Extension of Dupire's Formula to a General Martingale
                 in $\mathcal{M}_+^{0,c}$ .......................................... 246
          A.3.3  A Formula Relative to Lévy Processes Without Positive
                 Jumps.............................................. 248

**B    Bessel Functions and Bessel Processes** ........................... 251
     B.1  Bessel Functions (see [46], p. 108-136) ...................... 251
     B.2  Squared Bessel Processes (see [70] Chapter XI, or [26]) ......... 253
          B.2.1  Definition of Squared Bessel Processes ................. 253
          B.2.2  BESQ as a Diffusion ............................... 254
          B.2.3  Brownian Local Times and BESQ Processes ............. 254
     B.3  Bessel Processes (see [70] Chapter XI, or [26]).................. 255
          B.3.1  Definition ........................................ 255
          B.3.2  An Implicit Representation in Terms of Geometric
                 Brownian Motions ................................. 256

**References**...................................................... 259

**Further Readings** ............................................... 265

**Index** ......................................................... 269

# List of Notation

## About Brownian motion

$(B_t, t \geq 0, \mathscr{F}_t, t \geq 0, \mathbb{P})$ — a standard Brownian motion started at 0 and equipped with its natural filtration

$\mathscr{F}_\infty := \bigvee_{t \geq 0} \mathscr{F}_t$ — smallest $\sigma$-algebra generated by $(\mathscr{F}_t)_{t \geq 0}$

$\mathbb{P}_{|\mathscr{F}_t}$ — restriction of the probability $\mathbb{P}$ to the $\sigma$-algebra $\mathscr{F}_t$

$(B_t^{(\nu)} := B_t + \nu t, t \geq 0)$ — Brownian motion with drift $\nu$ ($\nu \in \mathbb{R}$)

$\left( \mathscr{E}_t^{(\nu)} := \exp\left( \nu B_t - \frac{\nu^2}{2} t \right), t \geq 0 \right)$ — exponential martingale associated to $\nu B$

$T_a^{(\nu)} := \inf\{t \geq 0; B_t^{(\nu)} = a\}$ — first hitting time of level $a$ of a Brownian motion with drift $\nu$

$G_a^{(\nu)} := \sup\{t \geq 0; B_t^{(\nu)} = a\}$ — last passage time at level $a$ of a Brownian motion with drift $\nu$

$G_a^{(\nu)}(t) := \sup\{u \leq t; B_u^{(\nu)} = a\}$ — last passage time at level $a$ before time $t$ of a Brownian motion with drift $\nu$

$\Pi_K(t) := \mathbb{E}\left[ (K - \mathscr{E}_t)^+ \right]$ — European put associated to $(\mathscr{E}_t, t \geq 0)$

$C_K(t) := \mathbb{E}\left[ (\mathscr{E}_t - K)^+ \right]$ — European call associated to $(\mathscr{E}_t, t \geq 0)$

$\mathscr{N}(x) = \frac{1}{\sqrt{2\pi}} \int_{-\infty}^{x} e^{-\frac{y^2}{2}} dy$ — cumulative distribution function of a reduced Gaussian r.v.

$\Sigma_\mu^\pm := \int_0^\infty (\mathscr{E}_t - 1)^\pm \mu(dt)$ — "Black-Scholes" perpetuities

$\mathscr{F}_{s,t} = \sigma(B_u, u \leq s; B_h, h \geq t), (s \leq t)$ — the past-future Brownian filtration

$\Delta_{s,t}, (s \leq t)$ — a past-future martingale

PFH-function — past-future harmonic function

## About martingales and semimartingales

$(M_t, t \geq 0)$ — a positive (local) martingale

$\mathbb{P}^{(M)} ; \mathbb{P}^{(M)}_{|\mathscr{F}_t} = M_t \cdot \mathbb{P}_{|\mathscr{F}_t}$ — probability on $\mathscr{F}_\infty$

$(L_t^K, t \geq 0)$ — local time process at level $K$

$\Pi_M(K,t) := \mathbb{E}\left[(K - M_t)^+\right]$     European put associated to $(M_t, t \geq 0)$

$C_M(K,t) := \mathbb{E}\left[(M_t - K)^+\right]$     European call associated to $(M_t, t \geq 0)$

$(\langle M \rangle_t, t \geq 0)$     increasing process of $(M_t, t \geq 0)$

$(m_t(x), x \geq 0)$     density function of the r.v. $M_t$

$(\sigma_t^2, t \geq 0)$     density of $(\langle M \rangle_t, t \geq 0)$ with respect to the Lebesgue measure: $d\langle M \rangle_t = \sigma_t^2 dt$

$\theta_t(x) := \mathbb{E}\left[\sigma_t^2 | M_t = x\right]$     conditional expectation of $\sigma_t^2$, given $M_t = x$

$\mathcal{G}_K = \mathcal{G}_K^{(M)} := \sup\{t \geq 0; M_t = K\}$     last passage time at level $K$ of the local martingale $M$

$(\mathcal{F}_t^K, t \geq 0)$     smallest filtration containing $(\mathcal{F}_t)_{t \geq 0}$ and making $\mathcal{G}_K$ a $(\mathcal{F}_t^K)_{t \geq 0}$ stopping time

$\left(Z_t^{(K)} := \mathbb{P}(\mathcal{G}_K > t | \mathcal{F}_t), t \geq 0\right)$     Azéma's supermartingale associated to $\mathcal{G}_K$

**About some random variables**

$U$     a uniform r.v. on $[0,1]$

$\mathbf{e}$     a standard exponential r.v.

$\gamma_t$     a gamma r.v. with parameter $t$

$\beta(a,b)$     a beta r.v. with parameters $a$ and $b$

$\delta_0$     Dirac measure at 0

$f_Y$     density function of the r.v. $Y$

**About Bessel functions and Bessel processes**

$I_\nu, K_\nu$     modified Bessel and McDonald functions

$J_\nu, Y_\nu$     Bessel functions with index $\nu$

$(j_{\nu,n}, n \geq 1)$     increasing sequence of the positive zeroes of $J_\nu$

$\left((R_t, \mathcal{F}_t)_{t \geq 0}, \mathbb{P}_x^{(\nu)}, x \geq 0\right)$     a Bessel process of index $\nu$ $(\nu \geq -1)$

$p^{(\nu)}(t,x,y)$     density of the r.v. $R_t$ under $\mathbb{P}_x^{(\nu)}$

$u_\lambda^{(\nu)}(x,y)$     density of potential kernel under $\mathbb{P}^{(\nu)}$

$\left(\mathbb{P}_x^{(\nu,a)}, x \geq 0\right)$     law of the Bessel process with index $\nu$ and drift $a$

$\left(\mathbb{Q}_x^{(\nu,\beta)}, x \geq 0\right)$     law of the generalized squared Ornstein-Uhlenbeck process

**About diffusions**

$(X_t, t \geq 0)$     a linear diffusion

$m(dx) = \rho(x)dx$     its speed measure

$s(x)$     its scale function

$p(t,x,y)$     density of the r.v. $X_t$ under $\mathbb{P}_x$ with respect to the Lebesgue measure $dy$

$q(t,x,y)$     density of the r.v. $X_t$ under $\mathbb{P}_x$ with respect to the speed measure $m(dy)$

$u_\lambda(x,y)$     density of resolvent kernel of $(X_t, t \geq 0)$

$(X_t^{(\alpha)}, t \geq 0)$        a family of diffusions

$TF(x) := \dfrac{e^{2F(x)}}{\int_0^x e^{2F(y)} dy} - F'(x)$        an operator $T$ acting on $F$

$\overline{\rho}(x) = (m[0,x])^2 s'(x)$        density of the speed measure associated with Biane's transformation

$\overline{s}(x) = \dfrac{1}{m[0,+\infty[} - \dfrac{1}{m[0,x]}$        scale function associated with Biane's transformation

$(Y_{x,y}, y > x)$ (resp. $(Y_{x,y}, y < x)$)        an increasing (resp. decreasing) pseudo-inverse

$E_{x,y}^{(g)} := \mathbb{E}_x \left[ \exp \int_0^t g(X_s) ds \Big| X_t = y \right]$        Feynman-Kac conditional expectation

# Chapter 1
# Reading the Black-Scholes Formula in Terms of First and Last Passage Times

**Abstract** We first recall the classical Black-Scholes formula (Theorem 1.1), and then give two new formulations of it:

- the first one in terms of first and last passage times of a Brownian motion with drift (Theorem 1.2 and Theorem 1.3),
- the second one as an expectation with respect to the law of $B_1^2$ (Theorem 1.4).

## 1.1 Introduction and Notation

### 1.1.1 Basic Notation

We present some basic notation for the Brownian items we shall deal with throughout this Chapter, as well as classical results about the laws of the first and last passage times for Brownian motion with drift. For every $\nu \in \mathbb{R}$, we denote the Brownian motion with drift $\nu$ by $(B_t^{(\nu)}, t \geq 0)$:

$$(B_t^{(\nu)}, t \geq 0) := (B_t + \nu t, t \geq 0). \tag{1.1}$$

We denote by $(\mathscr{F}_t, t \geq 0)$ the natural filtration of $(B_t, t \geq 0)$:

$$(\mathscr{F}_t, t \geq 0) := (\sigma(B_s, s \leq t); t \geq 0) \tag{1.2}$$

and

$$\mathscr{F}_\infty := \bigvee_{t \geq 0} \mathscr{F}_t.$$

C. Profeta et al., *Option Prices as Probabilities*, Springer Finance,
DOI 10.1007/978-3-642-10395-7_1, © Springer-Verlag Berlin Heidelberg 2010

## 1.1.2 Exponential Martingales and the Cameron-Martin Formula

Let $\left(\mathscr{E}_t^{(\nu)}, t \geq 0\right)$ be the positive $(\mathscr{F}_t, t \geq 0)$-martingale defined by:

$$\left(\mathscr{E}_t^{(\nu)}, t \geq 0\right) := \left(\exp\left(\nu B_t - \frac{\nu^2}{2}t\right), t \geq 0\right). \tag{1.3}$$

Note that $\left(\mathscr{E}_t^{(\nu)}, t \geq 0\right)$ has, a priori, little to do with $(B_t^{(\nu)}, t \geq 0)$, although see the Cameron-Martin formula (1.5) below. For $\nu = 1$, we shall simply write $\mathscr{E}_t$ instead of $\mathscr{E}_t^{(1)}$. Throughout this chapter, many facts pertaining to $\left(\mathscr{E}_t^{(\nu)}, t \geq 0\right)$ may be reduced to $(\mathscr{E}_t, t \geq 0)$ since by scaling:

$$\left(\mathscr{E}_t^{(\nu)}, t \geq 0\right) \overset{(law)}{=} (\mathscr{E}_{\nu^2 t}, t \geq 0). \tag{1.4}$$

Moreover, the Cameron-Martin formula relates the laws of $B^{(\nu)}$ and $B$ as follows:

$$\mathbb{E}\left[F(B_s^{(\nu)}, s \leq t)\right] = \mathbb{E}\left[F(B_s, s \leq t)\mathscr{E}_t^{(\nu)}\right] =: \mathbb{E}^{(\mathscr{E}^\nu)}\left[F(B_s, s \leq t)\right] \tag{1.5}$$

for any positive functional $F$ on $\mathscr{C}([0,t], \mathbb{R})$.

## 1.1.3 First and Last Passage Times

Let us define for $a \in \mathbb{R}, \nu \in \mathbb{R}$:

$$T_a^{(\nu)} := \inf\{u \geq 0; B_u^{(\nu)} = a\} \tag{1.6}$$
$$(= +\infty \text{ if the set } \{u \geq 0; B_u^{(\nu)} = a\} \text{ is empty}).$$
$$G_a^{(\nu)} := \sup\{u \geq 0; B_u^{(\nu)} = a\} \tag{1.7}$$
$$(= 0 \text{ if the set } \{u \geq 0; B_u^{(\nu)} = a\} \text{ is empty}).$$

It is obvious by symmetry that:

$$T_a^{(\nu)} \overset{(law)}{=} T_{-a}^{(-\nu)} \quad ; \quad G_a^{(\nu)} \overset{(law)}{=} G_{-a}^{(-\nu)} \tag{1.8}$$

and, by time inversion, that:

$$\frac{1}{T_a^{(\nu)}} \overset{(law)}{=} G_\nu^{(a)} \quad ; \quad \frac{1}{G_a^{(\nu)}} \overset{(law)}{=} T_\nu^{(a)}. \tag{1.9}$$

We recall the classical formulae, for $\nu > 0$ and $a > 0$:

$$\mathbb{P}\left(G_a^{(-\nu)} > 0\right) = \mathbb{P}\left(T_a^{(-\nu)} < +\infty\right) = \exp\left(-2\nu a\right) \qquad (1.10)$$

and

$$\mathbb{P}\left(T_a^{(\nu)} \in dt\right) = \frac{a}{\sqrt{2\pi t^3}} \exp\left(-\frac{(a-\nu t)^2}{2t}\right) dt, \qquad (1.11)$$

$$\mathbb{P}\left(G_a^{(\nu)} \in dt\right) = \frac{\nu}{\sqrt{2\pi t}} \exp\left(-\frac{(a-\nu t)^2}{2t}\right) dt, \qquad (1.12)$$

whereas, for $a > 0$ and $\nu > 0$:

$$\mathbb{P}\left(T_a^{(-\nu)} \in dt\right) = \frac{a}{\sqrt{2\pi t^3}} \exp\left(-\frac{(a+\nu t)^2}{2t}\right) dt, \qquad (1.13)$$

$$\mathbb{P}\left(G_a^{(-\nu)} \in dt\right) = \frac{\nu}{\sqrt{2\pi t}} \exp\left(-\frac{(a+\nu t)^2}{2t}\right) dt. \qquad (1.14)$$

In agreement with equation (1.10), the measures given by formulae (1.13) and (1.14) are subprobabilities on $[0, +\infty[$ with common total mass $\exp(-2\nu a)$. Note that the proof of formula (1.11) may be reduced to the case $\nu = 0$ thanks to the Cameron-Martin formula (1.5).

### 1.1.4 The Classical Black-Scholes Formula

A reduced form of the celebrated Black-Scholes formula is the following:

**Theorem 1.1 ([11], [44]).** *For every* $K \geq 0$:

i) *The European put price equals:*

$$\mathbb{E}\left[(K - \mathscr{E}_t)^+\right] = K\mathscr{N}\left(\frac{\log(K)}{\sqrt{t}} + \frac{\sqrt{t}}{2}\right) - \mathscr{N}\left(\frac{\log(K)}{\sqrt{t}} - \frac{\sqrt{t}}{2}\right) \qquad (1.15)$$

*with* $\mathscr{N}(x) := \dfrac{1}{\sqrt{2\pi}} \displaystyle\int_{-\infty}^{x} e^{-\frac{y^2}{2}} dy.$

ii) *The European call price equals:*

$$\mathbb{E}\left[(\mathscr{E}_t - K)^+\right] = \mathscr{N}\left(-\frac{\log(K)}{\sqrt{t}} + \frac{\sqrt{t}}{2}\right) - K\mathscr{N}\left(-\frac{\log(K)}{\sqrt{t}} - \frac{\sqrt{t}}{2}\right). \qquad (1.16)$$

iii) *Formula (1.16) (or (1.15)) may be split into two parts:*

$$\mathbb{E}\left[\mathscr{E}_t 1_{\{\mathscr{E}_t > K\}}\right] = 1 - \mathbb{E}\left[\mathscr{E}_t 1_{\{\mathscr{E}_t < K\}}\right] = \mathscr{N}\left(-\frac{\log(K)}{\sqrt{t}} + \frac{\sqrt{t}}{2}\right), \qquad (1.17)$$

$$K\mathbb{P}(\mathscr{E}_t > K) = K\left(1 - \mathbb{P}(\mathscr{E}_t < K)\right) = K\mathscr{N}\left(-\frac{\log(K)}{\sqrt{t}} - \frac{\sqrt{t}}{2}\right). \qquad (1.18)$$

*iv) In the case $K = 1$, the equalities (1.15) and (1.16) reduce to:*

$$\mathbb{E}\left[|\mathcal{E}_t - 1|\right] = 2\mathbb{P}\left(B_1^2 \leq \frac{t}{4}\right), \qquad (1.19)$$

*or equivalently to:*

$$\mathbb{E}\left[(\mathcal{E}_t - 1)^+\right] = \mathbb{E}\left[(1 - \mathcal{E}_t)^+\right] = \mathbb{P}\left(B_1^2 \leq \frac{t}{4}\right).$$

This theorem can easily be proven thanks to the Cameron-Martin formula (1.5). Indeed:

$$\mathbb{E}\left[\mathcal{E}_t 1_{\{\mathcal{E}_t > K\}}\right] = \mathbb{P}\left(e^{B_t + \frac{t}{2}} > K\right) \quad \text{(from (1.5))}$$

$$= \mathbb{P}\left(B_t > \log(K) - \frac{t}{2}\right)$$

$$= \mathbb{P}\left(B_1 > \frac{\log(K)}{\sqrt{t}} - \frac{\sqrt{t}}{2}\right) \quad \text{(by scaling)}$$

$$= 1 - \mathcal{N}\left(\frac{\log(K)}{\sqrt{t}} - \frac{\sqrt{t}}{2}\right)$$

$$= \mathcal{N}\left(-\frac{\log(K)}{\sqrt{t}} + \frac{\sqrt{t}}{2}\right).$$

This is formula (1.17). The other formulae can be proven using similar arguments. Besides, formula (1.19) is a consequence of the following equalities:

$$\mathbb{E}\left[(\mathcal{E}_t - 1)^+\right] - \mathbb{E}\left[(\mathcal{E}_t - 1)^-\right] = \mathbb{E}\left[\mathcal{E}_t - 1\right] = 0$$

and

$$\mathbb{E}\left[|\mathcal{E}_t - 1|\right] = \mathbb{E}\left[(\mathcal{E}_t - 1)^+\right] + \mathbb{E}\left[(\mathcal{E}_t - 1)^-\right]$$

$$= 2\mathbb{E}\left[(\mathcal{E}_t - 1)^+\right]$$

$$= 2\left(\mathcal{N}\left(\frac{\sqrt{t}}{2}\right) - \mathcal{N}\left(-\frac{\sqrt{t}}{2}\right)\right) \quad \text{(from (1.16))}$$

$$= 2\mathbb{P}\left(B_1 \in \left[-\frac{\sqrt{t}}{2}, \frac{\sqrt{t}}{2}\right]\right)$$

$$= 2\mathbb{P}\left(B_1^2 \leq \frac{t}{4}\right).$$

## 1.2 The Black-Scholes Formula in Terms of First and Last Passage Times

### 1.2.1 A New Expression for the Black-Scholes Formula

The aim of this section is to give a new expression for the Black-Scholes formula making use of the first passage times $T_a^{(\nu)}$ and the last passage times $G_a^{(\nu)}$. More precisely, we have (see [47]):

**Theorem 1.2.** *For any* $K \geq 0$:

i) *The European put price admits the representation:*

$$\mathbb{E}\left[(K - \mathscr{E}_t)^+\right] = K\mathbb{P}\left(G_{\log(K)}^{(-1/2)} \leq t\right). \tag{1.20}$$

ii) *The European call price admits the representation:*

$$\mathbb{E}\left[(\mathscr{E}_t - K)^+\right] = \mathbb{E}\left[\mathscr{E}_t 1_{\{\mathscr{E}_t > K\}}\right] - K\mathbb{P}(\mathscr{E}_t > K)$$
$$= \mathbb{P}\left(G_{\log(K)}^{(1/2)} \leq t\right). \tag{1.21}$$

iii) *For* $K \geq 1$:

$$\mathbb{E}\left[\mathscr{E}_t 1_{\{\mathscr{E}_t > K\}}\right] + K\mathbb{P}(\mathscr{E}_t > K) = \mathbb{P}\left(T_{\log(K)}^{(1/2)} \leq t\right) \tag{1.22}$$

*while for* $K \leq 1$:

$$\mathbb{E}\left[\mathscr{E}_t 1_{\{\mathscr{E}_t < K\}}\right] + K\mathbb{P}(\mathscr{E}_t < K) = \mathbb{P}\left(T_{\log(K)}^{(1/2)} \leq t\right). \tag{1.23}$$

Of course, relations (1.21), (1.22) and (1.23) imply:
• For $K \geq 1$:

$$\mathbb{E}\left[\mathscr{E}_t 1_{\{\mathscr{E}_t > K\}}\right] = \frac{1}{2}\left(\mathbb{P}\left(T_{\log(K)}^{(1/2)} \leq t\right) + \mathbb{P}\left(G_{\log(K)}^{(1/2)} \leq t\right)\right) \tag{1.24}$$

and

$$K\mathbb{P}(\mathscr{E}_t > K) = \frac{1}{2}\mathbb{P}\left(T_{\log(K)}^{(1/2)} \leq t \leq G_{\log(K)}^{(1/2)}\right). \tag{1.25}$$

• For $0 \leq K \leq 1$:

$$\mathbb{E}\left[\mathscr{E}_t 1_{\{\mathscr{E}_t > K\}}\right] = \frac{1}{2}\left(1 + K - \mathbb{P}\left(T_{\log(K)}^{(1/2)} \leq t \leq G_{\log(K)}^{(1/2)}\right)\right) \tag{1.26}$$

and

$$K\mathbb{P}(\mathscr{E}_t > K) = \frac{1}{2}\left\{1 + K - \left(\mathbb{P}\left(T_{\log(K)}^{(1/2)} \leq t\right) + \mathbb{P}\left(G_{\log(K)}^{(1/2)} \leq t\right)\right)\right\}. \tag{1.27}$$

## 1.2.2 Comments

a) Formula (1.20) is the prototype of a more general formula that we shall develop in Chapter 2. More precisely, let $(M_t, t \geq 0)$ denote a positive, continuous local martingale, such that $M_0 = 1$ and $\lim_{t \to +\infty} M_t = 0$ a.s. We shall prove in Chapter 2, Theorem 2.1 that:

$$\mathbb{E}\left[(K - M_t)^+\right] = K\mathbb{P}\left(\mathscr{G}_K^{(M)} \leq t\right) \tag{1.28}$$

for all $K \geq 0$ and $t \geq 0$, with:

$$\mathscr{G}_K^{(M)} := \sup\{t \geq 0; M_t = K\}. \tag{1.29}$$

Hence, formula (1.20) is a particular case of (1.28) with $(M_t = \mathscr{E}_t, t \geq 0)$ since:

$$\mathscr{G}_K^{(\mathscr{E})} = G_{\log(K)}^{(-1/2)}. \tag{1.30}$$

b) Formula (1.21) can also be obtained from a more general formula that we shall prove in Chapter 2, Theorem 2.2. More precisely, let $(M_t, t \geq 0)$ a positive, continuous $(\Omega, (\mathscr{F}_t, t \geq 0), \mathscr{F}_\infty, \mathbb{P})$-martingale, such that $M_0 = 1$ and $\lim_{t \to +\infty} M_t = 0$ a.s. The relative absolute continuity formula

$$\mathbb{P}_{|\mathscr{F}_t}^{(M)} := M_t \cdot \mathbb{P}_{|\mathscr{F}_t} \tag{1.31}$$

induces a probability on $(\Omega, \mathscr{F}_\infty)$ (see Azéma-Jeulin [2] for some precisions) and, for every $K \geq 0$ and $t \geq 0$, the following relation:

$$\mathbb{E}\left[(M_t - K)^+\right] = \mathbb{P}^{(M)}\left(\mathscr{G}_K^{(M)} \leq t\right) \tag{1.32}$$

holds. Formula (1.21) is then a particular case of (1.32) since, from the Cameron-Martin formula, under the probability $\mathbb{P}^{(\mathscr{E})}$:

$$\mathscr{G}_K^{(\mathscr{E})} \stackrel{(law)}{=} G_{\log(K)}^{(1/2)}.$$

c) On the other hand, formulae (1.22) and (1.23) do not have a plain generalization to a larger class of martingales. Indeed, as we shall see in the proof below, formulae (1.22) and (1.23) rely on Désiré André's symmetry principle for Brownian motion, a principle which an "ordinary" martingale does not satisfy in general.[1]

d) When $(M_t, t \geq 0)$ is a martingale, the functions: $x \mapsto (K - x)^+$ and $x \mapsto (x - K)^+$ being convex, the processes $\left(\frac{1}{K}(K - M_t)^+, t \geq 0\right)$ and $((M_t - K)^+, t \geq 0)$ are sub-

---

[1] A class of martingales satisfying the reflection principle consists of the Ocone martingales, i.e. martingales whose Dambis-Dubins-Schwarz representation $M_t = \beta_{\langle M \rangle_t}$ features a Brownian motion $\beta$ independent of $\langle M \rangle$. Many examples are known, such as: the Winding number $(\theta_t)$ of a planar BM, Lévy's stochastic area $\mathscr{A}_t = \int_0^t X_s dY_s - Y_s dX_s \cdots$ (see [20]).

martingales. Hence, the functions $\psi_1$ and $\psi_2$ defined by: $\psi_1(t) := \frac{1}{K}\mathbb{E}\left[(K - M_t)^+\right]$ and $\psi_2(t) := \mathbb{E}\left[(M_t - K)^+\right]$ are increasing (and continuous) functions. Then, applying the Dominated Convergence Theorem, we obtain

$$\lim_{t \to \infty} \frac{1}{K}\mathbb{E}\left[(K - M_t)^+\right] = 1,$$

and, using the relation $1 - \frac{1}{K} = \psi_1(t) - \frac{1}{K}\psi_2(t)$,

$$\lim_{t \to \infty} \mathbb{E}\left[(M_t - K)^+\right] = 1.$$

Therefore, there exists two positive random variables $Z_1$ and $Z_2$ such that:

$$\frac{1}{K}\mathbb{E}\left[(K - M_t)^+\right] = \mathbb{P}(Z_1 \leq t) \quad \text{and} \quad \mathbb{E}\left[(M_t - K)^+\right] = \mathbb{P}(Z_2 \leq t).$$

In the case $M = \mathcal{E}$, formulae (1.20) and (1.21) (and more generally (1.28) and (1.32)) make it possible to identify the laws of $Z_1$ and $Z_2$:

$$Z_1 \overset{(law)}{=} G_{\log(K)}^{(-1/2)} \quad , \quad Z_2 \overset{(law)}{=} G_{\log(K)}^{(1/2)}.$$

More generally, we shall prove in Chapter 6 that the function $(\mathbb{E}\left[(K - \mathcal{E}_t)^+\right]; 0 \leq K \leq 1, t \geq 0)$ is the distribution function of a couple of r.v.'s taking values in $[0, 1] \times \mathbb{R}^+$, whose law will be explicitly described.

### 1.2.3 Proof of Theorem 1.2

We first prove (1.20) and (1.21)
• We shall show that, for any $a$ and $\mu$ in $\mathbb{R}$:

$$\mathbb{P}\left(G_a^{(\mu)} \geq t | \mathscr{F}_t\right) = \left(\exp\left\{2\mu\left(a - B_t^{(\mu)}\right)\right\}\right) \wedge 1. \qquad (1.33)$$

We now prove (1.33). For $\mu \geq 0$, we have, applying the Markov property to the process $\left(B_t^{(\mu)}, t \geq 0\right)$:

$$\mathbb{P}\left(G_a^{(\mu)} \geq t | \mathscr{F}_t\right) = \mathbb{P}\left(\inf_{s \geq 0}\left(x + B_s^{(\mu)}\right) \leq a\right) \quad \text{with } x = B_t^{(\mu)}$$

$$= \mathbb{P}\left(\inf_{s \geq 0} B_s^{(\mu)} \leq a - x\right)$$

$$= \left(\exp\left\{2\mu\left(a - B_t^{(\mu)}\right)\right\}\right) \wedge 1 \quad \text{(from (1.10))}.$$

The proof for $\mu \leq 0$ is similar.
• We apply (1.33) with $\mu = -1/2$ and $a = \log(K)$:

$$\mathbb{P}\left(G_{\log(K)}^{(-1/2)} \geq t|\mathscr{F}_t\right) = \left(\frac{1}{K}\exp\left\{B_t - \frac{t}{2}\right\}\right) \wedge 1, \tag{1.34}$$

which implies:

$$\mathbb{P}\left(G_{\log(K)}^{(-1/2)} \leq t\right) = \mathbb{E}\left[\left(1 - \frac{1}{K}\exp\left(B_t - \frac{t}{2}\right)\right)^+\right]$$

$$= \frac{1}{K}\mathbb{E}\left[(K - \mathscr{E}_t)^+\right].$$

This is relation (1.20).

- We apply (1.33) with $\mu = 1/2$ and $a = \log(K)$:

$$\mathbb{P}\left(G_{\log(K)}^{(1/2)} \geq t|\mathscr{F}_t\right) = \left\{K\exp\left(-B_t - \frac{t}{2}\right)\right\} \wedge 1, \tag{1.35}$$

which implies:

$$\mathbb{P}\left(G_{\log(K)}^{(1/2)} \leq t\right) = \mathbb{E}\left[\left(1 - K\exp\left(-B_t - \frac{t}{2}\right)\right)^+\right]$$

$$= \mathbb{E}\left[e^{B_t - \frac{t}{2}}\left(e^{-B_t + \frac{t}{2}} - K\right)^+\right] \quad (\text{since } B_t \overset{(law)}{=} -B_t)$$

$$= \mathbb{E}\left[\left(e^{-(B_t + t) + \frac{t}{2}} - K\right)^+\right] \quad (\text{from the Cameron-Martin formula})$$

$$= \mathbb{E}\left[(\mathscr{E}_t - K)^+\right] \quad (\text{since } B_t \overset{(law)}{=} -B_t).$$

This is formula (1.21).

We now prove (1.25)
Using again (1.33) we see that, for $K \geq 1$, (1.25) is equivalent to:

$$K\mathbb{P}\left(B_t - \frac{t}{2} > \log(K)\right) = \frac{1}{2}\mathbb{E}\left[1_{\left\{T_{\log(K)}^{(1/2)} \leq t\right\}}\left(\frac{K}{e^{B_t + t/2}} \wedge 1\right)\right]. \tag{1.36}$$

We now use the Cameron-Martin formula on both sides to reduce the statement of (1.36) to a statement about standard Brownian motion $(B_t, t \geq 0)$, for which we denote: $S_t := \sup_{s \leq t} B_s$. (1.36) is then equivalent to:

$$K\mathbb{E}\left[1_{\{B_t > \log(K)\}}e^{-\frac{B_t}{2}}\right] = \frac{1}{2}\mathbb{E}\left[1_{\{S_t > \log(K)\}}\left(Ke^{-B_t} \wedge 1\right)e^{\frac{B_t}{2}}\right]. \tag{1.37}$$

We now decompose the RHS of (1.37) in a sum of two quantities:

$$\frac{1}{2}\left(\mathbb{E}\left[1_{\{S_t > \log(K)\}}1_{\{B_t > \log(K)\}}Ke^{-\frac{B_t}{2}}\right] + \mathbb{E}\left[1_{\{S_t > \log(K)\}}1_{\{B_t < \log(K)\}}e^{\frac{B_t}{2}}\right]\right).$$

Thus, (1.37) gets simplified to the equivalent form:

$$KE\left[1_{\{B_t>\log(K)\}}e^{-\frac{B_t}{2}}\right] = E\left[1_{\{S_t>\log(K)\}}1_{\{B_t<\log(K)\}}e^{\frac{B_t}{2}}\right] \tag{1.38}$$

which, taking $x = \log(K)$, may be written as:

$$E\left[1_{\{B_t>x\}}\exp\left(x-\frac{B_t}{2}\right)\right] = E\left[1_{\{S_t>x>B_t\}}e^{\frac{B_t}{2}}\right]. \tag{1.39}$$

We now show (1.39) from the right to the left, as a consequence of the reflection principle of Désiré André. Conditionally on $\mathscr{F}_{T_x}$, and $T_x < t$, we have:

$$B_t - x = \widehat{B}_{(t-T_x)}, \text{ with } \widehat{B} \text{ independent from } \mathscr{F}_{T_x};$$

hence, under this conditioning, the reflection principle boils down to:

$$B_t - x \stackrel{(law)}{=} -(B_t - x). \tag{1.40}$$

Thus, the RHS of (1.39) rewrites:

$$E\left[1_{\{T_x<t\}}1_{\{B_t-x<0\}}\exp\left(\frac{1}{2}\{x+(B_t-x)\}\right)\right]$$

$$= E\left[1_{\{T_x<t\}}1_{\{B_t-x>0\}}\exp\left(\frac{1}{2}\{x-(B_t-x)\}\right)\right]$$

$$= E\left[1_{\{B_t>x\}}\exp\left(x-\frac{B_t}{2}\right)\right],$$

which is the LHS of (1.39). This proves (1.25), and with the help of (1.21), it also proves (1.22).

We now prove that (1.22) implies (1.23) (with $0 \le K \le 1$)

We introduce the probability $\mathbb{P}^{(\mathscr{E})}$ defined by:

$$\mathbb{P}^{(\mathscr{E})}_{|\mathscr{F}_t} = \mathscr{E}_t \cdot \mathbb{P}_{|\mathscr{F}_t}. \tag{1.41}$$

We note that from the Cameron-Martin formula, under $\mathbb{P}^{(\mathscr{E})}$, $(B_t, t \ge 0)$ is a Brownian motion with drift $+1$, so that $\frac{1}{\mathscr{E}_t} := \widehat{\mathscr{E}}_t = \exp\left(\widehat{B}_t - \frac{t}{2}\right)$ for a new Brownian motion $(\widehat{B}_t, t \ge 0)$. Thus, the LHS of (1.23) writes:

$$\mathbb{P}^{(\mathscr{E})}\left(\mathscr{E}_t < K\right) + K\mathbb{E}^{(\mathscr{E})}\left[\widehat{\mathscr{E}}_t 1_{\{\mathscr{E}_t < K\}}\right] = \mathbb{P}^{(\mathscr{E})}\left(\widehat{\mathscr{E}}_t > \frac{1}{K}\right) + K\mathbb{E}^{(\mathscr{E})}\left[\widehat{\mathscr{E}}_t 1_{\{\widehat{\mathscr{E}}_t > \frac{1}{K}\}}\right]$$

$$= K\left(\mathbb{E}^{(\mathscr{E})}\left[\widehat{\mathscr{E}}_t 1_{\{\widehat{\mathscr{E}}_t > \frac{1}{K}\}}\right] + \frac{1}{K}\mathbb{P}^{(\mathscr{E})}\left(\widehat{\mathscr{E}}_t > \frac{1}{K}\right)\right)$$

$$= K\mathbb{P}^{(\mathscr{E})}\left(\widehat{T}^{(1/2)}_{-\log(K)} \le t\right)$$

(applying (1.22) with $1/K$ instead of $K$)

$$= K\mathbb{P}^{(\mathscr{E})}\left(\widehat{T}^{(-1/2)}_{\log(K)} \le t\right) \quad \text{(by symmetry)}$$

$$= \mathbb{P}\left(T^{(1/2)}_{\log(K)} \le t\right)$$

since, from the Cameron-Martin absolute continuity relationship:

$$\mathbb{P}^{(-\nu)}_{|\mathscr{F}_{T_a} \cap (T_a < \infty)} = \exp(-2\nu a) \cdot \mathbb{P}^{(\nu)}_{|\mathscr{F}_{T_a}} \tag{1.42}$$

where $\mathbb{P}^{(\nu)}$ (resp. $\mathbb{P}^{(-\nu)}$) denotes the law of the Brownian motion with drift $\nu$ (resp. $-\nu$). $\qquad\square$

### 1.2.4 On the Agreement Between the Classical Black-Scholes Formula (Theorem 1.1) and our Result (Theorem 1.2)

We now check in an elementary manner the equality between the classical Black-Scholes formulae given by Theorem 1.1 and the formulae given by Theorem 1.2.

• We first prove that:

$$\mathscr{N}\left(-\frac{\log(K)}{\sqrt{t}} + \frac{\sqrt{t}}{2}\right) - K\mathscr{N}\left(-\frac{\log(K)}{\sqrt{t}} - \frac{\sqrt{t}}{2}\right) = \mathbb{P}\left(G^{(1/2)}_{\log(K)} \le t\right) \tag{1.43}$$

$$\left(= \mathbb{E}\left[(\mathscr{E}_t - K)^+\right]\right).$$

We assume $K \ge 1$. Since both sides of (1.43) are equal to 0 for $t = 0$, we only need to check that the derivatives in $t$ are equal. On the one hand, we have:

$$\frac{\partial}{\partial t}\left\{\mathscr{N}\left(-\frac{\log(K)}{\sqrt{t}} + \frac{\sqrt{t}}{2}\right) - K\mathscr{N}\left(-\frac{\log(K)}{\sqrt{t}} - \frac{\sqrt{t}}{2}\right)\right\}$$

$$= \frac{K}{2\sqrt{2\pi t}}\exp\left(-\frac{1}{2t}\left(\log(K) + \frac{t}{2}\right)^2\right)$$

and, on the other hand, from (1.12):

$$\frac{\partial}{\partial t}\mathbb{P}\left(G_{\log(K)}^{(1/2)} \leq t\right) = \frac{1}{2\sqrt{2\pi t}}\exp\left(-\frac{1}{2t}\left(\log(K)-\frac{t}{2}\right)^2\right)$$

$$= \frac{K}{2\sqrt{2\pi t}}\exp\left(-\frac{1}{2t}\left(\log(K)+\frac{t}{2}\right)^2\right). \qquad (1.44)$$

This shows (1.43) when $K \geq 1$. The case $K \leq 1$ can be proven in the same way.

- We now prove that, for $K \geq 1$:

$$\mathcal{N}\left(-\frac{\log(K)}{\sqrt{t}}+\frac{\sqrt{t}}{2}\right)+K\mathcal{N}\left(-\frac{\log(K)}{\sqrt{t}}-\frac{\sqrt{t}}{2}\right) = \mathbb{P}\left(T_{\log(K)}^{(1/2)} \leq t\right) \qquad (1.45)$$

$$(= \mathbb{E}\left[\mathcal{E}_t 1_{\{\mathcal{E}_t > K\}}\right] + K\mathbb{P}(\mathcal{E}_t > K)).$$

Once again, both sides of (1.45) equal 0 when $t = 0$, so we only need to check that the derivatives in $t$ are equal. We have:

$$\frac{\partial}{\partial t}\left\{\mathcal{N}\left(-\frac{\log(K)}{\sqrt{t}}+\frac{\sqrt{t}}{2}\right)+K\mathcal{N}\left(-\frac{\log(K)}{\sqrt{t}}-\frac{\sqrt{t}}{2}\right)\right\}$$

$$= \frac{K\log(K)}{\sqrt{2\pi t^3}}\exp\left(-\frac{1}{2t}\left(\log(K)+\frac{t}{2}\right)^2\right)$$

while, from (1.11):

$$\frac{\partial}{\partial t}\mathbb{P}\left(T_{\log(K)}^{(1/2)} \leq t\right) = \frac{\log(K)}{\sqrt{2\pi t^3}}\exp\left(-\frac{1}{2t}\left(\log(K)-\frac{t}{2}\right)^2\right)$$

$$= \frac{K\log(K)}{\sqrt{2\pi t^3}}\exp\left(-\frac{1}{2t}\left(\log(K)+\frac{t}{2}\right)^2\right). \qquad (1.46)$$

The agreement between the other formulae of Theorem 1.1 and Theorem 1.2 can be obtained by similar computations. In fact, these are the precise manipulations which led us to believe in the truth of Theorem 1.2.

## 1.2.5  A Remark on Theorem 1.2 and Time Inversion

We come back to the time inversion property of BM, in order to throw another light upon our key result (1.20), which relates the European call price with the cumulative function of last Brownian passage times. (This paragraph has been partly inspired from unpublished notes by Peter Carr [15].) Indeed, a variant of (1.20) is the following:

For every $t \geq 0$, $K \geq 0$ and $\phi : \mathcal{C}([0,t], \mathbb{R}^+)$, measurable,

$$\mathbb{E}\left[\phi(B_u, u \leq t)(K-\mathcal{E}_t)^+\right] = K\mathbb{E}\left[\phi(B_u, u \leq t)1_{\{\mathcal{G}_K^{(\mathcal{E})} \leq t\}}\right] \qquad (1.47)$$

where $\mathcal{G}_K^{(\mathcal{E})} := \sup\{u \geq 0; \ \mathcal{E}_t = K\}$.

Writing (1.47) in terms of the Brownian motion $(\widehat{B}_h, h \geq 0)$ such that: $B_u = u\widehat{B}_{(1/u)}$, and setting $s = 1/t$, it is clearly seen that (1.47) is equivalent to:

$$\left(K - \exp\left(\frac{1}{s}\widehat{B}_s - \frac{1}{2s}\right)\right)^+ = K\mathbb{P}\left(\widehat{T}_{1/2}^{(-\log(K))} \geq s|\widehat{B}_s\right),$$

where $\widehat{T}_a^{(\nu)} := \inf\{u \geq 0, \widehat{B}_u + \nu u = a\}$. Since hats are no longer necessary for our purpose, we drop them, and we now look for an independent proof of:

$$\left(K - \exp\left(\frac{x}{s} - \frac{1}{2s}\right)\right)^+ = K\mathbb{P}\left(T_{1/2}^{(-\log(K))} \geq s|B_s = x\right). \tag{1.48}$$

On the RHS of (1.48), we may replace $\{B_s = x\}$ by $\{B_s - s\log(K) = x - s\log(K)\}$. Now, as a consequence of the Cameron-Martin relationship, the conditional expectation:

$$\mathbb{E}\left[F(B_u - \nu u, u \leq s)|B_s - \nu s = y\right]$$

does not depend on $\nu$; hence (1.48) is equivalent to:

$$\left(1 - \frac{1}{K}\exp\left(\frac{x}{s} - \frac{1}{2s}\right)\right)^+ = \mathbb{P}\left(T_{1/2} \geq s|B_s = x - s\log(K)\right),$$

which simplifies to:

$$\left(1 - \exp\left(\frac{y - \frac{1}{2}}{s}\right)\right)^+ = \mathbb{P}\left(\sup_{u \leq s} B_u < \frac{1}{2}|B_s = y\right),$$

or, by scaling:

$$\left(1 - \exp\left\{\frac{1}{\sqrt{s}}\left(\frac{y}{\sqrt{s}} - \frac{1}{2\sqrt{s}}\right)\right\}\right)^+ = \mathbb{P}\left(\sup_{u \leq 1} B_u < \frac{1}{2\sqrt{s}}|B_1 = \frac{y}{\sqrt{s}}\right).$$

This is equivalent to:

$$\left(1 - e^{2\sigma(y-\sigma)}\right)^+ = \mathbb{P}\left(\sup_{u \leq 1} B_u < \sigma|B_1 = y\right) \tag{1.49}$$

for $\sigma \geq 0$ and $y \in \mathbb{R}$. This formula is trivial for $\sigma < y$, and, for $\sigma \geq y$, it follows from the classical formula:

$$\mathbb{P}\left(\sup_{u \leq 1} B_u \in d\sigma, B_1 \in da\right) = \frac{da\,d\sigma}{\sqrt{2\pi}}2(2\sigma - a)e^{-\frac{(2\sigma-a)^2}{2}}1_{\{a < \sigma, \sigma \geq 0\}}. \tag{1.50}$$

The interested reader may refer to Chapter 5, proof of Point (i) of Theorem 5.1, for similar computations. Let us also mention that formula (1.49) plays an important role in numerical computations, see [61].

**Exercise 1.1 (On the law of $(S_t, B_t)$).**

Let $(B_t, t \geq 0)$ be a standard Brownian motion and define $S_t := \sup_{s \leq t} B_s$.

**1)** Prove that, for $\sigma \geq 0$, $\sigma \geq a$:

$$\int_0^\infty e^{-\lambda t} \mathbb{P}(S_t > \sigma, B_t < a)\, dt = \mathbb{E}\left[e^{-\lambda T_\sigma}\right] \int_0^\infty e^{-\lambda u} \mathbb{P}(B_u \leq a - \sigma)\, du$$

where $T_\sigma := \inf\{t \geq 0, X_t = \sigma\}$.

**2)** Compute $\mathbb{E}\left[e^{-\lambda T_\sigma}\right]$.

<u>Hint:</u> Consider the martingale $\left(\exp\left(\lambda B_t - \frac{\lambda^2}{2}t\right), t \geq 0\right)$.

**3)** Deduce then that the joint density of $(S_t, B_t)$, $f_t(\sigma, a)$, satisfies:

$$\int_0^\infty e^{-\lambda u} f_u(\sigma, a)\, du = 2\mathbb{E}\left[e^{-\lambda T_{2\sigma - a}}\right].$$

**4)** Finally, recover formula (1.50).

## 1.3 Extension of Theorem 1.2 to an Arbitrary Index $\nu$

### 1.3.1 Statement of the Main Result

So far, we have focused on the Black-Scholes formula relative to the martingale $\left(\mathscr{E}_t = e^{B_t - \frac{t}{2}}, t \geq 0\right)$. We shall now extend these results to the family of martingales $\left(\mathscr{E}_t^{(\nu)} := \exp\left(\nu B_t - \frac{\nu^2 t}{2}\right), t \geq 0\right)_{\nu \in \mathbb{R}}$. Of course, we have, by scaling and symmetry, for the three following processes:

$$\mathscr{E}_t^{(\nu)} \overset{(law)}{=} \mathscr{E}_t^{(-\nu)} \overset{(law)}{=} \mathscr{E}_{\nu^2 t} \qquad (\nu \neq 0) \tag{1.51}$$

hence,

$$\begin{cases} \dfrac{1}{4\nu^2} G_{2a\nu}^{(1/2)} \overset{(law)}{=} G_a^{(\nu)} & (\nu \neq 0) \\[3mm] \dfrac{1}{4\nu^2} T_{2a\nu}^{(1/2)} \overset{(law)}{=} T_a^{(\nu)} & (\nu \neq 0). \end{cases} \tag{1.52}$$

The counterpart of Theorem 1.2 in this new framework writes:

**Theorem 1.3.** *For all $\nu$ and $K, t > 0$:*

*i)* *The European put price associated to the martingale $(\mathscr{E}_t^{(-2\nu)}, t \geq 0)$ equals:*

$$\mathbb{E}\left[\left(K^{-2\nu} - \mathscr{E}_t^{(-2\nu)}\right)^+\right] = K^{-2\nu} \mathbb{P}\left(G_{\log(K)}^{(\nu)} \leq t\right). \tag{1.53}$$

ii) *The European call price associated to the martingale* $(\mathscr{E}_t^{(2\nu)}, t \geq 0)$ *equals:*

$$\mathbb{E}\left[\left(\mathscr{E}_t^{(2\nu)} - K^{2\nu}\right)^+\right] = \mathbb{P}\left(G_{\log(K)}^{(\nu)} \leq t\right). \tag{1.54}$$

iii) *For every* $K \geq 0$, *if* $\nu \log(K) > 0$:

$$\mathbb{E}\left[\mathscr{E}_t^{(2\nu)} 1_{\{\mathscr{E}_t^{(2\nu)} > K^{2\nu}\}}\right] + K^{2\nu}\mathbb{P}\left(\mathscr{E}_t^{(2\nu)} > K^{2\nu}\right) = \mathbb{P}\left(T_{\log(K)}^{(\nu)} \leq t\right), \tag{1.55}$$

*while, if* $\nu \log(K) < 0$:

$$\mathbb{E}\left[\mathscr{E}_t^{(2\nu)} 1_{\{\mathscr{E}_t^{(2\nu)} < K^{2\nu}\}}\right] + K^{2\nu}\mathbb{P}\left(\mathscr{E}_t^{(2\nu)} < K^{2\nu}\right) = \mathbb{P}\left(T_{\log(K)}^{(\nu)} \leq t\right). \tag{1.56}$$

*Furthermore, if* $K = 1$ *and* $\nu \neq 0$:

$$\begin{aligned}
\mathbb{P}\left(T_0^{(\nu)} \leq t\right) &= \mathbb{E}\left[\mathscr{E}_t^{(2\nu)} 1_{\{\mathscr{E}_t^{(2\nu)} > 1\}}\right] + \mathbb{P}(\mathscr{E}_t^{(2\nu)} > 1) \\
&= \mathbb{E}\left[\mathscr{E}_t^{(2\nu)} 1_{\{\mathscr{E}_t^{(2\nu)} < 1\}}\right] + \mathbb{P}(\mathscr{E}_t^{(2\nu)} < 1) \qquad (1.57) \\
&= 1.
\end{aligned}$$

### 1.3.2 Some Comments on Theorem 1.3

• Theorem 1.2 may be recovered from Theorem 1.3 by taking $\nu = -1/2$ or $\nu = 1/2$ (recall that $\left(\mathscr{E}_t^{(\nu)}, t \geq 0\right) \overset{(law)}{=} \left(\mathscr{E}_t^{(-\nu)}, t \geq 0\right)$ ). Besides, formulae (1.53) and (1.54) are seen to coincide with (1.20) and (1.21). Taking limits as $t$ tends to 0 in (1.53) and to $+\infty$ in (1.55) and (1.56), we obtain:

- if $\nu \log(K) > 0$:

$$\mathbb{P}\left(G_{\log(K)}^{(\nu)} = 0\right) = 0 = \mathbb{P}\left(T_{\log(K)}^{(\nu)} = +\infty\right),$$

- if $\nu \log(K) < 0$:

$$\mathbb{P}\left(G_{\log(K)}^{(\nu)} > 0\right) = \mathbb{P}\left(T_{\log(K)}^{(\nu)} < +\infty\right) = K^{2\nu}.$$

These results coincide with (1.10).

• The relations (1.55) and (1.56) for the case $\nu = 0$ $\left(\text{i.e. } \mathscr{E}_t^{(-\nu)} = \mathscr{E}_t^{(0)} = 1\right)$ may be obtained by passing to the limit as $\nu$ tends to 0 on both sides. For example, we have from (1.55), for $\nu > 0$ and $K \geq 1$:

$$\mathbb{P}\left(T^{(\nu)}_{\log(K)} < s\right) = \mathbb{E}\left[e^{2\nu(B_s - \nu s)} 1_{\{e^{2\nu(B_s - \nu s)} > K^{2\nu}\}}\right] + \mathbb{P}(e^{2\nu(B_s - \nu s)} > K^{2\nu})$$

$$= \mathbb{E}\left[e^{2\nu(B_s - \nu s)} 1_{\{B_s - \nu s > \log(K)\}}\right] + \mathbb{P}(B_s - \nu s > \log(K))$$

$$\xrightarrow[\nu \downarrow 0]{} 2\mathbb{P}(B_s > \log(K)).$$

On the other hand, since $T_a^{(0)} \stackrel{(law)}{=} \dfrac{a^2}{B_1^2}$:

$$\mathbb{P}\left(T^{(0)}_{\log(K)} < s\right) = \mathbb{P}\left(\frac{(\log(K))^2}{B_1^2} < s\right) = \mathbb{P}(|B_s| > \log(K)) = 2\mathbb{P}(B_s > \log(K)).$$

(Note that this is once again a form of the reflection principle.) A similar analysis holds for (1.56) as we let $\nu$ tend to 0.

### 1.3.3 A Short Proof of Theorem 1.3

We prove (1.54)
We have:

$$\mathbb{P}\left(G^{(\nu)}_{\log(K)} < t\right) = \mathbb{P}\left(\frac{1}{4\nu^2} G^{(1/2)}_{\log(K^{2\nu})} < t\right) \quad \text{(from (1.52))}$$

$$= \mathbb{P}\left(G^{(1/2)}_{\log(K^{2\nu})} < 4\nu^2 t\right)$$

$$= \mathbb{E}\left[(\mathscr{E}_{4\nu^2 t} - K^{2\nu})^+\right] \quad \text{(from Theorem 1.2)}$$

$$= \mathbb{E}\left[(\mathscr{E}_t^{(2\nu)} - K^{2\nu})^+\right] \quad \text{(from (1.51)).}$$

Similar arguments establish (1.53), (1.55) and (1.56).

We now prove (1.57)
From the Cameron-Martin formula (1.5), it holds:

$$\mathbb{E}\left[\mathscr{E}_t^{(2\nu)} 1_{\{\mathscr{E}_t^{(2\nu)} < 1\}}\right] = \mathbb{P}\left(e^{2\nu(B_t + 2\nu t) - 2\nu^2 t} < 1\right)$$

$$= \mathbb{P}\left(e^{-2\nu B_t - 2\nu^2 t} > 1\right)$$

$$= \mathbb{P}\left(\mathscr{E}_t^{(2\nu)} > 1\right).$$

Hence:

$$\mathbb{E}\left[\mathscr{E}_t^{(2\nu)} 1_{\{\mathscr{E}_t^{(2\nu)} > 1\}}\right] = \mathbb{P}\left(\mathscr{E}_t^{(2\nu)} < 1\right).$$

Consequently, we obtain:

$$\mathbb{E}\left[\mathscr{E}_t^{(2\nu)} 1_{\{\mathscr{E}_t^{(2\nu)}>1\}}\right] + \mathbb{P}\left(\mathscr{E}_t^{(2\nu)} > 1\right) = \mathbb{E}\left[\mathscr{E}_t^{(2\nu)} 1_{\{\mathscr{E}_t^{(2\nu)}<1\}}\right] + \mathbb{P}\left(\mathscr{E}_t^{(2\nu)} < 1\right),$$

and so:

$$\mathbb{E}\left[\mathscr{E}_t^{(2\nu)} 1_{\{\mathscr{E}_t^{(2\nu)}>1\}}\right] + \mathbb{E}\left[\mathscr{E}_t^{(2\nu)} 1_{\{\mathscr{E}_t^{(2\nu)}<1\}}\right] + \mathbb{P}\left(\mathscr{E}_t^{(2\nu)} < 1\right) + \mathbb{P}\left(\mathscr{E}_t^{(2\nu)} > 1\right) = 2,$$

which establishes (1.57).

## 1.4 Another Formulation of the Black-Scholes Formula

### 1.4.1 Statement of the Result

We shall now give yet another formulation of the Black-Scholes formula.

**Theorem 1.4.** *For all $K \geq 0$ and $\nu \in \mathbb{R}$:*

$$\mathbb{E}\left[(\mathscr{E}_t^{(2\nu)} - K)^{\pm}\right] = (1 - K)^{\pm} + \sqrt{K}\mathbb{E}\left[1_{\{B_1^2 \leq \nu^2 t\}} \exp\left(-\frac{(\log(K))^2}{8B_1^2}\right)\right]. \quad (1.58)$$

Of course, (1.58) is a generalization of (1.19), which is (1.58) with $K = 1$ and $\nu = 1/2$.

### 1.4.2 First Proof of Theorem 1.4

Since $\left(\mathscr{E}_t^{(2\nu)}, t \geq 0\right) \overset{(law)}{=} \left(\mathscr{E}_{4\nu^2 t}, t \geq 0\right)$, it suffices to show (1.58) for $\nu = 1/2$. The proof will hinge upon the use of the local time of the martingale $(\mathscr{E}_t, t \geq 0)$. Let $\left(L_t^K, t \geq 0\right)$ denote the local time at level $K$ of the martingale $(\mathscr{E}_t, t \geq 0)$. For any positive Borel function $f$, the occupation density formula writes:

$$\int_0^t f(\mathscr{E}_s)d\langle\mathscr{E}\rangle_s = \int_0^\infty f(K)L_t^K dK \quad (1.59)$$

where $(\langle\mathscr{E}\rangle_t, t \geq 0)$ denotes the increasing process of the martingale $(\mathscr{E}_t, t \geq 0)$. Since, from Itô's formula:

$$\mathscr{E}_t = 1 + \int_0^t \mathscr{E}_s dB_s, \quad (1.60)$$

we have $d\langle\mathscr{E}\rangle_s = (\mathscr{E}_s)^2 ds$. Hence:

$$\int_0^t \mathbb{E}\left[f(\mathscr{E}_s)(\mathscr{E}_s)^2\right] ds = \int_0^\infty f(K)\mathbb{E}\left[L_t^K\right] dK. \quad (1.61)$$

As (1.61) holds for all positive Borel functions $f$, it follows that:

$$\mathbb{E}\left[L_t^K\right] = K \int_0^t \frac{ds}{\sqrt{2\pi s}} \exp\left(-\frac{1}{2s}\left(\log(K) + \frac{s}{2}\right)^2\right) \tag{1.62}$$

where we have used for the density of $\mathcal{E}_s$:

$$f_{\mathcal{E}_s}(z) = \frac{1}{z}\frac{1}{\sqrt{2\pi s}} e^{-\frac{1}{2s}\left(\log(z) + \frac{s}{2}\right)^2} 1_{\{z \geq 0\}}. \tag{1.63}$$

On developing the square in the exponential in (1.62), we get:

$$\mathbb{E}\left[L_t^K\right] = \frac{\sqrt{K}}{\sqrt{2\pi}} \int_0^t \frac{ds}{\sqrt{s}} \exp\left(-\frac{(\log(K))^2}{2s} - \frac{s}{8}\right). \tag{1.64}$$

On the other hand, the Itô-Tanaka formula yields:

$$\mathbb{E}\left[(\mathcal{E}_t - K)^{\pm}\right] = (1 - K)^{\pm} + \frac{1}{2}\mathbb{E}\left[L_t^K\right]$$

$$= (1 - K)^{\pm} + \frac{\sqrt{K}}{2\sqrt{2\pi}} \int_0^t \frac{ds}{\sqrt{s}} \exp\left(-\frac{(\log(K))^2}{2s} - \frac{s}{8}\right) \quad \text{(from (1.64))}$$

$$= (1 - K)^{\pm} + \frac{\sqrt{K}}{2\sqrt{2\pi}} 4\sqrt{\frac{\pi}{2}} \mathbb{E}\left[1_{\{4B_1^2 \leq t\}} \exp\left(-\frac{(\log(K))^2}{8B_1^2}\right)\right], \tag{1.65}$$

where we have used the density of $B_1^2$:

$$f_{B_1^2}(z) = \frac{1}{\sqrt{2\pi z}} e^{-\frac{z}{2}} 1_{\{z \geq 0\}}.$$

Formula (1.58) has thus been proven. $\qquad\square$

## 1.4.3 A Second Proof of Theorem 1.4

The result is plainly true for $\nu = 0$ (since $\mathcal{E}_t^{(0)} = 1$), and, as both sides only depend on the absolute value of $\nu$, it is enough to consider the case $\nu > 0$. The derivative, with respect to $\nu$, of the RHS of (1.58) equals:

$$\frac{\partial}{\partial \nu}\left(\sqrt{K}\mathbb{E}\left[1_{\{B_1^2 \leq \nu^2 t\}} \exp\left(-\frac{(\log(K))^2}{8B_1^2}\right)\right]\right)$$

$$= 2\sqrt{K}\frac{\partial}{\partial \nu} \int_0^{\nu\sqrt{t}} \exp\left(-\frac{(\log(K))^2}{8x^2} - \frac{x^2}{2}\right) \frac{dx}{\sqrt{2\pi}} \tag{1.66}$$

$$= \frac{2\sqrt{Kt}}{\sqrt{2\pi}} \exp\left(-\frac{(\log(K))^2}{8\nu^2 t} - \frac{\nu^2 t}{2}\right). \tag{1.67}$$

On the other hand, we may directly evaluate the LHS of (1.58) using the Black-Scholes formula as:

$$\mathbb{E}\left[\left(\mathcal{E}_t^{(2\nu)} - K\right)^+\right] = \mathcal{N}\left(-\frac{\log(K)}{2\nu\sqrt{t}} + \nu\sqrt{t}\right) - K\mathcal{N}\left(-\frac{\log(K)}{2\nu\sqrt{t}} - \nu\sqrt{t}\right).$$

Taking partial derivatives with respect to $\nu$, we get:

$$
\frac{\partial}{\partial\nu}\mathbb{E}\left[\left(\mathcal{E}_t^{(2\nu)} - K\right)^+\right] = n\left(-\frac{\log(K)}{2\nu\sqrt{t}} + \nu\sqrt{t}\right)\left(\frac{\log(K)}{2\nu^2\sqrt{t}} + \sqrt{t}\right)
$$
$$
- Kn\left(-\frac{\log(K)}{2\nu\sqrt{t}} - \nu\sqrt{t}\right)\left(\frac{\log(K)}{2\nu^2\sqrt{t}} - \sqrt{t}\right) \qquad (1.68)
$$

where $n(x) = \mathcal{N}'(x) = \frac{1}{\sqrt{2\pi}}e^{-\frac{x^2}{2}}$. We now recognize:

$$
\begin{cases}
n\left(-\dfrac{\log(K)}{2\nu\sqrt{t}} + \nu\sqrt{t}\right) = \exp\left(-\dfrac{(\log(K))^2}{8\nu^2 t} - \dfrac{\nu^2 t}{2}\right)\dfrac{\sqrt{K}}{\sqrt{2\pi}}, \\[4mm]
n\left(-\dfrac{\log(K)}{2\nu\sqrt{t}} - \nu\sqrt{t}\right) = \exp\left(-\dfrac{(\log(K))^2}{8\nu^2 t} - \dfrac{\nu^2 t}{2}\right)\dfrac{1}{\sqrt{K}\sqrt{2\pi}}.
\end{cases}
$$

Substituting back into (1.68), we obtain:

$$\frac{\partial}{\partial\nu}\mathbb{E}\left[\left(\mathcal{E}_t^{(2\nu)} - K\right)^+\right] = \exp\left(-\frac{(\log(K))^2}{8\nu^2 t} - \frac{\nu^2 t}{2}\right)\frac{2\sqrt{Kt}}{\sqrt{2\pi}},$$

which matches (1.67), the derivative of the RHS of (1.58) with respect to $\nu$.  □

**Problem 1.1 (Computation of the laws of $T_a^{(\nu)}$ and $G_a^{(\nu)}$).**
Let $(B_t, t \geq 0)$ be a standard Brownian motion started at 0, and denote $S_t := \sup_{s \leq t} B_s$.
**1)** Prove, by using for instance the reflection principle, that for every $t \geq 0$, $S_t \overset{(law)}{=} |B_t|$.
**2)** Let $a \in \mathbb{R}$ and denote $T_a := \inf\{s \geq 0; B_s = a\}$. Deduce from **1)** that:

$$\mathbb{P}(T_a \in dt) = \frac{|a|}{\sqrt{2\pi t^3}}\exp\left(-\frac{a^2}{2t}\right)dt \qquad (t \geq 0).$$

**3)** Prove that $\mathbb{E}\left[e^{-\lambda T_a}\right] = e^{-|a|\sqrt{2\lambda}}$, $(\lambda \geq 0)$ and that $T_a \xrightarrow[a\to 0]{} 0$. Show that $(T_a, a > 0)$ is a stable subordinator of index 1/2.
**4)** We recall (see Appendix B.1) that:

$$K_\nu(x) = \frac{1}{2}\left(\frac{x}{2}\right)^\nu \int_0^\infty t^{-\nu-1}e^{-t-\frac{x^2}{4t}}\,dt \qquad (\nu \in \mathbb{R}), \qquad (1)$$

$$K_\nu = K_{-\nu} \qquad \text{and} \qquad K_{1/2}(x) = \sqrt{\frac{\pi}{2x}}e^{-x}. \qquad (2)$$

Prove that:

$$\int_0^\infty \frac{a}{\sqrt{2\pi t^3}} \exp\left(-\frac{a^2}{2t} - \lambda t\right) dt = e^{-a\sqrt{2\lambda}} \qquad (a, \lambda \geq 0)$$

which yields another proof of **2)**.

**5)** Let $a$ and $\nu > 0$. Using the martingale $\left(\mathscr{E}_t^{(2\nu)} := \exp\left(2\nu B_t - 2\nu^2 t\right), t \geq 0\right)$ and Lemma 2.1, prove formula (1.10):

$$\mathbb{P}\left(G_a^{(-\nu)} > 0\right) = \mathbb{P}\left(T_a^{(-\nu)} < +\infty\right) = \exp(-2\nu a).$$

**6)** By using Cameron-Martin formula (1.5), prove (1.11):

$$\mathbb{P}\left(T_a^{(\nu)} \in dt\right) = \frac{a}{\sqrt{2\pi t^3}} \exp\left(-\frac{(a - \nu t)^2}{2t}\right) dt \qquad (t > 0, \, a > 0).$$

**7)** Deduce then that:

$$\int_0^\infty \frac{a}{\sqrt{2\pi t^3}} \exp\left(-\frac{(a + \nu t)^2}{2t}\right) dt = e^{-2a\nu}.$$

This formula could also have been obtained from (1) and (2).

**8)** Let $\widetilde{B}_t := tB_{1/t}$ if $t > 0$, and $\widetilde{B}_0 = 0$. Prove that $(\widetilde{B}_t, t \geq 0)$ is a standard Brownian motion, and deduce then formula (1.9):

$$\frac{1}{T_a^{(\nu)}} \overset{(law)}{=} G_\nu^{(a)} \qquad (\nu, a \text{ reals}).$$

**9)** Deduce finally from **8)** and **6)** that:

$$\mathbb{P}\left(G_a^{(\nu)} \in dt\right) = \frac{|\nu|}{\sqrt{2\pi t}} \exp\left(-\frac{(a - \nu t)^2}{2t}\right) dt \qquad (t > 0, \, a > 0, \, \nu \in \mathbb{R}).$$

## 1.5 Notes and Comments

As explained in a number of preprints (Bentata-Yor [5], [6] and [5, F], [6, F], [7, F], Madan-Roynette-Yor [47], [48], [49], [50] and [38, F]), the motivation for the search of an expression such as:

$$\mathbb{E}\left[(\mathscr{E}_t - 1)^{\pm}\right] = \mathbb{P}(X \leq t) \tag{1.69}$$

came from a question by M. Qian, in August 2007, to the third author of this monograph, precisely, to give a closed form expression for:

$$\int_0^\infty e^{-\lambda t} \mathbb{E}\left[(\mathscr{E}_t - 1)^{\pm}\right] dt$$

and more generally:

$$\int_0^\infty h(t)\mathbb{E}\left[(\mathscr{E}_t - 1)^\pm\right]dt.$$

Since, from Theorem 1.1, $X \overset{(law)}{=} 4B_1^2$ in (1.69), we obtain:

$$\int_0^\infty e^{-\lambda t}\mathbb{E}\left[(\mathscr{E}_t - 1)^\pm\right]dt = \frac{1}{\lambda}\mathbb{E}\left[e^{-\lambda(4B_1^2)}\right] = \frac{1}{\lambda\sqrt{1+8\lambda}},$$

and more generally, there is the identity:

$$\int_0^\infty h(t)\mathbb{E}\left[(\mathscr{E}_t - 1)^\pm\right]dt = \mathbb{E}\left[H(4B_1^2)\right],$$

where $H(x) = \int_x^\infty h(t)dt$.

More general quantities will be studied in Chapter 4.

Theorems 1.3 and 1.4 are taken from [48], and Theorem 1.2 – which is a particular case of Theorem 1.3 for $\nu = 1$ – was first obtained in [50].

# Chapter 2
# Generalized Black-Scholes Formulae for Martingales, in Terms of Last Passage Times

**Abstract**  Let $(M_t, t \geq 0)$ be a positive, continuous local martingale such that $M_t \xrightarrow[t \to \infty]{} M_\infty = 0$ a.s. In Section 2.1, we express the European put $\Pi(K,t) := \mathbb{E}\left[(K - M_t)^+\right]$ in terms of the last passage time $\mathcal{G}_K^{(M)} := \sup\{t \geq 0; M_t = K\}$. In Section 2.2, under the extra assumption that $(M_t, t \geq 0)$ is a true martingale, we express the European call $C(K,t) := \mathbb{E}\left[(M_t - K)^+\right]$ still in terms of the last passage time $\mathcal{G}_K^{(M)}$. In Section 2.3, we shall give several examples of explicit computations of the law of $\mathcal{G}_K^{(M)}$, and Section 2.4 will be devoted to the proof of a more general formula for this law. In Section 2.5, we recover, using the results of Section 2.1, Pitman-Yor's formula for the law of $\mathcal{G}_K$ in the framework of transient diffusions. The next sections shall extend these results in different ways:

- In Section 2.6, we present an example where $(M_t, t \geq 0)$ is no longer continuous, but only càdlàg without positive jumps,
- In Section 2.7, we remove the assumption $M_\infty = 0$,
- Finally, in Section 2.8, we consider the framework of several orthogonal local martingales.

## 2.1 Expression of the European Put Price in Terms of Last Passage Times

### 2.1.1 Hypotheses and Notation

Let $(M_t, t \geq 0)$ be a local martingale defined on a filtered probability space $(\Omega, (\mathcal{F}_t, t \geq 0), \mathcal{F}_\infty, \mathbb{P})$. We assume that $(\mathcal{F}_t := \sigma(M_s; s \leq t), t \geq 0)$ is the natural filtration of $(M_t, t \geq 0)$ and that $\mathcal{F}_\infty := \bigvee_{t \geq 0} \mathcal{F}_t$. Let $\mathcal{M}_+^{0,c}$ denote the set of local martingales such that:

C. Profeta et al., *Option Prices as Probabilities*, Springer Finance,
DOI 10.1007/978-3-642-10395-7_2, © Springer-Verlag Berlin Heidelberg 2010

- for all $t \geq 0$, $M_t \geq 0$ a.s. $\qquad (2.1)$
- $(M_t, t \geq 0)$ is a.s. continuous $\qquad (2.2)$
- $M_0$ is a.s. constant and $\lim_{t \to \infty} M_t = 0$ a.s. $\qquad (2.3)$

Hence, a local martingale which belongs to $\mathcal{M}_+^{0,c}$ is a supermartingale.

For all $K, t \geq 0$, we define the put quantity $\Pi(K,t)$ associated to $M$:

$$\Pi(K,t) := \mathbb{E}\left[(K - M_t)^+\right].\qquad (2.4)$$

We note that, since $x \mapsto (K - x)^+$ is a bounded convex function, $((K - M_t)^+, t \geq 0)$ is a submartingale, and therefore $\mathbb{E}\left[(K - M_t)^+\right]$ is an increasing function of $t$, converging to $K$ as $t \to \infty$. It is thus natural to try to express $\Pi(K,t)$ as $K$ times the distribution function of a positive random variable. This is the purpose of the next paragraph, which is a generalization of Point $(i)$ of Theorem 1.2.

## 2.1.2 Expression of $\Pi(K,t)$ in Terms of $\mathcal{G}_K^{(M)}$

Let $(M_t, t \geq 0) \in \mathcal{M}_+^{0,c}$ and define $\mathcal{G}_K^{(M)}$ by:

$$\mathcal{G}_K^{(M)} := \sup\{t \geq 0; M_t = K\},\qquad (2.5)$$
$$(= 0 \text{ if the set } \{t \geq 0; M_t = K\} \text{ is empty.})$$

We shall often write $\mathcal{G}_K$ instead of $\mathcal{G}_K^{(M)}$ when there is no ambiguity.

**Theorem 2.1.** *Let $K > 0$:*

i) *For any $\mathscr{F}_t$-stopping time $T$:*

$$\left(1 - \frac{M_T}{K}\right)^+ = \mathbb{P}(\mathcal{G}_K \leq T | \mathscr{F}_T).\qquad (2.6)$$

ii) *Consequently:*

$$\Pi(K,T) = \mathbb{E}\left[(K - M_T)^+\right] = K\mathbb{P}(\mathcal{G}_K \leq T).\qquad (2.7)$$

## 2.1.3 Proof of Theorem 2.1

It hinges upon the following (classical) Lemma.

**Lemma 2.1 (Doob's maximal identity).**
*If $(N_t, t \geq 0) \in \mathcal{M}_+^{0,c}$, then:*

$$\sup_{t \geq 0} N_t \overset{(law)}{=} \frac{N_0}{U} \tag{2.8}$$

*where $U$ is a uniform r.v. on $[0,1]$ independent of $\mathcal{F}_0$.*

*Proof.* Let $a > N_0$ and $T_a := \inf\{t \geq 0; N_t = a\}$ $(= +\infty$ if this set is empty). We use Doob's optional stopping theorem to obtain:

$$\mathbb{E}[N_{T_a} | \mathcal{F}_0] = a\mathbb{P}(T_a < \infty | \mathcal{F}_0) = N_0$$

since $N_{T_a} = 0$ if $T_a = +\infty$ and $N_{T_a} = a$ if $T_a < +\infty$. Thus:

$$\mathbb{P}\left(\sup_{t \geq 0} N_t > a \Big| \mathcal{F}_0\right) = \frac{N_0}{a}$$

since $\{T_a < \infty\} = \left\{\sup_{t \geq 0} N_t > a\right\}$.

$\square$

**We now prove Theorem 2.1**
We note that, since $\lim_{t \to \infty} M_t = 0$ a.s.:

$$\{\mathcal{G}_K < T\} = \left\{\sup_{s \geq T} M_s < K\right\}.$$

We apply Lemma 2.1 to the local martingale $(M_{T+u}, u \geq 0)$, in the filtration $(\mathcal{F}_{T+u}, u \geq 0)$. Conditionally on $\mathcal{F}_T$:

$$\sup_{s \geq T} M_s \overset{(law)}{=} \frac{M_T}{U} \tag{2.9}$$

where $U$ is uniform on $[0,1]$, and independent of $\mathcal{F}_T$. Consequently:

$$\mathbb{P}\left(\sup_{s \geq T} M_s < K \Big| \mathcal{F}_T\right) = \mathbb{P}\left(\frac{M_T}{U} < K \Big| \mathcal{F}_T\right) = \left(1 - \frac{M_T}{K}\right)^+. \tag{2.10}$$

$\square$

*Remark 2.1.* Theorem 2.1 and Lemma 2.1 remain valid if we replace the hypothesis: $(M_t, t \geq 0)$ is a.s. continuous by the weaker one: $(M_t, t \geq 0)$ has càdlàg paths and no positive jumps. This relies on the fact that, in this new framework, we still have $M_{T_a} = a$ on the event $\{T_a < \infty\}$.

Of course, Theorem 2.1 has a practical interest only if we can explicitly compute the law of $\mathcal{G}_K$. We shall tackle this computation in Section 2.3 below, but, before that, we study the way Theorem 2.1 is modified when we replace the Put price by a Call price.

**Exercise 2.1 (Another proof of Doob's maximal identity).**
Let $(B_t, t \geq 0)$ denote a Brownian motion started at $x$, and denote by
$T_k := \inf\{t \geq 0; B_t = k\}$ its first hitting time of level $k$.
**1)** Let $a < x < b$. Prove that:

$$\mathbb{P}_x (T_a < T_b) = \frac{x - b}{a - b}.$$

**2)** We assume that $x = 1$. Deduce then that:

$$\sup_{t \geq 0} B_{t \wedge T_0} = \frac{1}{U}$$

where $U$ is a uniform r.v. on $[0, 1]$ independent from $(B_t, t \geq 0)$.
**3)** Let $(M_t, t \geq 0) \in \mathcal{M}_+^{0,c}$ such that $M_0 = 1$ a.s. Apply question **2)** and the Dambis, Dubins, Schwarz's Theorem to recover Doob's maximal identity. (Note that $\langle M \rangle_\infty < \infty$ a.s. since $(M_t, t \geq 0)$ converges a.s.)

## 2.2 Expression of the European Call Price in Terms of Last Passage Times

### 2.2.1 Hypotheses

Let $(M_t, t \geq 0) \in \mathcal{M}_+^{0,c}$. For all $K, t \geq 0$, we defined the call quantity $C(K,t)$ associated to $M$ by:

$$C(K,t) := \mathbb{E}\left[(M_t - K)^+\right]. \tag{2.11}$$

In order to state the counterpart of Theorem 2.1 for the call price, we add the extra assumption:

$(T)$ $\quad\quad\quad\quad\quad (M_t, t \geq 0)$ is a (true) martingale such that $M_0 = 1$.

In particular, $\mathbb{E}[M_t] = 1$ for all $t \geq 0$.

Let $\mathbb{P}^{(M)}$ be the probability on $(\Omega, \mathcal{F}_\infty)$ such that, for all $t \geq 0$:

$$\mathbb{P}_{|\mathcal{F}_t}^{(M)} = M_t \cdot \mathbb{P}_{|\mathcal{F}_t}. \tag{2.12}$$

In the framework of financial mathematics, this probability is called a change of numéraire probability.

We note that:
- $\mathbb{P}^{(M)}(T_0 < +\infty) = 0$, where $T_0 := \inf\{t \geq 0; M_t = 0\}$. Indeed,

$$\mathbb{P}^{(M)}(T_0 \leq t) = \mathbb{E}\left[M_t 1_{\{T_0 \leq t\}}\right] = \mathbb{E}\left[M_{T_0} 1_{\{T_0 \leq t\}}\right] = 0. \tag{2.13}$$

Hence, $T_0 = +\infty$ $\mathbb{P}^{(M)}$-a.s.

• It is easily seen that, under $\mathbb{P}^{(M)}$, $\left(\dfrac{1}{M_t}, t \geq 0\right)$ is a positive local martingale. It is thus a supermartingale, which converges a.s. In fact:

$$\frac{1}{M_t} \xrightarrow[t \to \infty]{} 0 \quad \mathbb{P}^{(M)}\text{-a.s.} \tag{2.14}$$

Indeed, for every $\varepsilon > 0$:

$$\mathbb{P}^{(M)}\left(\frac{1}{M_t} > \varepsilon\right) = \mathbb{P}^{(M)}\left(M_t < \frac{1}{\varepsilon}\right) = \mathbb{E}\left[M_t \mathbf{1}_{\{M_t < 1/\varepsilon\}}\right] \xrightarrow[t \to \infty]{} 0$$

from the dominated convergence theorem. Thus (2.14) holds.

## 2.2.2 Price of a European Call in Terms of Last Passage Times

We state the counterpart of Theorem 2.1 for the call price.

**Theorem 2.2.** *Let $(M_t, t \geq 0)$ be a positive, continuous martingale which converges to 0 a.s and such that $M_0 = 1$ a.s. Then:*

i) *For every bounded, $\mathscr{F}_t$-measurable r.v. $F_t$, and all $K \geq 0$:*

$$\mathbb{E}\left[F_t \left(M_t - K\right)^+\right] = \mathbb{E}^{(M)}\left[F_t \mathbf{1}_{\left\{\mathscr{G}_K^{(M)} \leq t\right\}}\right]. \tag{2.15}$$

*In particular:*

$$\mathbb{E}\left[(M_t - K)^+\right] = \mathbb{P}^{(M)}\left(\mathscr{G}_K^{(M)} \leq t\right). \tag{2.16}$$

ii) *For every bounded, $\mathscr{F}_t$-measurable r.v. $F_t$:*

$$\mathbb{E}\left[F_t |M_t - K|\right] = K\mathbb{E}\left[F_t \mathbf{1}_{\left\{\mathscr{G}_K^{(M)} \leq t\right\}}\right] + \mathbb{E}^{(M)}\left[F_t \mathbf{1}_{\left\{\mathscr{G}_K^{(M)} \leq t\right\}}\right] \tag{2.17}$$

iii) *For every $K, t \geq 0$:*

$$\mathbb{P}^{(M)}\left(\mathscr{G}_K^{(M)} \leq t\right) = 1 - K + K\mathbb{P}\left(\mathscr{G}_K^{(M)} \leq t\right) \tag{2.18}$$

$$= 1 - K\mathbb{P}(\mathscr{G}_K^{(M)} > t)$$

*and*

$$\mathbb{P}^{(M)}\left(\mathscr{G}_K^{(M)} > t\right) = K\mathbb{P}(\mathscr{G}_K^{(M)} > t).$$

*Therefore, if $K \geq 1$:*

$$\mathbb{P}^{(M)}\left(\mathscr{G}_K^{(M)} \geq t \mid \mathscr{G}_K^{(M)} > 0\right) = \mathbb{P}(\mathscr{G}_K^{(M)} \geq t),$$

*while if $K \leq 1$:*

$$\mathbb{P}\left(\mathscr{G}_K^{(M)} \geq t | \mathscr{G}_K^{(M)} > 0\right) = \mathbb{P}^{(M)}(\mathscr{G}_K^{(M)} \geq t).$$

*In particular, for $K = 1$:*

$$\mathscr{G}_1^{(M)} \text{ has the same distribution under } \mathbb{P} \text{ and under } \mathbb{P}^{(M)}. \tag{2.19}$$

### 2.2.3 Proof of Theorem 2.2

We first prove Point $(i)$
We have:

$$\mathbb{E}\left[F_t\left(M_t - K\right)^+\right] = \mathbb{E}\left[F_t\left(M_t - K\right)^+ 1_{\{M_t > 0\}}\right] \quad \text{(since } (M_t - K)^+ = 0 \text{ on } \{M_t = 0\})$$

$$= \mathbb{E}\left[F_t M_t \left(1 - \frac{K}{M_t}\right)^+ 1_{\{M_t > 0\}}\right]$$

$$= \mathbb{E}^{(M)}\left[F_t \left(1 - \frac{K}{M_t}\right)^+ 1_{\{M_t > 0\}}\right] \quad \text{(from the definition of } \mathbb{P}^{(M)})$$

$$= \mathbb{E}^{(M)}\left[F_t 1_{\{\mathscr{G}_{1/K}^{(1/M)} \leq t\}}\right]$$

by applying Theorem 2.1 with $1/M$ and $1/K$ instead of $M$ and $K$, and since $\mathbb{P}^{(M)}(M_t > 0) = 1$ from (2.13). But, by its very definition, $\mathscr{G}_{1/K}^{(1/M)} = \mathscr{G}_K^{(M)}$, and therefore:

$$\mathbb{E}\left[F_t\left(M_t - K\right)^+\right] = \mathbb{E}^{(M)}\left[F_t 1_{\{\mathscr{G}_K^{(M)} \leq t\}}\right].$$

This is Point $(i)$.

We now prove Point $(iii)$
We have,

$$\mathbb{E}\left[\left(M_t - K\right)^+\right] - \mathbb{E}\left[\left(K - M_t\right)^+\right] = \mathbb{E}[M_t - K] = 1 - K.$$

Hence, from Point $(ii)$ of Theorem 2.1 and Point $(i)$ of Theorem 2.2:

$$\mathbb{P}^{(M)}\left(\mathscr{G}_K^{(M)} \leq t\right) - K\mathbb{P}\left(\mathscr{G}_K^{(M)} \leq t\right) = 1 - K.$$

This is Point $(iii)$.
Finally, Point $(ii)$ is an easy consequence of the identity:

$$|M_t - K| = \left(M_t - K\right)^+ + \left(K - M_t\right)^+,$$

applying once again Point $(ii)$ of Theorem 2.1 and Point $(i)$ of Theorem 2.2.

$\square$

In fact, formula (2.16) can be improved in the following way: let $\varphi : \mathbb{R}^+ \to \mathbb{R}^+$ a Borel and locally integrable function, and $\Phi(x) := \int_0^x \varphi(y)dy$. Then, for any $t \geq 0$ and $F_t \in \mathscr{F}_t$, one has:

$$\mathbb{E}\left[F_t \Phi(M_t)\right] = \mathbb{E}^{(M)}\left[F_t \varphi\left(\inf_{s \geq t} M_s\right)\right]. \tag{2.20}$$

$\left(\text{Of course, (2.16) is also a particular case of (2.20) with } \Phi(x) = (x-K)^+ \text{ and } \varphi(x) = 1_{\{x>K\}}.\right)$

We prove (2.20)

We have:

$$\mathbb{E}\left[F_t \Phi(M_t)\right] = \mathbb{E}\left[F_t \int_0^{M_t} \varphi(y)dy\right]$$

$$= \mathbb{E}\left[F_t M_t \varphi(U M_t)\right]$$

(where $U$ is a uniform r.v. on $[0,1]$ independent from $\mathscr{F}_t$)

$$= \mathbb{E}^{(M)}\left[F_t \varphi(U M_t)\right]$$

$$= \mathbb{E}^{(M)}\left[F_t \varphi\left(\frac{1}{\frac{1}{U M_t}}\right)\right]$$

$$= \mathbb{E}^{(M)}\left[F_t \varphi\left(\frac{1}{\sup_{s \geq t} \frac{1}{M_s}}\right)\right]$$

(applying Lemma 2.1 to the $\mathbb{P}^{(M)}$-local martingale $(1/M_t, t \geq 0)$)

$$= \mathbb{E}^{(M)}\left[F_t \varphi\left(\inf_{s \geq t} M_s\right)\right].$$

## 2.3 Some Examples of Computations of the Law of $\mathscr{G}_K^{(M)}$

*Example 2.3.a.* We get back to the classical Black-Scholes formula, with $\mathscr{E}_t := \exp\left(B_t - \frac{t}{2}\right)$ where $B$ is a Brownian motion started from 0. From (2.7), the identity:

$$\mathbb{E}\left[\left(1 - \frac{\mathscr{E}_t}{K}\right)^+\right] = \mathbb{P}\left(\mathscr{G}_K^{(\mathscr{E})} \leq t\right)$$

holds. Taking $K = 1$, it suffices to obtain the identity:

$$\mathscr{G}_1 \overset{(law)}{=} 4B_1^2 \tag{2.21}$$

to recover formula (1.19). In fact, this identity (2.21) may be obtained simply by time inversion, since:

$$\mathcal{G}_1 := \sup\{t \geq 0; \, \mathcal{E}_t = 1\} = \sup\{t \geq 0; \, B_t - \frac{t}{2} = 0\},$$

hence, with the notation of Subsection 1.1.3:

$$\mathcal{G}_1 = G_0^{(-1/2)} \overset{(law)}{=} \frac{1}{T_{-1/2}^{(0)}} \quad \text{(from (1.9))}$$

$$\overset{(law)}{=} \frac{4}{T_1^{(0)}} \quad \text{(by scaling)}$$

$$\overset{(law)}{=} 4B_1^2.$$

*Example 2.3.b.* Here $(M_t, t \geq 0)$ is the martingale defined by: $\left(M_t = B_{t \wedge T_0}, t \geq 0\right)$ where $(B_t, t \geq 0)$ is a Brownian motion started from 1, and $T_0 := \inf\{t \geq 0; B_t = 0\}$. Then, for $K \leq 1$:

$$\mathcal{G}_K^{(M)} \overset{(law)}{=} \frac{(U_K)^2}{N^2} \overset{(law)}{=} T_{U_K} \tag{2.22}$$

where $(T_x, x \geq 0)$ is the first hitting time process of a Brownian motion $(\beta_t, t \geq 0)$ starting from 0, $N$ is a standard Gaussian r.v. and $U_K$ is uniform on $[1 - K, 1 + K]$, independent from $T$ and $N$.

Proof of (2.22)
Applying Williams' time reversal Theorem (see [91]), we have:

$$\left(B_{T_0 - u}, u \leq T_0\right) \overset{(law)}{=} \left(R_u, u \leq G_1(R)\right) \tag{2.23}$$

where $(R_u, u \geq 0)$ is a Bessel process of dimension 3 starting from 0 and $G_1(R) := \sup\{u \geq 0; R_u = 1\}$. Hence:

$$T_0 \overset{(law)}{=} T_K(R) + \mathcal{G}_K^{(M)}, \tag{2.24}$$

where on the RHS, $T_K(R)$ and $\mathcal{G}_K^{(M)}$ are independent and

$$T_K(R) := \inf\{u \geq 0; R_u = K\}.$$

Taking the Laplace transform in $\frac{\lambda^2}{2}$ on both sides of (2.24) gives:

$$e^{-\lambda} = \frac{\lambda K}{\sinh(\lambda K)} \mathbb{E}\left[e^{-\frac{\lambda^2}{2}\mathcal{G}_K^{(M)}}\right], \tag{2.25}$$

since $\mathbb{E}\left[e^{-\frac{\lambda^2}{2}T_0}\right] = e^{-\lambda}$ and $\mathbb{E}\left[e^{-\frac{\lambda^2}{2}T_K(R)}\right] = \frac{\lambda K}{\sinh(\lambda K)}.$

Then, formula (2.25) becomes:

$$\mathbb{E}\left[e^{-\frac{\lambda^2}{2}\mathscr{G}_K^{(M)}}\right] = \frac{e^{-\lambda(1-K)} - e^{-\lambda(1+K)}}{2\lambda K} = \frac{1}{2K}\int_{1-K}^{1+K} e^{-\lambda x} dx$$

$$= \frac{1}{2K}\int_{1-K}^{1+K}\mathbb{E}\left[e^{-\frac{\lambda^2}{2}T_x}\right] dx.$$

Thus:

$$\mathbb{E}\left[e^{-\frac{\lambda^2}{2}\mathscr{G}_K^{(M)}}\right] = \frac{1}{2K}\int_{1-K}^{1+K}\mathbb{E}\left[e^{-\frac{\lambda^2}{2}\frac{x^2}{N^2}}\right] dx = \mathbb{E}\left[e^{-\frac{\lambda^2}{2}\frac{(U_K)^2}{N^2}}\right]$$

since $T_x \overset{(law)}{=} \dfrac{x^2}{N^2}$. Hence,

$$\mathscr{G}_K^{(M)} \overset{(law)}{=} \frac{(U_K)^2}{N^2} \overset{(law)}{=} T_{U_K}.$$

$\square$

*Example 2.3.c.*  $(M_t, t \geq 0)$ is the (strict) local martingale defined by:
$\left(M_t := \dfrac{1}{R_t}, t \geq 0\right)$ where $(R_t, t \geq 0)$ is a 3-dimensional Bessel process starting from 1. Then, for every $K < 1$:

$$\mathscr{G}_K^{(M)} \overset{(law)}{=} \frac{\left(\widetilde{U}_K\right)^2}{N^2} \overset{(law)}{=} T_{\widetilde{U}_K} \tag{2.26}$$

where $(T_x, x \geq 0)$ is the first hitting time process of a Brownian motion $(\beta_t, t \geq 0)$ starting from 0, $N$ is a standard Gaussian r.v. and $\widetilde{U}_K$ is a uniform r.v. on $\left[\frac{1}{K} - 1, \frac{1}{K} + 1\right]$, independent from the process $T$ and $N$.

Proof of (2.26)

We observe that $\mathscr{G}_K^{(M)} := \sup\left\{u \geq 0; \dfrac{1}{R_u} = K\right\} = \sup\left\{u \geq 0; R_u = \dfrac{1}{K}\right\}$. We consider the process $R$ as obtained by time reversal from the Brownian motion $(B_t, t \geq 0)$ starting from $1/K$ and killed when it first hits 0. Hence, with the same notation as in Example 2.3.b, we have:

$$T_0 \overset{(law)}{=} \mathscr{G}_K^{(M)} + T_1(R), \tag{2.27}$$

where on the RHS, $\mathscr{G}_K^{(M)}$ and $T_1(R)$ are assumed to be independent. Taking once again the Laplace transform in $\frac{\lambda^2}{2}$ of both sides, one obtains:

$$e^{-\lambda/K} = \frac{\lambda}{\sinh(\lambda)}\mathbb{E}\left[e^{-\frac{\lambda^2}{2}\mathscr{G}_K^{(M)}}\right], \tag{2.28}$$

which can be rewritten:

$$\mathbb{E}\left[e^{-\frac{\lambda^2}{2}\mathscr{G}_K^{(M)}}\right] = \frac{e^{-\lambda(\frac{1}{K}-1)} - e^{-\lambda(\frac{1}{K}+1)}}{2\lambda} = \frac{1}{2}\int_{\frac{1}{K}-1}^{\frac{1}{K}+1} e^{-\lambda x}dx$$

$$= \frac{1}{2}\int_{\frac{1}{K}-1}^{\frac{1}{K}+1} \mathbb{E}\left[e^{-\frac{\lambda^2}{2}T_x}\right]dx$$

$$= \frac{1}{2}\int_{\frac{1}{K}-1}^{\frac{1}{K}+1} \mathbb{E}\left[e^{-\frac{\lambda^2}{2}\frac{x^2}{N^2}}\right]dx = \mathbb{E}\left[e^{-\frac{\lambda^2}{2}\frac{(\tilde{U}_K)^2}{N^2}}\right].$$

$\square$

*Example 2.3.d.* $(M_t, t \geq 0)$ is the martingale defined by:

$\left(M_t = |B_t|h(L_t) + \int_{L_t}^{\infty} h(x)dx, t \geq 0\right)$ where $(B_t, t \geq 0)$ is a Brownian motion start-

ing from 0, $(L_t, t \geq 0)$ its local time at level 0 and $h : \mathbb{R}^+ \to \mathbb{R}^+$ a strictly positive Borel function such that $\int_0^{\infty} h(x)dx = 1$. $(M_t, t \geq 0)$ is the Azéma-Yor martingale associated with $h$ (see [4]). We then have:

$$\mathscr{G}_1^{(M)} \overset{(law)}{=} \left(H^{-1}(U) + \frac{U}{h \circ H^{-1}(U)}\right)^2 \cdot T_1 \tag{2.29}$$

where, on the RHS, $T_1$ and $U$ are assumed to be independent, $U$ is uniform on $[0,1]$, $T_1$ is the first hitting time of 1 by a Brownian motion starting from 0, and $H(u) := \int_0^u h(y)dy$. In particular, if $h(l) = \frac{\lambda}{2}e^{-\frac{\lambda l}{2}}$, we have:

$$\mathscr{G}_1^{(M)} \overset{(law)}{=} T_{\frac{2}{\lambda}\left(\log(\frac{1}{U})+\frac{1}{U}-1\right)} \tag{2.30}$$

where $T_x$ is the first hitting time of $x$ by a Brownian motion starting from 0, and $U$ is uniform on $[0,1]$, independent from $(T_x, x \geq 0)$.

Proof of (2.29)

We use the fact that $\mathscr{G}_1^{(M)}$ has the same distribution under $\mathbb{P}$ and $\mathbb{P}^{(M)}$ (see (2.19)). The law of the canonical process $(X_t, t \geq 0)$ under $\mathbb{P}^{(M)}$ is fully described in [72]. In particular, under $\mathbb{P}^{(M)}$:

- $L_\infty < \infty$ $\mathbb{P}^{(M)}$-a.s., and admits $h$ as density function,
- Conditionally on $\{L_\infty = l\}$, $(X_t, t \leq \tau_l)$ is a Brownian motion stopped at $\tau_l$ (with $\tau_l := \inf\{t \geq 0; L_t > l\}$), independent from the process $\left(X_{\tau_l+t}, t \geq 0\right)$, and $\left(|X_{\tau_l+t}|, t \geq 0\right)$ is a 3-dimensional Bessel process started at 0.

Then, under $\mathbb{P}^{(M)}$, conditionally on $\{L_\infty = l\}$:

$$\mathscr{G}_1^{(M)} = \tau_l + \sup\left\{t \geq 0; \int_l^{\infty} h(x)dx + h(l)|X_t| = 1\right\}$$

$$= \tau_l + \sup\left\{t \geq 0; |X_t| = \frac{\int_0^l h(x)dx}{h(l)}\right\}.$$

Denoting $k(l) = \frac{1}{h(l)}H(l)$ we have, by time reversal (see Example 2.3.b and 2.3.c above):

$$\mathscr{G}_1^{(M)} \overset{(law)}{=} \tau_l + T_{k(l)} \tag{2.31}$$

where $T_x$ is the first hitting time at level $x$ of a Brownian motion started from 0, and where, on the RHS, $\tau_l$ and $T_{k(l)}$ are assumed to be independent. But, $(\tau_l, l \geq 0)$ and $(T_l, l \geq 0)$ being independent and having the same law, we have:

$$\mathscr{G}_1^{(M)} \overset{(law)}{=} T_{l+k(l)} \tag{2.32}$$

Then, since $L_\infty$ admits $h$ as density, $H(L_\infty)$ is uniformly distributed on $[0,1]$. Hence:

$$\mathscr{G}_1^{(M)} \overset{(law)}{=} T_{\left(H^{-1}(U) + \frac{U}{h \circ H^{-1}(U)}\right)}$$

$$\overset{(law)}{=} \left(H^{-1}(U) + \frac{U}{h \circ H^{-1}(U)}\right)^2 \cdot T_1 \qquad \text{(by scaling)}.$$

$\square$

*Example 2.3.e.* We end this series of examples by examining a situation which is no longer in the scope of Theorem 2.1 or Theorem 2.2: we shall compute the price of a call where $(M_t, t \geq 0)$ is only a local martingale, and not a (true) martingale. (In other words, we remove assumption (T)). More precisely, let $\left(X_t, t \geq 0, \mathbb{P}_a^{(3)}, a > 0\right)$ the canonical Bessel process of dimension 3, defined on the space $\Omega := \mathscr{C}(\mathbb{R}^+, \mathbb{R}^+)$. Let $(M_t, t \geq 0)$ be the local martingale $\left(M_t = \frac{1}{X_t}, t \geq 0\right)$. Then:

$$\mathbb{E}_1^{(3)}\left[F_t\left(\frac{1}{X_t} - 1\right)^+\right] = W_1\left[F_t 1_{\{\gamma \leq t \leq T_0\}}\right] \tag{2.33}$$

where $F_t$ is a generic bounded $\mathscr{F}_t$-measurable r.v., $W_1$ is the Wiener measure (with $W_1(X_0 = 1) = 1$), $T_0 := \inf\{t \geq 0; X_t = 0\}$ and $\gamma := \sup\{t < T_0; X_t = 1\}$. In particular:

$$\mathbb{E}_1^{(3)}\left[\left(\frac{1}{X_t} - 1\right)^+\right] = W_1\left(\gamma \leq t \leq T_0\right). \tag{2.34}$$

Proof of (2.33)
From the well-known Doob's $h$-transform relationship:

$$\mathbb{P}_{a|\mathscr{F}_t}^{(3)} = \frac{X_{t \wedge T_0}}{a} \cdot W_{a|\mathscr{F}_t} \tag{2.35}$$

we have:

$$\mathbb{E}_1^{(3)}\left[F_t\left(\frac{1}{X_t}-1\right)^+\right] = W_1\left[F_t\left(\frac{1}{X_t}-1\right)^+ X_{t\wedge T_0}1_{\{t\leq T_0\}}\right]$$

$$= W_1\left[F_t\left(1-X_{t\wedge T_0}\right)^+ 1_{\{t\leq T_0\}}\right]$$

$$= W_1\left[F_t 1_{\{\gamma\leq t\leq T_0\}}\right]$$

by applying Theorem 2.1, relation (2.6), with $K=1$ and $M_t = X_{t\wedge T_0}$.

$\square$

We observe that the LHS of (2.34) is not an increasing function of $t$. Indeed, the RHS converges to 0 as $t\to\infty$ as a consequence of Lebesgue's dominated convergence Theorem. In fact, we can compute explicitly this RHS, which equals:

$$r(t) = W_1(T_0\geq t) - W_1(\gamma\geq t) \quad \text{(since, by definition, } T_0\geq\gamma).$$

Recall that, under $W_1$: $T_0 \overset{(law)}{=} \dfrac{1}{B_1^2}$ and, from Example 2.3.b with $K=1$, $\gamma \overset{(law)}{=} \dfrac{(U_1)^2}{B_1^2}$ where $U_1$ is uniform on $[0,2]$ and independent of $B_1^2$. Therefore:

$$r(t) = \mathbb{P}\left(|B_1|\leq\frac{1}{\sqrt{t}}\right) - \mathbb{P}\left(|B_1|\leq\frac{U_1}{\sqrt{t}}\right), \tag{2.36}$$

that is:

$$r(t) = \sqrt{\frac{2}{\pi}}\int_0^{1/\sqrt{t}}e^{-x^2/2}dx - \sqrt{\frac{2}{\pi}}\int_0^\infty e^{-x^2/2}\left(1-\frac{x\sqrt{t}}{2}\right)^+ dx. \tag{2.37}$$

In particular, $r$ starts to increase, reaches its overall maximum, and then decreases. Moreover, it easily follows from (2.37) that:

$$r(t) \underset{t\to\infty}{\sim} \frac{1}{6}\sqrt{\frac{2}{\pi t^3}}. \tag{2.38}$$

It is also easily proven that:

$$r(t) \underset{t\to 0}{\sim} \sqrt{\frac{t}{2\pi}}. \tag{2.39}$$

This example will be taken up and developed in Section A.1 of the Complements.

## 2.4 A More General Formula for the Computation of the Law of $\mathscr{G}_K^{(M)}$

### 2.4.1 Hypotheses

Let $(M_t, t\geq 0)\in\mathscr{M}_+^{0,c}$. (See Section 2.1). Our purpose in this Section is to give a general formula for the law of $\mathscr{G}_K^{(M)}$. To proceed, we add extra hypotheses on $M$:

i)  For every $t > 0$, the law of the r.v. $M_t$ admits a density $(m_t(x), x \geq 0)$, and $(t,x) \mapsto m_t(x)$ may be chosen continuous on $(]0, +\infty[)^2$.

ii) Let us denote by $(\langle M \rangle_t, t \geq 0)$ the increasing process of $(M_t, t \geq 0)$. We suppose that $d\langle M \rangle_t = \sigma_t^2 dt$, and that there exists a jointly continuous function:

$$(t,x) \mapsto \theta_t(x) := \mathbb{E}\left[\sigma_t^2 | M_t = x\right] \quad \text{on } (]0, +\infty[)^2. \tag{2.40}$$

## 2.4.2 Description of the Law of $\mathscr{G}_K^{(M)}$

**Theorem 2.3.** *Under the preceding hypotheses, the law of $\mathscr{G}_K^{(M)}$ is given by:*

$$\mathbb{P}\left(\mathscr{G}_K^{(M)} \in dt\right) = \left(1 - \frac{a}{K}\right)^+ \delta_0(dt) + \frac{1}{2K}\theta_t(K)m_t(K)1_{\{t>0\}}dt \tag{2.41}$$

*where, in (2.41), $a = M_0$ and $\delta_0$ denotes the Dirac measure at 0.*

*Proof.* Using Tanaka's formula, one obtains:

$$\mathbb{E}\left[(K - M_t)^+\right] = (K - a)^+ + \frac{1}{2}\mathbb{E}\left[L_t^K(M)\right] \tag{2.42}$$

where $(L_t^K(M), t \geq 0, K \geq 0)$ denotes the bicontinuous family of local times of the martingale $(M_t, t \geq 0)$. Thus, from Theorem 2.1, there is the relationship:

$$\mathbb{P}\left(\mathscr{G}_K^{(M)} \in dt\right) = \left(1 - \frac{a}{K}\right)^+ \delta_0(dt) + \frac{1_{\{t>0\}}}{2K}d_t\mathbb{E}\left[L_t^K(M)\right] \tag{2.43}$$

and formula (2.41) is now equivalent to the following expression for $d_t\mathbb{E}\left[L_t^K(M)\right]$:

$$d_t\mathbb{E}\left[L_t^K(M)\right] = \theta_t(K)m_t(K)dt \quad (t > 0). \tag{2.44}$$

We now prove (2.44).
For every $f : \mathbb{R}^+ \to \mathbb{R}^+$ Borel, the density of occupation formula

$$\int_0^t f(M_s)d\langle M \rangle_s = \int_0^\infty f(K)L_t^K dK \tag{2.45}$$

becomes, under hypothesis (2.40),

$$\int_0^t f(M_s)\sigma_s^2 ds = \int_0^\infty f(K)L_t^K dK. \tag{2.46}$$

Thus, taking expectation on both sides of (2.46), we obtain:

$$\mathbb{E}\left[\int_0^t f(M_s)\sigma_s^2 ds\right] = \int_0^\infty f(K)\mathbb{E}\left[L_t^K\right] dK \tag{2.47}$$

and, the LHS of (2.47), thanks to (2.40), equals:

$$\int_0^t \mathbb{E}\left[f(M_s)\sigma_s^2\right]ds = \int_0^t \mathbb{E}\left[f(M_s)\mathbb{E}\left[\sigma_s^2|M_s\right]\right]ds$$

$$= \int_0^\infty f(K)dK \int_0^t \theta_s(K)m_s(K)ds. \qquad (2.48)$$

Comparing (2.47) and (2.48), we obtain:

$$\mathbb{E}\left[L_t^K(M)\right] = \int_0^t \theta_s(K)m_s(K)ds,$$

and Theorem 2.3 is proven.

$\square$

### 2.4.3  Some Examples of Applications of Theorem 2.3

*Example 2.4.3.a.* Here, $\left(M_t := \mathcal{E}_t = \exp\left(B_t - \frac{t}{2}\right), t \geq 0\right)$ where $(B_t, t \geq 0)$ is a Brownian motion started at 0. From Itô's formula, $\mathcal{E}_t = 1 + \int_0^t \mathcal{E}_s dB_s$, thus $d\langle \mathcal{E} \rangle_t = \mathcal{E}_t^2 dt$ and we may apply Theorem 2.3 with $\theta_t(x) = x^2$ and $m_t(x) = \frac{1}{x\sqrt{2\pi t}} \exp\left(-\frac{1}{2t}\left(\log(x) + \frac{t}{2}\right)^2\right)$. We obtain:

$$\mathbb{P}\left(\mathcal{G}_K^{(\mathcal{E})} \in dt\right) = \left(1 - \frac{1}{K}\right)^+ \delta_0(dt) + \frac{1_{\{t>0\}}}{2\sqrt{2\pi t}}\exp\left(-\frac{1}{2t}\left(\log(K) + \frac{t}{2}\right)^2\right)dt \qquad (2.49)$$

This formula (2.49) agrees with formulae (1.10) and (1.14) since $\mathcal{G}_K^{(\mathcal{E})} \overset{(law)}{=} G_{\log(K)}^{(-1/2)}$.

*Example 2.4.3.b.* Let $(M_t, 0 \leq t < 1)$ be the martingale defined by:

$$\left(M_t = \frac{1}{\sqrt{1-t}}\exp\left(-\frac{B_t^2}{2(1-t)}\right), t < 1\right).$$

This martingale is the Girsanov density of the law of the Brownian bridge $(b_u, 0 \leq u < 1)$ with respect to the Wiener measure on the $\sigma$-field $(\mathcal{F}_t)$ (see Exercise 2.2). We have here:

$$m_t(x) = \frac{1}{\sqrt{2\pi t}}\frac{2(1-t)}{x}\frac{1}{\sqrt{\Delta(x)}}e^{-\frac{\Delta(x)}{2t}}1_{\left\{x < \frac{1}{\sqrt{1-t}}\right\}} \qquad (2.50)$$

with $\Delta(x) := -2(1-t)\log\left(x\sqrt{1-t}\right)$ and:

$$\theta_t(x) = \frac{x^2}{(1-t)^2}\Delta(x).\tag{2.51}$$

Hence, from Theorem 2.3:

$$\mathbb{P}\left(\mathscr{G}_K^{(M)} \in dt\right)$$

$$= \left(1 - \frac{1}{K}\right)^{+}\delta_0(dt) + \frac{\sqrt{\Delta(K)}}{(1-t)\sqrt{2\pi t}}e^{-\frac{\Delta(K)}{2t}}1_{\left\{K<\frac{1}{\sqrt{1-t}}\right\}}1_{\{t<1\}}dt\tag{2.52}$$

$$= \left(1 - \frac{1}{K}\right)^{+}\delta_0(dt) + \left(K\sqrt{1-t}\right)^{\frac{1}{t}-1}\sqrt{\frac{-\log(K\sqrt{1-t})}{\pi t(1-t)}}1_{\left\{K<\frac{1}{\sqrt{1-t}}\right\}}1_{\{t<1\}}dt.$$

*Example 2.4.3.c.* Here, $(M_t, t \geq 0)$ is the martingale defined by:
$\left(M_t = \cosh(B_t)e^{-t/2}, t \geq 0\right)$. We have from Itô's formula: $\sigma_t = \sinh(B_t)e^{-t/2}$, hence
$\sigma_t^2 = M_t^2 - e^{-t}$ and $\theta_t(x) = x^2 - e^{-t}$. On the other hand:

$$m_t(x) = \sqrt{\frac{2}{\pi t}}\exp\left(-\frac{1}{2t}\left(\text{Argcosh}(xe^{t/2})\right)^2\right)\frac{1}{\sqrt{x^2 - e^{-t}}}1_{\{x>e^{-t/2}\}}.\tag{2.53}$$

Hence, from Theorem 2.3:

$$\mathbb{P}\left(\mathscr{G}_K^{(M)} \in dt\right) = \left(1 - \frac{1}{K}\right)^{+}\delta_0(dt) + \frac{1}{2K}(K^2 - e^{-t})m_t(K)1_{\{t>0\}}dt,$$

where $m_t(K)$ is given by (2.53).

*Example 2.4.3.d.* We consider Feller's martingale, i.e. the solution of the stochastic equation:

$$M_t = l + 2\int_0^t \sqrt{M_s}dB_s.$$

$(M_t, t \geq 0)$ is a square Bessel process of dimension 0 started from $l$. From Itô's formula, $\theta_t(x) = 4x$, and it is known that the law of the r.v. $M_t$ is given by:

$$q_t^0(l, dK) = \exp(-l/2t)\delta_0(dK) + \frac{1}{2t}\sqrt{\frac{l}{K}}\exp\left(-\frac{l+K}{2t}\right)I_1\left(\sqrt{\frac{lK}{t}}\right)dK,$$

where $I_1$ is the modified Bessel function of index 1 (see Appendix B.1). Hence, from Theorem 2.3:

$$\mathbb{P}\left(\mathscr{G}_K^{(M)} \in dt\right) = \left(1 - \frac{l}{K}\right)^{+}\delta_0(dt) + \frac{1}{t}\sqrt{\frac{l}{K}}\exp\left(-\frac{l+K}{2t}\right)I_1\left(\sqrt{\frac{lK}{t}}\right)1_{\{t>0\}}dt.$$

$$\tag{2.54}$$

**Exercise 2.2 (Girsanov's density of the Brownian bridge with respect to the Brownian motion).**
Let $(B_t, 0 \le t \le 1)$ denote the canonical Brownian motion started at 0, $\mathbb{P}$ the Wiener measure, and $(b_t, 0 \le t \le 1)$ the standard Brownian bridge (with $b_0 = b_1 = 0$). $(b_t, 0 \le t < 1)$ is the strong solution of the SDE:

$$\begin{cases} db_t &= \dfrac{-b_t}{1-t} dt + dB_t \\ b_0 &= 0 \end{cases} \qquad (0 \le t < 1).$$

Let $\Pi$ be the law on $\mathscr{C}([0,1], \mathbb{R})$ of $(b_t, 0 \le t \le 1)$.
**1)** Prove, applying Girsanov's Theorem, that:

$$\Pi_{|\mathscr{F}_t} = M_t \cdot \mathbb{P}_{|\mathscr{F}_t} \qquad (0 \le t < 1) \tag{1}$$

with

$$M_t := \exp\left( -\int_0^t \frac{B_s dB_s}{1-s} - \frac{1}{2} \int_0^t \frac{B_s^2 ds}{(1-s)^2} \right), \quad 0 \le t < 1.$$

**2)** Use Itô's formula to show that:

$$M_t = \frac{1}{\sqrt{1-t}} \exp\left( -\frac{B_t^2}{2(1-t)} \right), \quad 0 \le t < 1.$$

Show that $M_1 = 0$ a.s.
**3)** In the case of the $n$-dimensional Brownian motion in $\mathbb{R}^n$, prove that the analogous martingale writes:

$$M_t^{(n)} = \frac{1}{(1-t)^{n/2}} \exp\left( -\frac{\|B_t\|^2}{2(1-t)} \right), \quad 0 \le t < 1$$

where $\|B_t\|^2 = \left(B_t^1\right)^2 + \ldots + \left(B_t^n\right)^2$.
**4)** More generally, let $(X_t, t \ge 0; \mathscr{F}_t, t \ge 0; \mathbb{P}_x, x \in \mathbb{R})$ be the canonical realization of a regular diffusion on $\mathbb{R}$. We denote by $(P_t, t \ge 0)$ and $(p_t(x,y); t > 0, x, y \in \mathbb{R})$ the associated semi-group and its density kernel (with respect to the Lebesgue measure), which we assumed to be regular.

Let $l > 0$. We denote by $\Pi_{x \to y}^{(l)}$ the law, on $\mathscr{C}([0,l], \mathbb{R})$ of the bridge of length $l$ $(x_u, 0 \le u \le l)$ such that $x_0 = x, x_l = y$. Let $0 \le t < l$. For every $F : \mathscr{C}([0,l], \mathbb{R}) \to \mathbb{R}$ bounded and measurable and every $f : \mathbb{R} \to \mathbb{R}$ Borel and bounded, we have:

$$\mathbb{E}_x \left[ F(X_u, u \le t) f(X_l) \right] = \int_{\mathbb{R}} \mathbb{E}_x \left[ F(X_u, u \le t) | X_l = y \right] f(y) p_l(x,y) dy.$$

*i)* Prove that:

$$\mathbb{E}_x \left[ F(X_u, u \le t) f(X_l) \right] = \mathbb{E}_x \left[ F(X_u, u \le t) P_{l-t} f(X_t) \right].$$

*ii*) Deduce that:

$$\mathbb{E}_x\left[F(X_u, u \leq t)|X_l = y\right] = \mathbb{E}_x\left[F(X_u, u \leq t)\frac{p_{l-t}(X_t, y)}{p_l(x, y)}\right]$$

and

$$\Pi^{(l)}_{x \to y|\mathscr{F}_t} = M_t \cdot \mathbb{P}_{x|\mathscr{F}_t} \qquad \text{with } M_t := \frac{p_{l-t}(X_t, y)}{p_l(x, y)}. \tag{2}$$

Prove that (1) is a particular case of (2).

Comment: Relation (2) makes it possible to derive the expression of the bridge of a diffusion as the solution of a SDE, thanks to Girsanov's Theorem.

## 2.5 Computation of the Law of $\mathscr{G}_K$ in the Framework of Transient Diffusions

### 2.5.1 General Framework

Theorem 2.1 and Theorem 2.2 cast some light on our ability to compute explicitly the law of $\mathscr{G}_K^{(M)}$ when $M$ is a positive (local) martingale. We temporarily leave this framework and give (following Pitman-Yor, [65]) a general formula for the law of $\mathscr{G}_K^{(X)}$ when $(X_t, t \geq 0)$ is a transient diffusion taking values in $\mathbb{R}^+$.

We consider the canonical realization of a transient diffusion $(X_t, t \geq 0; \mathbb{P}_x, x > 0)$ on $\mathscr{C}(\mathbb{R}^+, \mathbb{R}^+)$ (See [12]). For simplicity, we assume that:

i) $\mathbb{P}_x(T_0 < \infty) = 0$ for every $x > 0$, with $T_0 := \inf\{t \geq 0; X_t = 0\}$.

ii) $\mathbb{P}_x\left(\lim_{t \to \infty} X_t = +\infty\right) = 1, x > 0$.

As a consequence of (*i*) and (*ii*), there exists a scale function $s$ for this diffusion which satisfies:

$$s(0^+) = -\infty \quad , \quad s(+\infty) = 0 \quad (s \text{ is increasing}). \tag{2.55}$$

Let $\Gamma$ be the infinitesimal generator of $(X_t, t \geq 0)$, and take the speed measure $m$ to be such that:

$$\Gamma = \frac{\partial}{\partial m}\frac{\partial}{\partial s}. \tag{2.56}$$

Let, for $K > 0$:

$$\mathscr{G}_K^{(X)} := \sup\{t \geq 0; X_t = K\}. \tag{2.57}$$

Let us also denote by $q(t, x, y)$ $(= q(t, y, x))$ the density of the r.v. $X_t$ under $\mathbb{P}_x$, with respect to $m$; thus

$$\mathbb{P}_x(X_t \in A) = \int_A q(t, x, y) m(dy), \tag{2.58}$$

for every Borel set $A$.

## 2.5.2 A General Formula for the Law of $\mathscr{G}_K^{(X)}$

**Theorem 2.4 (Pitman-Yor, [65]).** *For all $x, K > 0$:*

$$\mathbb{P}_x\left(\mathscr{G}_K^{(X)} \in dt\right) = \left(1 - \frac{s(x)}{s(K)}\right)^+ \delta_0(dt) - \frac{1}{s(K)}q(t,x,K)dt \qquad (2.59)$$

*where $\delta_0(dt)$ denotes the Dirac measure at 0.*

In particular, if $K \geq x$, since $s$ is increasing and negative, formula (2.59) reduces to:

$$\mathbb{P}_x\left(\mathscr{G}_K^{(X)} \in dt\right) = -\frac{1}{s(K)}q(t,x,K)dt.$$

*Proof.* We apply Theorem 2.1 to the local martingale $(M_t = -s(X_t), t \geq 0)$. This leads to:

$$\mathbb{P}_x\left(\mathscr{G}_K^{(X)} \leq t\right) = \mathbb{P}_x\left(\mathscr{G}_{-s(K)}^{-s(X)} \leq t\right) = \mathbb{E}\left[\left(1 - \frac{-s(X_t)}{-s(K)}\right)^+\right]. \qquad (2.60)$$

As we apply Tanaka's formula to the submartingale $\left(1 - \frac{s(X_t)}{s(K)}\right)^+$, we obtain:

$$\mathbb{P}_x\left(\mathscr{G}_K^{(X)} \leq t\right) = \left(1 - \frac{s(x)}{s(K)}\right)^+ - \frac{1}{2s(K)}\mathbb{E}_x\left[L_t^{-s(K)}(M)\right] \qquad (2.61)$$

where $(L_t^a(M), t \geq 0)$ denotes the local time of the local martingale $(M_t, t \geq 0)$ at level $a$.
We now prove:

$$\frac{\partial}{\partial t}\mathbb{E}_x\left[L_t^{-s(K)}(M)\right] = 2q(t,x,K) \qquad (2.62)$$

which obviously, together with (2.61), implies Theorem 2.4. In fact, (2.62) follows from the density of occupation formula for our diffusion $(X_t, t \geq 0)$:
for any $f : \mathbb{R}^+ \to \mathbb{R}^+$, Borel,

$$\int_0^t f(X_s)ds = \int_0^\infty f(K)l_t^K m(dK) \qquad (2.63)$$

where $(l_t^a, t \geq 0, a > 0)$ is the family of diffusion local times. (See [12]). On the LHS of (2.63), taking the expectation, we have:

$$\mathbb{E}_x\left[\int_0^t f(X_s)ds\right] = \int_0^\infty f(K)\left(\int_0^t q(s,x,K)ds\right)m(dK). \qquad (2.64)$$

Thus, (2.63) and (2.64) imply:

$$\mathbb{E}_x\left[l_t^K\right] = \int_0^t q(s,x,K)ds. \tag{2.65}$$

On the other hand, there is the following relationship between the diffusion and martingale local times:

$$2l_t^K = L_t^{-s(K)}(M). \tag{2.66}$$

Hence, from (2.61), (2.65) and (2.66):

$$\frac{\partial}{\partial t}\mathbb{P}_x\left(\mathscr{G}_K^{(X)} \leq t\right) = \left(1 - \frac{s(x)}{s(K)}\right)^+ \delta_0 - \frac{1}{2s(K)}\frac{\partial}{\partial t}\left(\mathbb{E}_x\left[L_t^{-s(K)}(M)\right]\right)$$

$$= \left(1 - \frac{s(x)}{s(K)}\right)^+ \delta_0 - \frac{1}{s(K)}\frac{\partial}{\partial t}\left(\mathbb{E}_x\left[l_t^K\right]\right)$$

$$= \left(1 - \frac{s(x)}{s(K)}\right)^+ \delta_0 - \frac{1}{s(K)}q(t,x,K),$$

where $\delta_0$ is the derivative of the Heaviside step function, i.e. the Dirac measure at 0. $\qquad\square$

*Remark 2.2.* Formula (2.59) still holds in the more general framework of a transient diffusion $(X_t, t \geq 0)$ taking values in $\mathbb{R}$, such that for example $X_t \xrightarrow[t\to\infty]{} +\infty$ a.s. Indeed, introducing $(\overline{X}_t := \exp(X_t), t \geq 0)$, it is easily seen that $\overline{X}$ satisfies the hypotheses of Subsection 2.5.1, and therefore Theorem 2.4 applies (with obvious notation):

$$\overline{\mathbb{P}}_{\overline{x}}\left(\mathscr{G}_K^{(\overline{X})} \in dt\right) = \left(1 - \frac{\overline{s}(\overline{x})}{\overline{s}(K)}\right)^+ \delta_0(dt) - \frac{1}{\overline{s}(K)}\overline{q}(t,\overline{x},K)dt.$$

Then, using the identities: $\mathscr{G}_K^{(\overline{X})} = \mathscr{G}_{\log(K)}^{(X)}$, $\overline{s}(\overline{x}) = s(\log(\overline{x})) = s(x)$ and $\overline{q}(t,\overline{x},\overline{y}) = q(t,\log(\overline{x}),\log(\overline{y})) = q(t,x,y)$, we obtain:

$$\mathbb{P}_x\left(\mathscr{G}_{\log(K)}^{(X)} \in dt\right) = \left(1 - \frac{s(x)}{s(\log(K))}\right)^+ \delta_0(dt) - \frac{1}{s(\log(K))}q(t,x,\log(K))dt$$

which is (2.59).

### 2.5.3 Case Where the Infinitesimal Generator is Given by its Diffusion Coefficient and its Drift

In practice, it may be useful to write formula (2.59) in terms of the density $p(t,x,y)$ of the r.v. $X_t$ with respect to the Lebesgue measure $dy$ (and not $m(dy)$). We shall give this new expression in the following. Let us assume that the infinitesimal generator $\Gamma$ is of the form:

$$\Gamma = \frac{1}{2}a(x)\frac{\partial^2}{\partial x^2} + b(x)\frac{\partial}{\partial x}. \tag{2.67}$$

Consequently:

$$\frac{dm(y)}{dy} = \frac{2}{s'(y)a(y)} \tag{2.68}$$

and

$$q(t,x,y) = \frac{1}{2}p(t,x,y)s'(y)a(y) \tag{2.69}$$

so that formula (2.59) becomes:

$$\mathbb{P}_x\left(\mathcal{G}_K^{(X)} \in dt\right) = \left(1 - \frac{s(x)}{s(K)}\right)^+ \delta_0(dt) - \frac{s'(K)a(K)}{2s(K)}p(t,x,K)dt. \tag{2.70}$$

We now give several examples of application of Theorem 2.4, and of relation (2.70).

• Let us go back to Example 2.4.3.d where we computed $\mathbb{P}\left(\mathcal{G}_K^{(M)} \in dt\right)$ for $(M_t, t \geq 0)$ a square Bessel process of dimension 0 started from $l$. We then define the diffusion $\left(X_t, t \leq \mathcal{G}_l^{(X)}\right)$ to be the time reversed process of the martingale $(M_t, t \leq T_0)$. $(X_t, t \geq 0)$ is a square Bessel process of dimension 4 started from 0 and therefore satisfies the hypotheses of Theorem 2.4. In this set-up, we have:

$$s(x) = -\frac{1}{x}, \qquad a(x) = 4x,$$

and

$$p(t,x,y) = \frac{1}{2t}\sqrt{\frac{y}{x}}\exp\left(-\frac{x+y}{2t}\right)I_1\left(\sqrt{\frac{xy}{t}}\right).$$

Now, applying the Markov property, we see that the law of $\mathcal{G}_K^{(M)}$ under $\mathbb{P}$ is the same as the law of $\mathcal{G}_l^{(X)}$ under $\mathbb{Q}_K^{(4)}$, where $\mathbb{Q}_K^{(4)}$ denotes the law of a square Bessel process of dimension 4 started at $K$. Then, relation (2.70) yields:

$$\mathbb{P}\left(\mathcal{G}_K^{(M)} \in dt\right) = \mathbb{Q}_K^{(4)}\left(\mathcal{G}_l^{(X)} \in dt\right)$$

$$= \left(1 - \frac{s(K)}{s(l)}\right)^+ \delta_0(dt) - \frac{s'(l)a(l)}{2s(l)}p(t,K,l)dt$$

$$= \left(1 - \frac{l}{K}\right)^+ \delta_0(dt) + \frac{1}{t}\sqrt{\frac{l}{K}}\exp\left(-\frac{l+K}{2t}\right)I_1\left(\sqrt{\frac{Kl}{t}}\right)dt,$$

which is (2.54).

- If $X_t = B_t + \nu t$ (i.e. $a = 1$ and $b = \nu$), we have: $s'(x) = e^{-2\nu x}$,

$$p(t, 0, K) = \frac{1}{\sqrt{2\pi t}} \exp\left(-\frac{1}{2t}(K - \nu t)^2\right) \text{ and, from (2.70), for } \nu \text{ and } K > 0:$$

$$\mathbb{P}_0\left(\mathcal{G}_K^{(X)} \in dt\right) = \mathbb{P}\left(G_K^{(\nu)} \in dt\right) = \frac{\nu}{\sqrt{2\pi t}} \exp\left(-\frac{1}{2t}(K - \nu t)^2\right) dt,$$

and this formula agrees with (1.12).

- If $(X_t, t \geq 0)$ is a transient Bessel process, i.e. if $a = 1$ and $b(x) = \dfrac{2\nu + 1}{2x}$, with index $\nu > 0$ (i.e. with dimension $d = 2\nu + 2 > 2$), we have: $s(x) = -x^{-2\nu}$ and

$$p(t, 0, K) = \frac{2^{-\nu}}{\Gamma(\nu + 1)} t^{-(\nu+1)} K^{2\nu+1} \exp\left(-\frac{K^2}{2t}\right). \text{ Hence, from (2.70):}$$

$$\mathbb{P}_0^{(\nu)}\left(\mathcal{G}_K^{(X)} \in dt\right) = \frac{\nu 2^{-\nu}}{\Gamma(\nu + 1)} \frac{1}{K} \frac{K^{2\nu+1}}{t^{\nu+1}} \exp\left(-\frac{K^2}{2t}\right) dt$$

$$= \frac{2^{-\nu}}{\Gamma(\nu)} \frac{K^{2\nu}}{t^{\nu+1}} \exp\left(-\frac{K^2}{2t}\right) dt. \tag{2.71}$$

Note that, by time reversal, we recover Getoor's result (see [25]):

$$\mathbb{P}_0^{(\nu)}\left(\mathcal{G}_K^{(X)} \in dt\right) = \mathbb{P}_K^{(-\nu)}(T_0 \in dt) = \mathbb{P}\left(\frac{K^2}{2\gamma_\nu} \in dt\right) \tag{2.72}$$

where $\mathbb{P}_K^{(-\nu)}$ denotes the law of a Bessel process of index $(-\nu)$ for $0 < \nu < 1$ (i.e. of dimension $\delta = 4 - d$ for $2 < d < 4$) started at $K$ and where $\gamma_\nu$ is a gamma r.v. with parameter $\nu$. See also Problem 4.1 in the present monograph, and e.g., [98], Paper # 1, for some closely related computations and references.

## 2.6 Computation of the Put Associated to a Càdlàg Martingale Without Positive Jumps

### 2.6.1 Notation

Let $(B_t, t \geq 0)$ be a Brownian motion started from 0, and $\nu > 0$. As in Chapter 1, we denote by $(B_t^{(\nu)} := B_t + \nu t, t \geq 0)$ the Brownian motion with drift $\nu$ and $T_a^{(\nu)} := \inf\{t \geq 0; B_t + \nu t = a\}$. We have:

$$\mathbb{E}\left[e^{-\frac{\lambda^2}{2} T_a^{(\nu)}}\right] = \exp\left(-a\left(\sqrt{\nu^2 + \lambda^2} - \nu\right)\right) \tag{2.73}$$

and in particular, for $\lambda^2 = 1 + 2\nu$:

$$\mathbb{E}\left[e^{-\left(\frac{1}{2}+\nu\right)T_a^{(\nu)}}\right] = \exp(-a) \qquad (a \geq 0). \tag{2.74}$$

It follows from the fact that $(B_t^{(\nu)}, t \geq 0)$ is a Lévy process (which implies $T_{a+b}^{(\nu)} \overset{(law)}{=} T_a^{(\nu)} + T_b^{(\nu)}$, with $T_a^{(\nu)}$ and $T_b^{(\nu)}$ independent), together with (2.74) that:

$$\left(M_a^{(\nu)} := \exp\left\{a - \left(\frac{1}{2} + \nu\right)T_a^{(\nu)}\right\}, a \geq 0\right) \text{ is a martingale.} \tag{2.75}$$

In fact, this is a positive martingale, without positive jumps, and such that

$$\lim_{a \to \infty} M_a^{(\nu)} = 0 \text{ a.s.} \tag{2.76}$$

Indeed, from the law of large numbers:

$$\frac{B_t + \nu t}{\left(\frac{1}{2} + \nu\right)t} \xrightarrow[t \to \infty]{} \frac{\nu}{\frac{1}{2} + \nu} \quad \text{a.s.}$$

Hence, since $T_a^{(\nu)} \xrightarrow[a \to \infty]{} \infty$ a.s.:

$$\frac{B_{T_a^{(\nu)}} + \nu T_a^{(\nu)}}{\left(\frac{1}{2} + \nu\right)T_a^{(\nu)}} = \frac{a}{\left(\frac{1}{2} + \nu\right)T_a^{(\nu)}} \xrightarrow[a \to \infty]{} \frac{\nu}{\frac{1}{2} + \nu} < 1 \quad \text{a.s.}$$

From (2.75), this implies that $M_a^{(\nu)} \xrightarrow[a \to \infty]{} 0$ a.s.

## 2.6.2 Computation of the Put Associated to the Martingale $\left(M_a^{(\nu)}, a \geq 0\right)$

**Proposition 2.1.** *Let $K > 0$.*

*i) If $K < e^a$:*

$$\mathbb{E}\left[\left(K - M_a^{(\nu)}\right)^+\right] = K\mathbb{E}\left[\left(\mathcal{E}_t - e^{2a\nu}\right)^+\right] - \mathbb{E}\left[\left(\mathcal{E}_t - e^{2a(\nu+1)}\right)^+\right], \tag{2.77}$$

$$\text{with } t = t(a, \nu, K) = \frac{2a^2(2\nu + 1)}{a - \log(K)}. \tag{2.78}$$

*ii) If $K \geq e^a$:*

$$\mathbb{E}\left[\left(K - M_a^{(\nu)}\right)^+\right] = K - 1. \tag{2.79}$$

*Proof.* (2.79) is obvious. We prove (2.77):

$$\mathbb{E}\left[\left(K-M_a^{(\nu)}\right)^+\right] = \int_{\frac{a-\log(K)}{\frac{1}{2}+\nu}}^{\infty} \left(K-e^{a-\left(\frac{1}{2}+\nu\right)u}\right)\frac{a}{\sqrt{2\pi u^3}}\exp\left(-\frac{(a-\nu u)^2}{2u}\right)du$$

(from the explicit formula for the density of $T_a^{(\nu)}$ given by (1.11))

$$= e^{a\nu}\int_0^{\frac{a^2(2\nu+1)}{a-\log(K)}}\left(K-e^{a-\left(\frac{1}{2}+\nu\right)\frac{a^2}{s}}\right)\frac{e^{a\nu}}{\sqrt{2\pi s}}\exp\left(-\frac{s}{2}-\frac{a^2\nu^2}{2s}\right)ds$$

(after the change of variable $\dfrac{a^2}{u}=s$)

$$= Ke^{a\nu}\mathbb{E}\left[1_{\left\{B_1^2\le\frac{a^2(2\nu+1)}{2(a-\log(K))}\right\}}e^{-\frac{a^2\nu^2}{2B_1^2}}\right]$$

$$- e^{a(\nu+1)}\mathbb{E}\left[1_{\left\{B_1^2\le\frac{a^2(2\nu+1)}{2(a-\log(K))}\right\}}e^{-\frac{a^2(\nu+1)^2}{2B_1^2}}\right], \tag{2.80}$$

i.e., with $t=\dfrac{2a^2(2\nu+1)}{a-\log(K)}$, $A=e^{2a\nu}$ and $B=e^{2a(\nu+1)}$:

$$\mathbb{E}\left[\left(K-M_a^{(\nu)}\right)^+\right] = K\sqrt{A}\,\mathbb{E}\left[1_{\{4B_1^2\le t\}}e^{-\frac{\log^2(A)}{8B_1^2}}\right] - \sqrt{B}\,\mathbb{E}\left[1_{\{4B_1^2\le t\}}e^{-\frac{\log^2(B)}{8B_1^2}}\right].$$
$$\tag{2.81}$$

We now apply formula (1.58), Theorem 1.4:

$$\mathbb{E}[(\mathscr{E}_t-K)^\pm] = (1-K)^\pm + \sqrt{K}\,\mathbb{E}\left[1_{\{4B_1^2\le t\}}e^{-\frac{\log^2(K)}{8B_1^2}}\right]$$

successively with $K=e^{2a\nu}$ and $K=e^{2a(\nu+1)}$. We obtain:

$$\mathbb{E}\left[\left(\mathscr{E}_t-e^{2a\nu}\right)^+\right] = e^{a\nu}\mathbb{E}\left[1_{\{4B_1^2\le t\}}e^{-\frac{a^2\nu^2}{2B_1^2}}\right], \tag{2.82}$$

$$\mathbb{E}\left[\left(\mathscr{E}_t-e^{2a(\nu+1)}\right)^+\right] = e^{a(\nu+1)}\mathbb{E}\left[1_{\{4B_1^2\le t\}}e^{-\frac{a^2(\nu+1)^2}{2B_1^2}}\right]. \tag{2.83}$$

Gathering (2.80), (2.82) and (2.83) ends the proof of (2.77) and of Proposition 2.1.

$\square$

## 2.6.3 Computation of the Law of $\mathscr{G}_K^{(M^{(\nu)})}$

We shall apply Proposition 2.1 to get the law of the r.v. $\mathscr{G}_K^{(M^{(\nu)})} := \sup\{a \geq 0;\ M_a^{(\nu)} = K\}$.

**Proposition 2.2.** *The r.v.* $\mathscr{G}_K^{(M^{(\nu)})}$ *admits as probability density the function* $f_{\mathscr{G}_K^{(M^{(\nu)})}}$ *given, for $K < e^a$, by:*

$$f_{\mathscr{G}_K^{(M^{(\nu)})}}(a) = \nu \left\{ \mathbb{P}\left( G_{2a\nu}^{(1/2)} \leq t \right) - \mathbb{P}\left( T_{2a\nu}^{(1/2)} < t \right) \right\}$$

$$+ \frac{\nu+1}{K} \left\{ \mathbb{P}\left( T_{2a(\nu+1)}^{(1/2)} \leq t \right) - \mathbb{P}\left( G_{2a(\nu+1)}^{(1/2)} < t \right) \right\} \tag{2.84}$$

$$= 2\frac{\nu+1}{K} e^{2a(\nu+1)} \mathbb{P}\left( B_t^{(-1/2)} > 2a(\nu+1) \right) - 2\nu e^{2a\nu} \mathbb{P}\left( B_t^{(-1/2)} > 2a\nu \right) \tag{2.85}$$

*with the notations of Chapter 1 and where $t = \dfrac{2a^2(2\nu+1)}{a - \log(K)}$.*

*Proof.* We first prove (2.84).

Since $\left( M_a^{(\nu)}, a \geq 0 \right)$ has no positive jumps, we may apply Theorem 2.1 (see Remark 2.1) to obtain:

$$\mathbb{P}\left( \mathscr{G}_K^{(M^{(\nu)})} \leq a \right) = \frac{1}{K} \mathbb{E}\left[ \left( K - M_a^{(\nu)} \right)^+ \right]$$

$$= \frac{1}{K} \left( \mathbb{E}\left[ K \left( \mathscr{E}_t - e^{2a\nu} \right)^+ \right] - \mathbb{E}\left[ \left( \mathscr{E}_t - e^{2a(\nu+1)} \right)^+ \right] \right)$$

$$\text{(from Proposition 2.1, with } t = \frac{2a^2(2\nu+1)}{a - \log(K)} \text{)}$$

$$= \mathbb{P}\left( G_{2a\nu}^{(1/2)} \leq t \right) - \frac{1}{K} \mathbb{P}\left( G_{2a(\nu+1)}^{(1/2)} \leq t \right) \quad \text{(from Theorem 1.2)} \tag{2.86}$$

$$= \int_0^{\frac{2a^2(2\nu+1)}{a-\log(K)}} \frac{1}{2\sqrt{2\pi u}} \exp\left( -\frac{1}{2u} \left( 2a\nu - \frac{u}{2} \right)^2 \right) du \tag{2.87}$$

$$- \frac{1}{K} \int_0^{\frac{2a^2(2\nu+1)}{a-\log(K)}} \frac{1}{2\sqrt{2\pi u}} \exp\left( -\frac{1}{2u} \left( 2a(\nu+1) - \frac{u}{2} \right)^2 \right) du.$$

We then differentiate both sides of (2.87) with respect to $a$. The terms coming from the differentiation of the upper bound in both integrals on the RHS cancel, and it finally remains:

$$\frac{\partial}{\partial a}\mathbb{P}\left(\mathscr{G}_K^{(M^{(\nu)})}\leq a\right)$$

$$=\int_0^t\frac{1}{2\sqrt{2\pi u}}\left(\nu-\frac{4a\nu^2}{u}\right)\exp\left(-\frac{1}{2u}\left(2a\nu-\frac{u}{2}\right)^2\right)du$$

$$-\frac{1}{K}\int_0^t\frac{1}{2\sqrt{2\pi u}}\left(\nu+1-\frac{4a(\nu+1)^2}{u}\right)\exp\left(-\frac{1}{2u}\left(2a(\nu+1)-\frac{u}{2}\right)^2\right)du$$

$$=\nu\left\{\mathbb{P}\left(G_{2a\nu}^{(1/2)}\leq t\right)-\mathbb{P}\left(T_{2a\nu}^{(1/2)}\leq t\right)\right\}$$

$$-\frac{\nu+1}{K}\left\{\mathbb{P}\left(G_{2a(\nu+1)}^{(1/2)}\leq t\right)-\mathbb{P}\left(T_{2a(\nu+1)}^{(1/2)}\leq t\right)\right\}.$$

This is relation (2.84).

We now prove (2.85)

From Theorem 1.3, identities (1.54) and (1.55) for $K>1$ and $\nu>0$, we deduce:

$$\mathbb{P}\left(T_{\log(K)}^{(\nu)}\leq t\right)=\mathbb{E}\left[\mathscr{E}_t^{(2\nu)}1_{\{\mathscr{E}_t^{(2\nu)}>K^{2\nu}\}}\right]+K^{2\nu}\mathbb{P}\left(\mathscr{E}_t^{(2\nu)}>K^{2\nu}\right)$$

$$\mathbb{P}\left(G_{\log(K)}^{(\nu)}\leq t\right)=\mathbb{E}\left[\mathscr{E}_t^{(2\nu)}1_{\{\mathscr{E}_t^{(2\nu)}>K^{2\nu}\}}\right]-K^{2\nu}\mathbb{P}\left(\mathscr{E}_t^{(2\nu)}>K^{2\nu}\right).$$

By subtracting:

$$\mathbb{P}\left(T_{\log(K)}^{(\nu)}\leq t\right)-\mathbb{P}\left(G_{\log(K)}^{(\nu)}\leq t\right)=2K^{2\nu}\mathbb{P}\left(\mathscr{E}_t^{(2\nu)}>K^{2\nu}\right)$$

$$=2K^{2\nu}\mathbb{P}\left(B_t^{(\nu)}>\log(K)\right).\qquad(2.88)$$

Finally, we obtain (2.85) from (2.84) by applying (2.88) first with $\nu=1/2$ and $K=e^{2a\nu}$, and then with $K=e^{2a(\nu+1)}$ noting that this is allowed since $e^{2a\nu}$ and $e^{2a(\nu+1)}$ are larger than 1.

$\square$

### 2.6.4 A More Probabilistic Approach of Proposition 2.2

We shall now prove again Proposition 2.2 via a more probabilistic method, when $\nu=0$. More precisely, we will show:

**Proposition 2.3 ($\nu=0$).**

i) $\mathbb{P}\left(\mathscr{G}_K^{(M^{(0)})}\leq a\right)=\mathbb{P}\left(G_{\log(K)}^{(-1/2)}\leq2(a-\log(K))\right)$

$$-\mathbb{P}\left(G_{2a-\log(K)}^{(1/2)}\leq2(a-\log(K))\right).\qquad(2.89)$$

*ii)* $f_{\mathscr{G}_K^{(M^{(0)})}}(a) = \mathbb{P}\left(T_{2a-\log(K)}^{(1/2)} \le 2(a-\log(K))\right) - \mathbb{P}\left(G_{2a-\log(K)}^{(1/2)} \le 2(a-\log(K))\right)$

$$= 2\frac{e^{2a}}{K}\mathbb{P}\left(B_{2(a-\log(K))} > 3a - 2\log(K)\right). \tag{2.90}$$

In the following, we shall first prove Proposition 2.3, and then, we will check that relations (2.89) and (2.90) coincide with those of Proposition 2.2.

*Proof.* We first give a geometric representation of the two events $\left\{\mathscr{G}_K^{(M^{(0)})} \le a\right\}$ and $\left\{G_{\log(K)}^{(-1/2)} \le 2(a-\log(K))\right\}$. We have,

$$\mathscr{G}_K(M^{(0)}) := \sup\left\{a \ge 0;\ e^{a-\frac{T_a}{2}} \ge K\right\} = \sup\left\{a \ge 0;\ a \ge \frac{1}{2}T_a + \log(K)\right\},$$

hence the event $\left\{\mathscr{G}_K^{(M^{(0)})} \le a\right\}$ is composed of the Brownian paths which do not leave the hatched area of Fig. 1:

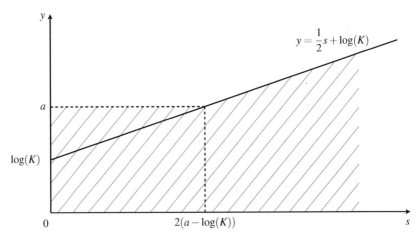

**Fig. 1** $\left\{\mathscr{G}_K^{(M^{(0)})} \le a\right\}$

Similarly:

$$\mathscr{G}_{\log(K)}^{(-1/2)} := \sup\left\{s \ge 0;\ B_s^{(-1/2)} = \log(K)\right\} = \sup\left\{s \ge 0;\ B_s = \frac{1}{2}s + \log(K)\right\}$$

and the event $\left\{G_{\log(K)}^{(-1/2)} \le 2(a-\log(K))\right\}$ is composed of the Brownian paths which do not leave the hatched area of Fig. 2:

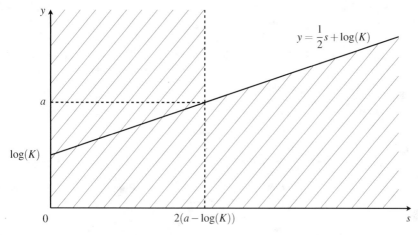

**Fig. 2** $\left\{ \mathcal{G}_{\log(K)}^{(-1/2)} \le 2(a - \log(K)) \right\}$

Consequently:

$$\mathbb{P}\left( \mathcal{G}_K(M^{(0)}) \le a \right) = \mathbb{P}\left( \left\{ G_{\log(K)}^{(-1/2)} \le 2(a - \log(K)) \right\} \cap \left\{ \sup_{s \le 2(a - \log(K))} B_s \le a \right\} \right).$$

We put $T = 2(a - \log(K))$, and this identity becomes:

$$\mathbb{P}\left( \mathcal{G}_K(M^{(0)}) \le a \right) = \mathbb{P}\left( G_{\log(K)}^{(-1/2)} \le T \right) - \mathbb{P}\left( \left\{ G_{\log(K)}^{(-1/2)} \le T \right\} \cap \left\{ \sup_{s \le T} B_s > a \right\} \right).$$

$$(2.91)$$

Let us now study the event:

$$\left\{ G_{\log(K)}^{(-1/2)} \le T \right\} \cap \left\{ \sup_{s \le T} B_s > a \right\} = \left\{ \inf_{s \ge T} \left( \frac{s}{2} + \log(K) - B_s \right) > 0 \right\} \cap \left\{ \sup_{s \le T} B_s > a \right\}.$$

On this event, the Brownian paths hit a.s. level $a$ before time $T$, hence, applying Désiré André's reflection principle, we can replace the piece of path after $T_a(B)$ by its symmetric with respect to the horizontal line of $y$-coordinate $a$. We then obtain:

$$\mathbb{P}\left( \left\{ \inf_{s \ge T} \frac{s}{2} + \log(K) - B_s > 0 \right\} \cap \left\{ \sup_{s \le T} B_s > a \right\} \right)$$

$$= \mathbb{P}\left( \left\{ \inf_{s \ge T} B_s - \left( 2a - \log(K) - \frac{s}{2} \right) > 0 \right\} \cap \left\{ \sup_{s \le T} B_s > a \right\} \right)$$

$$= \mathbb{P}\left( \inf_{s \ge T} B_s - \left( 2a - \log(K) - \frac{s}{2} \right) > 0 \right)$$

$$= \mathbb{P}\left( G_{2a - \log(K)}^{(1/2)} \le T \right).$$

Plugging back this expression in (2.91), we finally get:

$$\mathbb{P}\left(\mathscr{G}_K^{(M^{(0)})} \le a\right) = \mathbb{P}\left(G_{\log(K)}^{(-1/2)} \le 2(a - \log(K))\right) - \mathbb{P}\left(G_{2a-\log(K)}^{(1/2)} \le 2(a - \log(K))\right),$$

which is Point ($i$) of Proposition 2.3.

Let us now compute the density of $\mathscr{G}_K^{(M^{(0)})}$. We have:

$$\mathbb{P}\left(\mathscr{G}_K^{(M^{(0)})} \le a\right) = \int_0^{2(a-\log(K))} \frac{1}{2\sqrt{2\pi u}} \left(e^{-\frac{1}{2u}\left(\log(K)+\frac{u}{2}\right)^2} - e^{-\frac{1}{2u}\left(2a-\log(K)-\frac{u}{2}\right)^2}\right) du.$$

We differentiate this identity with respect to $a$. Once again, the terms coming from the differentiation of the upper bound in both integrals on the RHS cancel, and it finally remains:

$$
\begin{aligned}
f_{\mathscr{G}_K^{(M^{(0)})}}(a) &= \int_0^{2(a-\log(K))} \frac{1}{2\sqrt{2\pi u}} \frac{1}{2u} 4\left(2a - \log(K) - \frac{u}{2}\right) e^{-\frac{1}{2u}\left(2a-\log(K)-\frac{u}{2}\right)^2} du \\
&= \int_0^{2(a-\log(K))} \left(\frac{2a - \log(K)}{\sqrt{2\pi u^3}} - \frac{1}{2\sqrt{2\pi u}}\right) e^{-\frac{1}{2u}\left(2a-\log(K)-\frac{u}{2}\right)^2} du \\
&= \mathbb{P}\left(T_{2a-\log(K)}^{(1/2)} \le 2(a - \log(K))\right) - \mathbb{P}\left(G_{2a-\log(K)}^{(1/2)} \le 2(a - \log(K))\right) \\
&= 2\frac{e^{2a}}{K} \mathbb{P}\left(B_{2(a-\log(K))} > 3a - 2\log(K)\right),
\end{aligned}
$$

the last equality coming from (2.88) with $\nu = 1/2$, and where we have replaced $K$ by $\frac{e^{2a}}{K}$ $\left(\frac{e^{2a}}{K} > 1 \text{ since } K < e^a\right)$.

$\square$

We now show that the two expressions (2.84) and (2.90) of the density of $\mathscr{G}_K^{(M^{(0)})}$ coincide. With $\nu = 0$, (2.84) becomes:

$$
\begin{aligned}
f_{\mathscr{G}_K^{(M^{(0)})}}(a) &= \frac{1}{K}\left\{\mathbb{P}\left(T_{2a}^{1/2} \le \frac{2a^2}{a - \log(K)}\right) - \mathbb{P}\left(G_{2a}^{1/2} \le \frac{2a^2}{a - \log(K)}\right)\right\} \\
&= \frac{1}{K}\int_0^{\frac{2a^2}{a-\log(K)}} \left(\frac{2a}{\sqrt{2\pi u^3}} - \frac{1}{2\sqrt{2\pi u}}\right) e^{-\frac{1}{2u}\left(2a-\frac{u}{2}\right)^2} du \\
&= \frac{1}{K}\int_{2(a-\log(K))}^{\infty} \left(\frac{1}{\sqrt{2\pi s}} - \frac{a}{2\sqrt{2\pi s^3}}\right) e^{-\frac{1}{2s}(a-s)^2} ds \\
&\quad \text{(after the change of variable } s = \frac{4a^2}{u}\text{)} \\
&= \frac{1}{K}\left\{\mathbb{P}\left(G_a^{(1)} \ge 2(a - \log(K))\right) - \mathbb{P}\left(T_a^{(1)} \ge 2(a - \log(K))\right)\right\} \\
&= \frac{1}{K}\left\{\mathbb{P}\left(T_a^{(1)} \le 2(a - \log(K))\right) - \mathbb{P}\left(G_a^{(1)} \le 2(a - \log(K))\right)\right\}.
\end{aligned}
$$

$$(2.92)$$

Then, from (2.88), one obtains:

$$\mathbb{P}\left(T_a^{(1)} \le 2(a - \log(K))\right) - \mathbb{P}\left(G_a^{(1)} \le 2(a - \log(K))\right)$$

$$= 2\left(e^a\right)^2 \mathbb{P}\left(B_{2(a-\log(K))}^{(-1)} > a\right)$$

$$= 2e^{2a}\mathbb{P}\left(B_{2(a-\log(K))} - 2(a - \log(K)) > a\right)$$

$$= 2e^{2a}\mathbb{P}\left(B_{2(a-\log(K))} > 3a - 2\log(K)\right). \tag{2.93}$$

Hence, plugging (2.93) in (2.92):

$$f_{\mathscr{G}_K^{(M^{(0)})}}(a) = 2\frac{e^{2a}}{K}\mathbb{P}\left(B_{2(a-\log(K))} > 3a - 2\log(K)\right),$$

which is (2.84).

### 2.6.5 An application of Proposition 2.1 to the Local Times of the Martingale $(\mathscr{E}_t, t \ge 0)$

We end this Section 2.6 by an application of Proposition 2.1 to the local times $(L_t^K, t \ge 0, K \ge 0)$ of the martingale $(\mathscr{E}_t, t \ge 0)$.

**Proposition 2.4.** Let $(L_t^K, t \ge 0, K \ge 0)$ denote the family of local times of the martingale $(\mathscr{E}_t, t \ge 0)$. Then, for all $0 < K < e^a$ and $\nu > 0$:

$$(K-1)^+ + \frac{1}{2}\mathbb{E}\left[L_{T_a^{(\nu)}}^K\right] = \frac{K}{2}\mathbb{E}\left[L_t^{e^{2a\nu}}\right] - \frac{1}{2}\mathbb{E}\left[L_t^{e^{2a(\nu+1)}}\right]. \tag{2.94}$$

where $t = \dfrac{2a^2(2\nu + 1)}{a - \log(K)}$.

*Proof.* We start by applying Tanaka's formula to the martingale $(\mathscr{E}_t, t \ge 0)$ stopped at the time $s \wedge T_a^{(\nu)}$. Using the identity $B_t - \frac{t}{2} = B_t^{(\nu)} - \left(\frac{1}{2} + \nu\right)t$:

$$\mathbb{E}\left[\left(K - \mathscr{E}_{s\wedge T_a^{(\nu)}}\right)^+\right] = \mathbb{E}\left[\left(K - \exp\left\{B_{s\wedge T_a^{(\nu)}}^{(\nu)} - \left(\frac{1}{2} + \nu\right)(s \wedge T_a^{(\nu)})\right\}\right)^+\right]$$

$$= (K-1)^+ + \frac{1}{2}\mathbb{E}\left[L_{s\wedge T_a^{(\nu)}}^K\right].$$

We then let $s$ tend to $+\infty$, using the dominated convergence theorem for the term $\mathbb{E}\left[\left(K - \mathscr{E}_{s\wedge T_a^{(\nu)}}\right)^+\right]$, and the monotone convergence theorem for $\mathbb{E}\left[L_{s\wedge T_a^{(\nu)}}^K\right]$. This leads to:

$$(K-1)^+ + \frac{1}{2}\mathbb{E}\left[L^K_{T_a^{(\nu)}}\right] = \mathbb{E}\left[\left(K - \exp\left\{a - \left(\frac{1}{2}+\nu\right)T_a^{(\nu)}\right\}\right)^+\right]$$

$$= K\mathbb{E}\left[\left(\mathscr{E}_t - e^{2a\nu}\right)^+\right] - \mathbb{E}\left[\left(\mathscr{E}_t - e^{2a(\nu+1)}\right)^+\right],$$

where $t = \dfrac{2a^2(2\nu+1)}{a-\log(K)}$ from Proposition 2.1. It only remains to apply once again Tanaka's formula to obtain the result (2.94), since $e^{2a\nu}$ and $e^{2a(\nu+1)}$ are larger than $1 \ (= \mathscr{E}_0)$.

$\square$

*Remark 2.3.* Since the application $\varphi : x \mapsto e^x$ is a $\mathscr{C}^1$-diffeomorphism from $\mathbb{R}$ onto $]0,+\infty[$, we have, with obvious notation:

$$L^{\varphi(x)}_t(\varphi(B^{(-1/2)})) = \varphi'(x)L^x_t(B^{(-1/2)}).$$

Thus, denoting by $\left(L^x_t(B^{(-1/2)}), t \geq 0, x \in \mathbb{R}\right)$ the family of local times of the semimartingale $\left(B^{(-1/2)}_t, t \geq 0\right)$, Proposition 2.4 can be rewritten:

$$(K-1)^+ + \frac{K}{2}\mathbb{E}\left[L^{\log(K)}_{T_a^{(\nu)}}(B^{(-1/2)})\right]$$

$$= \frac{K}{2}e^{2a\nu}\mathbb{E}\left[L^{2a\nu}_t(B^{(-1/2)})\right] - \frac{e^{2a(\nu+1)}}{2}\mathbb{E}\left[L^{2a(\nu+1)}_t(B^{(-1/2)})\right]. \quad (2.95)$$

In the same spirit as that of Proposition 2.3, we could give, when $\nu = 0$, a probabilistic proof of identity (2.95). We leave this proof to the interested reader. Formula (2.95) writes then:

$$\mathbb{E}\left[L^{\log(K)}_{T_a}(B^{(-1/2)})\right] = \mathbb{E}\left[L^{\log(K)}_{2(a-\log(K))}(B^{(-1/2)})\right] - \mathbb{E}\left[L^{2a-\log(K)}_{2(a-\log(K))}(B^{(1/2)})\right].$$

$$(2.96)$$

## 2.7 The case $M_\infty \neq 0$

### 2.7.1 Hypotheses

Our aim in this Section is to give a generalization of Theorem 2.1 when we remove the assumption $M_\infty = 0$ a.s. More precisely, we still consider a positive and continuous local martingale $(M_t, t \geq 0)$ defined on a filtered probability space $(\Omega, (\mathscr{F}_t, t \geq 0), \mathscr{F}_\infty, \mathbb{P})$. We assume that $(\mathscr{F}_t := \sigma(M_s, s \leq t), t \geq 0)$ is the natural filtration of $(M_t, t \geq 0)$ and that $\mathscr{F}_\infty := \bigvee_{t \geq 0} \mathscr{F}_t$. As a positive local martingale,

$(M_t, t \geq 0)$ is a positive supermartingale, which therefore converges a.s. towards a r.v $M_\infty$ as $t \to \infty$. But, as opposed to the previous sections, we no longer assume that $M_\infty = 0$ a.s. Then, in this new framework, Theorem 2.1 extends in the following way:

### 2.7.2 A Generalization of Theorem 2.1

**Theorem 2.5.** *Let $(M_t, t \geq 0)$ a positive continuous local martingale. For every $K \geq 0$:*

$$\mathbb{E}\left[1_{\{\mathscr{G}_K^{(M)} \leq t\}} (K - M_\infty)^+ | \mathscr{F}_t\right] = (K - M_t)^+. \tag{2.97}$$

*Of course, if $M_\infty = 0$ a.s., we recover Theorem 2.1.*

### 2.7.3 First Proof of Theorem 2.5

It hinges on the balayage formula which we first recall (see [70], Chapter VI, p.260):

**Balayage formula:**
*Let $(Y_t, t \geq 0)$ be a continuous semi-martingale and $\mathscr{G}_Y(t) = \sup\{s \leq t; Y_s = 0\}$. Then, for any bounded predictable process $(\Phi_s, s \geq 0)$, we have:*

$$\Phi_{\mathscr{G}_Y(t)} Y_t = \Phi_0 Y_0 + \int_0^t \Phi_{\mathscr{G}_Y(s)} dY_s. \tag{2.98}$$

We note $\mathscr{G}_K(s) := \sup\{u \leq s; M_u = K\}$. The balayage formula, applied to $(Y_t = (K - M_t)^+, t \geq 0)$, becomes, for every bounded and predictable process $(\Phi_s, s \geq 0)$ and $t \geq 0$:

$$\Phi_{\mathscr{G}_K(t)} (K - M_t)^+ = \Phi_0 (K - M_0)^+ - \int_0^t \Phi_{\mathscr{G}_K(s)} 1_{\{M_s < K\}} dM_s + \frac{1}{2} \int_0^t \Phi_{\mathscr{G}_K(s)} dL_s^K$$

(where $(L_s^K, s \geq 0)$ denotes the local time at level $K$
of the local martingale $(M_s, s \geq 0)$),

$$= \Phi_0 (K - M_0)^+ - \int_0^t \Phi_{\mathscr{G}_K(s)} 1_{\{M_s < K\}} dM_s + \frac{1}{2} \int_0^t \Phi_s dL_s^K \tag{2.99}$$

since $dL_s^K$ charges only the set of times for which $M_s = K$ i.e. for which $\mathscr{G}_K(s) = s$. We now apply (2.99) between $t$ and $+\infty$ to obtain:

$$\mathbb{E}\left[\Phi_{\mathscr{G}_K^{(M)}} (K - M_\infty)^+ | \mathscr{F}_t\right] = \Phi_{\mathscr{G}_K(t)} (K - M_t)^+ + \frac{1}{2} \mathbb{E}\left[\int_t^\infty \Phi_s dL_s^K | \mathscr{F}_t\right] \tag{2.100}$$

since $\mathscr{G}_K(\infty) = \mathscr{G}_K^{(M)}$. Taking in (2.100) $\Phi_s = \varphi(s)1_{\{s \le t\}}$ for a bounded Borel function $\varphi : \mathbb{R}^+ \mapsto \mathbb{R}$ and observing that $\mathscr{G}_K(t) \le t$ and $\int_t^\infty \varphi(s)1_{\{s \le t\}}dL_s^K = 0$, we obtain:

$$\mathbb{E}\left[ \varphi\left(\mathscr{G}_K^{(M)}\right) 1_{\{\mathscr{G}_K^{(M)} \le t\}} (K - M_\infty)^+ \mid \mathscr{F}_t \right] = \varphi(\mathscr{G}_K(t))(K - M_t)^+ .$$

$\square$

### 2.7.4 A Second Proof of Theorem 2.5

Let $T$ be a $\mathscr{F}_t$-stopping time. It is clear that:

$$\mathbb{E}\left[ 1_{\{\mathscr{G}_K^{(M)} \le T\}} (K - M_\infty)^+ \right] = \mathbb{E}\left[ 1_{\{d_T^K = \infty\}} (K - M_\infty)^+ \right] \tag{2.101}$$

with $d_T^K := \inf\{t > T; M_t = K\}$, since $\{\mathscr{G}_K^{(M)} \le T\} = \{d_T^K = \infty\}$. Then:

$$\mathbb{E}\left[ 1_{\{d_T^K = \infty\}} (K - M_\infty)^+ \right] = \mathbb{E}\left[ 1_{\{d_T^K = \infty\}} \left(K - M_{d_T^K}\right)^+ \right] = \mathbb{E}\left[ \left(K - M_{d_T^K}\right)^+ \right] \tag{2.102}$$

since, on the set $(d_T^K < \infty)$, $\left(K - M_{d_T^K}\right)^+ = 0$. We now note that, for $t$ between $T$ and $d_T^K$, $L_t^K$ is constant since $M_t \ne K$. Hence, from Tanaka's formula, the RHS of (2.102) equals:

$$\mathbb{E}\left[ \left(K - M_{d_T^K}\right)^+ \right] = \mathbb{E}\left[ (K - M_T)^+ \right]. \tag{2.103}$$

Gathering (2.101), (2.102) and (2.103), we obtain:

$$\mathbb{E}\left[ 1_{\{\mathscr{G}_K^{(M)} \le T\}} (K - M_\infty)^+ \right] = \mathbb{E}\left[ (K - M_T)^+ \right]. \tag{2.104}$$

This identity may be reinforced as:

$$\mathbb{E}\left[ 1_{\{\mathscr{G}_K^{(M)} \le T\}} (K - M_\infty)^+ \mid \mathscr{F}_T \right] = (K - M_T)^+$$

by replacing in (2.104) $T$ by $T_\Lambda := \begin{cases} T \text{ on } \Lambda, \\ +\infty \text{ on } \Lambda^c \end{cases}$ for any generic set $\Lambda \in \mathscr{F}_T$.

$\square$

## 2.7.5 On the Law of $S_\infty := \sup_{t \geq 0} M_t$

We have seen in Section 2.1 that, when $M_t \xrightarrow[t \to \infty]{} 0$ a.s., then:

$$S_\infty \overset{(law)}{=} \frac{M_0}{U} \tag{2.105}$$

where $U$ is uniform on $[0,1]$ and independent from $M_0$. What can be said about the law of $S_\infty$ if we remove the assumption $M_\infty = 0$ ? Going back to the proof of Lemma 2.1 and applying Doob's optional stopping Theorem, we obtain, for $b > a = M_0$ and $T_b := \inf\{t \geq 0; M_t = b\}$:

$$\mathbb{E}\left[M_{T_b}\right] = a \tag{2.106}$$

i.e.

$$b\mathbb{P}\left(S_\infty \geq b\right) + \mathbb{E}\left[M_\infty 1_{\{S_\infty < b\}}\right] = a. \tag{2.107}$$

Let us remark that, applying the monotone convergence theorem:

$$\lim_{b \to \infty} b\mathbb{P}\left(S_\infty \geq b\right) = a - \mathbb{E}[M_\infty]. \tag{2.108}$$

Furthermore, relation (2.107) leads us to introduce the function $\Phi : \mathbb{R}^+ \to \mathbb{R}^+$ defined by:

$$\Phi(S_\infty) = \mathbb{E}\left[M_\infty | S_\infty\right]. \tag{2.109}$$

It is clear that $\Phi(x) \leq x$ and (2.107) becomes:

$$b\mathbb{P}\left(S_\infty \geq b\right) + \mathbb{E}\left[\Phi(S_\infty)1_{\{S_\infty < b\}}\right] = a. \tag{2.110}$$

Assuming $\Phi$ is given, we may consider (2.110) as an equation for the distribution of $S_\infty$, and we obtain:

**Proposition 2.5.** *For simplicity, we assume that for every $x > 0$, $\Phi(x) < x$. Then, the law of $S_\infty$ is given by:*

$$\mathbb{P}\left(S_\infty \geq b\right) = \exp\left(-\int_a^b \frac{dx}{x - \Phi(x)}\right). \tag{2.111}$$

Observe that, since $S_\infty < \infty$ a.s., it follows from (2.111) that:

$$\int_a^\infty \frac{dx}{x - \Phi(x)} = +\infty. \tag{2.112}$$

*Proof.* From formula (2.110), denoting $\overline{\mu}(b) := \mathbb{P}\left(S_\infty \geq b\right)$, we obtain:

$$b\overline{\mu}(b) - \int_a^b \Phi(x)d\overline{\mu}(x) = a. \tag{2.113}$$

Consequently, by differentiation:

$$bd\overline{\mu}(b) + \overline{\mu}(b)db - \Phi(b)d\overline{\mu}(b) = 0$$

i.e.

$$(b - \Phi(b))d\overline{\mu}(b) = -\overline{\mu}(b)db \tag{2.114}$$

The above equation yields:

$$\overline{\mu}(b) = C\exp\left(-\int_a^b \frac{dx}{x - \Phi(x)}\right) \tag{2.115}$$

which implies $C = 1$ by taking $b = a$, since $\overline{\mu}(a) = 1$.

□

*Example 2.7.5.* Let $(B_t, t \geq 0)$ be a Brownian motion started at $a > 0$, and $S_t^B :=$ $\sup_{s \leq t} B_s$. For $0 < \alpha < 1$, we define the stopping time:

$$T_a^{(\alpha)} := \inf\{t \geq 0; B_t = \alpha S_t^B\} \tag{2.116}$$

to which we associate the martingale $\left(M_t = B_{t \wedge T_a^{(\alpha)}}, t \geq 0\right)$. Then, $\Phi(x) = \alpha x$, and consequently, we have:

$$\int_a^b \frac{dx}{x - \Phi(x)} = \frac{1}{1 - \alpha}\log\left(\frac{b}{a}\right) \qquad (b \geq a) \tag{2.117}$$

and

$$\overline{\mu}(b) = \exp\left(-\frac{1}{1 - \alpha}\log\left(\frac{b}{a}\right)\right) = \left(\frac{a}{b}\right)^{\frac{1}{1-\alpha}} \qquad (b \geq a). \tag{2.118}$$

(Observe that $\overline{\mu}$ is the tail of a Pareto distribution.)

*Remark 2.4.*
**a)** Under which condition(s) is $(M_t, t \geq 0)$ uniformly integrable ? This question had of course a negative answer when $M_\infty = 0$ a.s. But now ? Uniform integrability is equivalent to $\mathbb{E}[M_\infty] = \mathbb{E}[M_0] = a$, which is satisfied if and only if (see (2.109)):

$$\mathbb{E}[\Phi(S_\infty)] = a. \tag{2.119}$$

From (2.108), uniform integrability of $(M_t, t \geq 0)$ is also equivalent to:

$$\lim_{b \to \infty} b\mathbb{P}(S_\infty \geq b) = 0. \tag{2.120}$$

From (2.119) and (2.111), uniform integrability of $(M_t, t \geq 0)$ is also equivalent to:

$$\int_a^\infty \frac{\Phi(x)}{x - \Phi(x)}\left(\exp\left(-\int_a^x \frac{dy}{y - \Phi(y)}\right)\right)dx = a. \tag{2.121}$$

Let us check that (2.120) and (2.121) coincide. Indeed, we have, from (2.121):

$$-a = \int_a^\infty \left( \frac{x - \Phi(x)}{x - \Phi(x)} - \frac{x}{x - \Phi(x)} \right) \left( \exp\left( -\int_a^x \frac{dy}{y - \Phi(y)} \right) \right) dx$$

$$= \int_a^\infty \overline{\mu}(x)dx - \int_a^\infty \frac{x}{x - \Phi(x)} \exp\left( -\int_a^x \frac{dy}{y - \Phi(y)} \right) dx$$

$$= \int_a^\infty \overline{\mu}(x)dx + [x\overline{\mu}(x)]_a^\infty - \int_a^\infty \overline{\mu}(x)dx$$

(after an integration by parts)

$$= \lim_{b \to \infty} b\mathbb{P}(S_\infty \geq b) - a.$$

Going back to Example 2.7.5, we have: $\Phi(x) = \alpha x$ and $\overline{\mu}(b) = \mathbb{P}(S_\infty \geq b) = \left( \frac{a}{b} \right)^{\frac{1}{1-\alpha}}$. Hence: $b\mathbb{P}(S_\infty \geq b) = a^{\frac{1}{1-\alpha}} b^{-\frac{\alpha}{1-\alpha}} \xrightarrow[b \to \infty]{} 0$. Then, Example 2.7.5 is a case of uniform integrability.

**b)** Can we describe all the laws of $(M_t, t \geq 0)$ which satisfy (2.109) for a given $\Phi$ ? See Rogers [71] where the law of $(S_\infty, M_\infty)$ is described in all generality. See also Vallois [87]. However, these authors assume a priori that $(M_t, t \geq 0)$ is uniformly integrable. We shall study later (see Chapter 3) the law of $\mathscr{G}_K^{(M)}$ when $M_\infty \neq 0$.

## 2.8 Extension of Theorem 2.1 to the Case of Orthogonal Local Martingales

### 2.8.1 Statement of the Main Result

In this section, we shall extend Theorem 2.1 to the case of orthogonal local martingales. Let $(M_t^{(i)}, t \geq 0; i = 1, \cdots, n)$ be a set of $n$ positive, continuous local martingales such that, for all $i = 1, \cdots, n$:

$$\lim_{t \to \infty} M_t^{(i)} = 0 \quad \text{a.s.} \tag{2.122}$$

We assume moreover that these martingales are orthogonal, i.e., for all $1 \leq i < j \leq n$:

$$\left\langle M^{(i)}, M^{(j)} \right\rangle_t = 0 \quad (0 \leq t < \infty) \tag{2.123}$$

where $\left( \left\langle M^{(i)}, M^{(j)} \right\rangle_t, t \geq 0 \right)$ denotes the bracket of the martingales $M^{(i)}$ and $M^{(j)}$. The counterpart of Theorem 2.1 writes then:

**Theorem 2.6.** *Under the preceding hypotheses:*

i)  *For every $K_i \geq 0$ and every bounded $\mathscr{F}_t$-measurable r.v. $F_t$:*

$$\mathbb{E}\left[F_t \prod_{i=1}^{n}\left(K_i - M_t^{(i)}\right)^+\right] = \mathbb{E}\left[F_t \left(\prod_{i=1}^{n}K_i\right) 1_{\left\{\bigvee_{i=1,\cdots,n} \mathscr{G}_{K_i}^{(i)} \leq t\right\}}\right] \tag{2.124}$$

*where*   $\displaystyle\bigvee_{i=1,\cdots,n} \mathscr{G}_{K_i}^{(i)} := \sup_{i=1,\cdots,n} \mathscr{G}_{K_i}^{(M^{(i)})}.$   $\tag{2.125}$

ii)  *In other words, the submartingale $\mathbb{P}\left(\displaystyle\bigvee_{i=1,\cdots,n} \mathscr{G}_{K_i}^{(i)} \leq t | \mathscr{F}_t\right)$ equals:*

$$\mathbb{P}\left(\bigvee_{i=1,\cdots,n} \mathscr{G}_{K_i}^{(i)} \leq t | \mathscr{F}_t\right) = \prod_{i=1}^{n}\left(1 - \frac{M_t^{(i)}}{K_i}\right)^+ \tag{2.126}$$

$$= \prod_{i=1}^{n}\mathbb{P}\left(\mathscr{G}_{K_i}^{(i)} \leq t | \mathscr{F}_t\right). \tag{2.127}$$

We shall discuss the law of $\displaystyle\bigvee_{i=1,\cdots,n} \mathscr{G}_{K_i}^{(i)}$ later (see Theorem 2.7). For the moment, we give two proofs of Theorem 2.6. For clarity's sake, we shall assume that $n = 2$, i.e. we are considering two orthogonal local martingales $(M_t, t \geq 0)$ and $(M_t', t \geq 0)$. Clearly, our arguments extend easily to the general case.

## 2.8.2 First Proof of Theorem 2.6, via Enlargement Theory

(2.124) writes:

$$\mathbb{E}\left[F_t\left(K - M_t\right)^+\left(K' - M_t'\right)^+\right] = \mathbb{E}\left[F_t K K' 1_{\{\mathscr{G}_K \vee \mathscr{G}_{K'}' \leq t\}}\right]. \tag{2.128}$$

To prove (2.128), we first use Point $(i)$ of Theorem 2.1 which allows to write the LHS of (2.128) as:

$$\mathbb{E}\left[F_t\left(K - M_t\right)^+\left(K' - M_t'\right)^+\right] = \mathbb{E}\left[F_t\left(K' - M_t'\right)^+ K 1_{\{\mathscr{G}_K \leq t\}}\right]. \tag{2.129}$$

Let us define $\left(\mathscr{F}_t^K, t \geq 0\right)$ the smallest filtration containing $(\mathscr{F}_t, t \geq 0)$ and making $\mathscr{G}_K$ a stopping time. From Lemma 2.2 below, $(M_t', t \geq 0)$ remains a local martingale in this new filtration. Then, we may use again Theorem 2.1, this time with respect to $(M_t', t \geq 0)$ which is a local martingale in $\left(\mathscr{F}_t^K, t \geq 0\right)$ to obtain:

$$\mathbb{E}\left[F_t\left(K-M_t\right)^+\left(K'-M_t'\right)^+\right] = \mathbb{E}\left[F_tK1_{\{\mathscr{G}_K\leq t\}}\left(K'-M_t'\right)^+\right]$$

$$= \mathbb{E}\left[F_tK1_{\{\mathscr{G}_K\leq t\}}K'1_{\{\mathscr{G}_{K'}'\leq t\}}\right]$$

$$= \mathbb{E}\left[F_tKK'1_{\{\mathscr{G}_K\vee\mathscr{G}_{K'}'\leq t\}}\right].$$

This is Theorem 2.6.

**Lemma 2.2.** *($M_t', t \geq 0$) is a local martingale in the enlarged filtration ($\mathscr{F}_t^K, t \geq 0$).*

*Proof.* For the reader's convenience, we recall some enlargement formulae (see Mansuy-Yor [51], and Subsection 3.2.1. Let $Z_t^{(K)} := \mathbb{P}(\mathscr{G}_K > t | \mathscr{F}_t)$ the Azéma supermartingale associated with $\mathscr{G}_K$. Then, if ($M_t', t \geq 0$) is a continuous ($\mathscr{F}_t, t \geq 0$)-local martingale, there exists ($\widetilde{M}_t', t \geq 0$) a ($\mathscr{F}_t^K, t \geq 0$) local martingale such that:

$$M_t' = \widetilde{M}_t' + \int_0^{t\wedge\mathscr{G}_K}\frac{d\left\langle M',Z^{(K)}\right\rangle_s}{Z_s^{(K)}} - \int_{t\wedge\mathscr{G}_K}^t\frac{d\left\langle M',Z^{(K)}\right\rangle_s}{1-Z_s^{(K)}}. \tag{2.130}$$

In our situation, we have, from Theorem 2.1:

$$Z_t^{(K)} = \left(1-\frac{M_t}{K}\right)^+. \tag{2.131}$$

This implies that:

$$d\left\langle M',Z^{(K)}\right\rangle_s = -1_{\{M_s<K\}}\frac{d\langle M',M\rangle_s}{K} = 0 \tag{2.132}$$

since $M$ and $M'$ are orthogonal. Hence, Lemma 2.2 and Theorem 2.6 are proven.

□

## 2.8.3  Second Proof of Theorem 2.6, via Knight's Representation of Orthogonal Continuous Martingales

It hinges upon the following:

**Lemma 2.3.** *For every $t \geq 0$, conditionally on $\mathscr{F}_t$:*

$$\left(\sup_{u\geq t}M_u, \sup_{u\geq t}M_u'\right) \overset{(law)}{=} \left(\frac{M_t}{U}, \frac{M_t'}{U'}\right) \tag{2.133}$$

*where, on the RHS of (2.133), $U$ and $U'$ are two independent r.v's which are uniform on $[0,1]$, and independent from $\mathscr{F}_t$.*

Let us admit for a moment Lemma 2.3, and let us prove Theorem 2.6. We have, since $M_t \underset{t\to\infty}{\longrightarrow} 0$ a.s. and $M_t' \underset{t\to\infty}{\longrightarrow} 0$ a.s.:

$$\mathbb{P}\left(\mathcal{G}_K \vee \mathcal{G}'_{K'} \leq t | \mathcal{F}_t\right) = \mathbb{P}\left(\left(\sup_{u \geq t} M_u < K\right) \cap \left(\sup_{u \geq t} M'_u < K'\right) | \mathcal{F}_t\right)$$

$$= \mathbb{P}\left(\left(\frac{M_t}{U} < K\right) \cap \left(\frac{M'_t}{U'} < K'\right) | \mathcal{F}_t\right) \quad \text{(from Lemma 2.3)}$$

$$= \mathbb{P}\left(\left(\frac{M_t}{K} < U\right) \cap \left(\frac{M'_t}{K'} < U'\right) | \mathcal{F}_t\right)$$

$$= \left(1 - \frac{M_t}{K}\right)^+ \left(1 - \frac{M'_t}{K'}\right)^+$$

which is Theorem 2.6.

We now give two proofs of Lemma 2.3.

**First proof of Lemma 2.3**
Replacing $(M_s, s \geq 0)$ and $(M'_s, s \geq 0)$ by $(M_{t+s}, s \geq 0)$ and $(M'_{t+s}, s \geq 0)$, it is enough to prove Lemma 2.3 for $t = 0$. Let $b \geq a$, $b' \geq a'$, $T_b := \inf\{t \geq 0; M_t = b\}$ and $T'_{b'} := \inf\{t \geq 0; M'_t = b'\}$. Since $M$ and $M'$ are orthogonal, $\left(M_{t \wedge T_b} \cdot M'_{t \wedge T'_{b'}}, t \geq 0\right)$ is a bounded martingale. By Doob's optional stopping Theorem, we have:

$$aa' = \mathbb{E}\left[M_{T_b} M'_{T'_{b'}}\right]$$

$$= bb' \mathbb{P}\left(T_b < \infty, T'_{b'} < \infty\right)$$

since $M_{T_b} = 0$ (resp. $M'_{T'_{b'}} = 0$) on the set $\{T_b = +\infty\}$ (resp. $\{T'_{b'} = +\infty\}$). Thus:

$$\mathbb{P}\left(\left\{\sup_{s \geq 0} M_s > b\right\} \cap \left\{\sup_{s \geq 0} M'_s > b'\right\}\right) = \mathbb{P}\left(T_b < \infty, T'_{b'} < \infty\right) = \frac{a}{b}\frac{a'}{b'}$$

which is Lemma 2.3.

**Second proof of Lemma 2.3**
From the Dambis, Dubins, Schwarz's Theorem, there exist two Brownian motions $(\beta_u, u \leq T_0(\beta))$ and $(\beta'_u, u \leq T_0(\beta'))$ with $T_0(\beta) := \inf\{t \geq 0; \beta_t = 0\}$ (resp. $T_0(\beta') := \inf\{t \geq 0; \beta'_t = 0\}$), started at $a$ and $a'$ such that:

$$M_u = \beta_{\langle M \rangle_u} \quad , \quad M'_u = \beta'_{\langle M' \rangle_u}, \tag{2.134}$$

Moreover:

$$\langle M \rangle_\infty = T_0(\beta) \quad , \quad \langle M' \rangle_\infty = T_0(\beta'). \tag{2.135}$$

Let us admit for a moment that the orthogonality of $M$ and $M'$ implies:

$$(\beta_u, u \leq T_0(\beta)) \quad \text{and} \quad (\beta'_u, u \leq T_0(\beta')) \text{ are independent,} \tag{2.136}$$

and let us show that (2.136) implies Lemma 2.3. As we have already mentioned it, it is sufficient to prove Lemma 2.3 for $t = 0$. But, since from Lemma 2.1:

$$\sup_{u \geq 0} M_u = \sup_{u \leq T_0(\beta)} \beta_u \overset{(law)}{=} \frac{a}{U},$$

(2.136) implies then:

$$\left( \sup_{u \geq 0} M_u , \sup_{u \geq 0} M'_u \right) = \left( \sup_{u \leq T_0(\beta)} \beta_u , \sup_{u \leq T_0(\beta')} \beta'_u \right) \overset{(law)}{=} \left( \frac{a}{U} , \frac{a'}{U'} \right) \quad (2.137)$$

where, on the RHS of (2.137), $U$ and $U'$ are two independent r.v.'s uniform on $[0,1]$.

It remains to prove (2.136)

Of course, when $\langle M \rangle_\infty = \langle M' \rangle_\infty = +\infty$, relation (2.136) is the celebrated Knight's Theorem on the representation of orthogonal martingales. The proof we shall present below is a variant of P.A. Meyer's proof [56] of Knight's Theorem. We consider two square integrable variables $H$ and $H'$ which are measurable with respect to $\sigma(\beta_u, u \leq T_0(\beta))$ and $\sigma(\beta'_u, u \leq T_0(\beta'))$. From Itô's representation Theorem, they may be written as:

$$H = \mathbb{E}[H] + \int_0^{T_0(\beta)} h_u d\beta_u \quad , \quad H' = \mathbb{E}[H'] + \int_0^{T_0(\beta')} h'_u d\beta'_u \quad (2.138)$$

where $(h_u, u \geq 0)$ and $(h'_u, u \geq 0)$ are two predictable processes with respect to the natural filtration of $\left( \beta_{u \wedge T_0(\beta)}, u \geq 0 \right)$ and $\left( \beta'_{u \wedge T_0(\beta')}, u \geq 0 \right)$ such that:

$$\mathbb{E}\left[ \int_0^{T_0(\beta)} (h_u)^2 du + \int_0^{T_0(\beta')} (h'_u)^2 du \right] < \infty.$$

Now, after the change of variable $u = \langle M \rangle_t$ (resp. $u = \langle M' \rangle_t$) in (2.138), we obtain:

$$H = \mathbb{E}[H] + \int_0^\infty h_{\langle M \rangle_t} dM_t \quad , \quad H' = \mathbb{E}[H'] + \int_0^\infty h'_{\langle M' \rangle_t} dM'_t, \quad (2.139)$$

and, because of the orthogonality of $M$ and $M'$, and Itô's formula:

$$\mathbb{E}[HH'] = \mathbb{E}[H]\mathbb{E}[H'],$$

which is Lemma 2.3.

$\square$

## 2.8.4 On the Law of $\bigvee_{i=1,\cdots,n} \mathscr{G}_{K_i}^{(i)}$

We now give the counterpart of Theorem 2.3 in the situation of Section 2.8, where we deal with several orthogonal local martingales. As previously, to simplify, we assume that $n = 2$ and we make the following hypotheses:

i)  The r.v. $M_t$ (resp. $M_t'$) admits a density $x \mapsto m_t(x)$ (resp. $x \mapsto m_t'(x)$) which is jointly continuous in $t$ and $x$.

ii) There exist two previsible processes $(\sigma_t, t \geq 0)$ and $(\sigma_t', t \geq 0)$ such that:

$$d \langle M \rangle_t = \sigma_t^2 dt \quad , \quad d \langle M' \rangle_t = \left( \sigma_t' \right)^2 dt.$$

Then, there is the following.

**Theorem 2.7.**

$$\mathbb{P} \left( \left( \mathscr{G}_K \vee \mathscr{G}_{K'}' \right) \in dt \right) = \left( 1 - \frac{M_0}{K} \right)^+ \left( 1 - \frac{M_0'}{K} \right)^+ \delta_0(dt) + \gamma_{K,K'}(t) dt \quad (t \geq 0) \tag{2.140}$$

with

$$\gamma_{K,K'}(t) = \frac{1}{2K} \mathbb{E} \left[ \sigma_t^2 \left( 1 - \frac{M_t'}{K'} \right)^+ \Big| M_t = K \right] m_t(K)$$

$$+ \frac{1}{2K'} \mathbb{E} \left[ \left( \sigma_t' \right)^2 \left( 1 - \frac{M_t}{K} \right)^+ \Big| M_t' = K' \right] m_t'(K') \tag{2.141}$$

where all the functions of several variables appearing in (2.141) admit jointly continuous versions.

The proof of (2.140) is essentially the same as that of Theorem 2.3. We start by writing, applying Theorem 2.6:

$$\mathbb{P} \left( \left( \mathscr{G}_K \vee \mathscr{G}_{K'}' \right) \leq t \right) = \mathbb{E} \left[ \left( 1 - \frac{M_t}{K} \right)^+ \left( 1 - \frac{M_t'}{K} \right)^+ \right] \tag{2.142}$$

and we develop the RHS of (2.142) thanks to Tanaka's formula to obtain:

$$\mathbb{P} \left( \left( \mathscr{G}_K \vee \mathscr{G}_{K'}' \right) \leq t \right) = \left( 1 - \frac{M_0}{K} \right)^+ \left( 1 - \frac{M_0'}{K} \right)^+$$

$$+ \frac{1}{2} \left( \mathbb{E} \left[ \int_0^t \left( 1 - \frac{M_s'}{K'} \right)^+ \frac{dL_s^K}{K} \right] + \mathbb{E} \left[ \int_0^t \left( 1 - \frac{M_s}{K} \right)^+ \frac{dL_s'^{K'}}{K'} \right] \right)$$

(since $M$ and $M'$ are orthogonal) where $(L_s^K, s \geq 0)$ (resp. $(L_s'^{K'}, s \geq 0)$) denotes the local time of $M$ at level $K$ (resp. of $M'$ at level $K'$). And, as for Theorem 2.3, the proof can now be ended by applying the occupation density formula.

**Problem 2.1 (On time inversion of a Lévy process).**

**1)** Let $X_1, X_2, \ldots, X_n, \ldots$ denote a sequence of i.i.d. integrable r.v.'s. Define:

$$S_0 = 0, \qquad S_n = \sum_{i=1}^{n} X_i \quad (n \geq 1).$$

i) Prove that, for every $1 \leq i \leq n$, $\mathbb{E}[X_i|S_n] = \dfrac{S_n}{n}$.

ii) Let $\mathscr{F}_{n+1}^+ := \sigma(S_{n+1}, S_{n+2}, \ldots)$. Prove that, for every $n \geq 1$:

$$\mathbb{E}\left[\frac{S_n}{n}\Big|\mathscr{F}_{n+1}^+\right] = \frac{S_{n+1}}{n+1}.$$

**2)** Let $(\Lambda_t, t \geq 0)$ be an integrable Lévy process, i.e. for any $t \geq 0$, $\mathbb{E}[|\Lambda_t|] < \infty$.

i) Prove that $\mathbb{E}[\Lambda_t] = t\mathbb{E}[\Lambda_1]$.

ii) Let $\mathscr{F}_t^+ := \sigma(\Lambda_s; s \geq t)$. Prove that $\left(\dfrac{1}{t}\Lambda_t, t \geq 0\right)$ is a $\mathscr{F}_t^+$-inverse martingale,

i.e, for every $s < t$:

$$\mathbb{E}\left[\frac{\Lambda_s}{s}\Big|\mathscr{F}_t^+\right] = \frac{\Lambda_t}{t}.$$

Hint: Use discretization and **1)** (see [36]).

iii) Define:

$$M_t = \begin{cases} t\Lambda_{\frac{1}{t}} & \text{if } t > 0, \\ \mathbb{E}[\Lambda_1] & \text{if } t = 0. \end{cases} \tag{1}$$

Prove that $(M_t, t \geq 0)$ is a martingale with respect to the filtration
$\mathscr{G} := \left\{\sigma\left(\Lambda_u, u \geq \frac{1}{t}\right), t > 0\right\}$, which is the natural filtration of $(M_t, t \geq 0)$.
(Note that $\Lambda$ being a càdlàg process, the martingale $M$ as defined by $(1)$ is left-continuous (and right-limited). To get a right-continuous process, one should take $M_t := t\Lambda_{\left(\frac{1}{t}\right)^-}$.)

iv) Prove that the process $(M_t, t \geq 0)$ is an (inhomogeneous) Markov process.

v) Identify $(M_t, t \geq 0)$ when $\Lambda_t$ is a Brownian motion with drift.

**3)** We now assume that $(\Lambda_t, t \geq 0)$ is an integrable subordinator, with Lévy measure $\nu$, and without drift term. Thus, there is the Lévy-Khintchine formula:

$$\mathbb{E}\left[e^{-\lambda\Lambda_t}\right] = \exp\left(-t\int_0^\infty (1 - e^{-\lambda x})\nu(dx)\right) \qquad \text{with } \int_0^\infty (x \wedge 1)\nu(dx) < \infty. \tag{2}$$

i) Prove that the integrability of $\Lambda_t$ is equivalent to $\int_0^\infty x\nu(dx) < \infty$.

ii) Observe that the martingale $(M_t, t \geq 0)$ has no positive jumps.

iii) Prove that $M_t \xrightarrow[t \to \infty]{} 0$ a.s.; to start with, one may show that $M_t \xrightarrow[t \to \infty]{} 0$ in law.

iv) Prove that, for every $K \leq \mathbb{E}[\Lambda_1]$ and every $t \geq 0$:

$$\frac{1}{K}\mathbb{E}\left[(K - M_t)^+\right] = \mathbb{P}\left(\mathscr{G}_K^{(M)} \leq t\right) \tag{3}$$

with $\mathscr{G}_K^{(M)} := \sup\{t \geq 0; M_t = K\}$. (Theorem 2.1 may be useful).

v) Let, for every $K < \mathbb{E}[\Lambda_1]$, $T_K^{(\Lambda)} := \inf\{u \geq 0; \Lambda_u \geq Ku\}$. Prove that $T_K^{(\Lambda)}$ is finite
a.s. and that $\mathscr{G}_K^{(M)} = \dfrac{1}{T_K^{(\Lambda)}}$ a.s. Deduce that, for every $t \geq 0$:

$$\mathbb{P}\left(T_K^{(\Lambda)} \geq t\right) = \frac{1}{K}\mathbb{E}\left[(K - M_t)^+\right].$$

vi) Compute explicitly the law of $T_K^{(\Lambda)}$ when $(\Lambda_t, t \geq 0)$ is the gamma subordinator. Recall that the density $f_{\Lambda_t}$ of $\Lambda_t$ is then equal to:

$$f_{\Lambda_t}(x) = \frac{1}{\Gamma(t)}e^{-x}x^{t-1}\mathbb{1}_{\{x>0\}}. \tag{4}$$

Solve the same question when $(\Lambda_t, t \geq 0)$ is the Poisson process with parameter $\lambda$.

**4)** We now complete the results of question **3)** when $(\Lambda_t, t \geq 0)$ is the gamma subordinator. We recall that the density of $\Lambda_t$ is given by (4), and that its Lévy measure $\nu$ is given by:

$$\nu(dx) = \frac{e^{-x}}{x}dx. \tag{5}$$

i)  Exploit the fact that, conditionally upon $\gamma_{\frac{1}{s}} = y$, the law of $\gamma_{\frac{1}{t}}$, for $s < t$, is given by:

$$\gamma_{\frac{1}{t}} \overset{(law)}{=} y\beta\left(\frac{1}{t}, \frac{1}{s} - \frac{1}{t}\right) \tag{6}$$

where $\beta\left(\frac{1}{t}, \frac{1}{s} - \frac{1}{t}\right)$ is a beta variable with parameters $\frac{1}{t}$ and $\frac{1}{s} - \frac{1}{t}$, in order to show that, for every $\alpha > 0$:

$$\left(M_t^{(\alpha)} = \frac{\Gamma\left(\frac{1}{t}\right)\left(\gamma_{\frac{1}{t}}\right)^{\alpha}}{\Gamma\left(\alpha + \frac{1}{t}\right)}, t \geq 0\right) \text{ is a } (\mathscr{G}_t, t \geq 0) \text{ martingale.} \tag{7}$$

ii) Using (7) with $\alpha = 2$, deduce that the bracket of $(M_t, t \geq 0)$ is given by:

$$\langle M \rangle_t = \int_0^t \frac{M_s^2}{1+s}ds.$$

iii) Prove more generally that:

$$d\langle M^{(\alpha)}, M^{(\beta)} \rangle_t = -M_t^{(\alpha)}M_t^{(\beta)}\frac{\theta'_{\alpha,\beta}(t)}{\theta_{\alpha,\beta}(t)}dt$$

where $\theta_{\alpha,\beta}(t) = \dfrac{\Gamma(\alpha + \frac{1}{t})\Gamma(\beta + \frac{1}{t})}{\Gamma(\frac{1}{t})\Gamma(\alpha + \beta + \frac{1}{t})}$.

iv) Prove that the infinitesimal generator $\widetilde{L}$ of the Markov process $(M_t, t \geq 0)$ satisfies:

$$\widetilde{L}f(s,x) = \frac{xf'(x)}{s} + \frac{1}{s^2}\int_0^1 (f(xz) - f(x))\frac{z^{\frac{1}{s}-1}}{1-z}dz \tag{8}$$

for every $\mathscr{C}^1$ function $f$. Check that, for $f_1(x) := x$, $\widetilde{L}f_1 = 0$.

v) Prove that, for every $\mathscr{C}^2$ function $f$, $\lim\limits_{s \to 0} \widetilde{L} f(s,x) = \dfrac{x^2 f''(x)}{2}$.

vi) Use (8) to give another proof of (7) when $\alpha \in \mathbb{N}^*$.

vii) Use relation (7) to obtain, for any integer $n$, the law of $T_K^{(n)}$ where:

$$T_K^{(n)} := \inf\left\{u \geq 0; \Lambda_u \geq [Ku(1+u)(1+2u)\ldots(1+(n-1)u)]^{\frac{1}{n}}\right\}$$

(and $K < \mathbb{E}[\Lambda_1] = 1$).

**5)** We now improve upon the results of question **3)** when $(\Lambda_t, t \geq 0)$ is the Poisson process of parameter 1 (and Lévy measure $\nu(dx) = \delta_1(dx)$).

i) Prove that, for $s < t$, conditionally upon $\Lambda_{\frac{1}{s}} = y$, $\Lambda_{\frac{1}{t}}$ follows a binomial distribution $\mathscr{B}\left(y, \frac{s}{t}\right)$ with parameters $y$ and $\left(\frac{s}{t}\right)$.

ii) Let $\widetilde{L}$ denote the infinitesimal generator of the Markov process $(M_t, t \geq 0)$. Prove that:

$$\widetilde{L} f(s,x) = \frac{x f'(x)}{s} + (f(x-s) - f(x)) \frac{x}{s^2} \tag{9}$$

for $f$ in $\mathscr{C}^1$. Check that, for $f_1(x) =: x$, $\widetilde{L} f_1 = 0$.

iii) Prove that, for every $\mathscr{C}^2$ function $f$, $\lim\limits_{s \to 0} \widetilde{L} f(s,x) = \dfrac{1}{2} x f''(x)$.

iv) Compute $\widetilde{L} f_2(s,x)$, with $f_2(x) = x^2$. Deduce that $(M_t^2 - t M_t, t \geq 0)$ is a $(\mathscr{G}_t, t \geq 0)$ martingale and that the bracket of $(M_t, t \geq 0)$ is given by:

$$\langle M \rangle_t = \int_0^t M_s ds.$$

v) Compute $\widetilde{L} f_n(s,x)$, with $f_n(x) = x^n$ $(n \in \mathbb{N})$.

$$\left(\text{Answer: } \widetilde{L} f(s,x) = \sum_{j=1}^{n-1} \binom{n}{j-1} x^j (-s)^{n-1-j}.\right)$$

## 2.9 Notes and Comments

The results of Sections 2.1, 2.2, 2.3, 2.4 and 2.7 are taken from the preprints of A. Bentata and M. Yor (see [5], [6] and [5, F], [6, F], [7, F]). The description of the European put and call in terms of last passage times also appears in D. Madan, B. Roynette and M. Yor (see [47] and [48]). The results of Section 2.5 are taken from J. Pitman and M. Yor ([65]). The study of the local martingale of Example 2.3.e, $\left(\dfrac{1}{X_t}, t \geq 0\right)$, where $(X_t, t \geq 0)$ is a Bessel process of dimension 3 is borrowed from Ju-Yi Yen and M. Yor ([93]) while the study of the càdlàg martingale in Section 2.6 is due to C. Profeta ([67]). The generalization of these results to the case of several orthogonal continuous local martingales is taken from B. Roynette, Y. Yano and M. Yor (see [74]).

# Chapter 3
# Representation of some particular Azéma supermartingales

**Abstract** We show how the formula obtained in Chapter 2 for the supermartingale associated with a last passage time at a given level fits with a more general representation of Azéma supermartingales.

We also recall progressive enlargement formulae, and particularize them to our framework.

Finally, we discuss a representation problem for Skorokhod submartingales.

## 3.1 A General Representation Theorem

### 3.1.1 Introduction

Let $(M_t, t \geq 0)$ be a positive, continuous local martingale which converges to 0 as $t \to \infty$, and, for $K \geq 0$:

$$\mathscr{G}_K := \mathscr{G}_K^{(M)} := \sup\{t \geq 0; M_t = K\}. \tag{3.1}$$

We have seen in Chapter 2 that $\mathscr{G}_K$ plays an important (although somewhat hidden!) role in option pricing. In particular, from Theorem 2.1, the formula

$$\mathbb{P}(\mathscr{G}_K \leq t | \mathscr{F}_t) = \left(1 - \frac{M_t}{K}\right)^+ \tag{3.2}$$

holds, or equivalently,

$$\mathbb{P}(\mathscr{G}_K > t | \mathscr{F}_t) = \left(\frac{M_t}{K}\right) \wedge 1. \tag{3.3}$$

Of course, $\mathscr{G}_K$ is a last passage time. That is why we start this chapter by recalling, in a general framework, a representation result for Azéma supermartingales associated with ends of predictable sets.

### 3.1.2 General Framework

Let $(\Omega, (\mathscr{F}_t, t \geq 0), (X_t, t \geq 0), \mathbb{P}_x, x \geq 0)$ be a transient diffusion taking values in $\mathbb{R}^+$, and $\Gamma$ a compact subset of $\mathbb{R}^+$. We define $L$ to be:

$$L := \sup\{t \geq 0; X_t \in \Gamma\} \tag{3.4}$$
$$(= 0 \text{ if the set } \{t \geq 0; X_t \in \Gamma\} \text{ is empty.})$$

We would like to describe the pre $L$-process $(X_t, t \leq L)$ and the post $L$-process $(X_{t+L}, t \geq 0)$. But, in order to apply the method of enlargement of filtration, we need first to study the Azéma supermartingale $(Z_t^{(L)}, t \geq 0)$ defined by:

$$\left(Z_t := Z_t^{(L)} = \mathbb{P}(L > t | \mathscr{F}_t), \; t \geq 0\right). \tag{3.5}$$

We shall start by stating a representation theorem of such a supermartingale. To this end, we make two extra assumptions:

| | | |
|---|---|---|
| (C) | All $(\mathscr{F}_t)$-martingales are continuous. | (3.6) |
| (A) | For any stopping time $T$, $\mathbb{P}(L = T) = 0$. | (3.7) |

((C) stands for continuous (martingales), and (A) for avoiding (stopping times)). Of course, assumption (C) is fulfilled in the Brownian set up.

### 3.1.3 Statement of the Representation Theorem

**Theorem 3.1.** *Under (C) and (A), there exists a unique positive local martingale* $(N_t, t \geq 0)$, *with* $N_0 = 1$ *such that:*

$$\mathbb{P}(L > t | \mathscr{F}_t) = \frac{N_t}{S_t} \tag{3.8}$$

*where*

$$S_t := \sup_{s \leq t} N_s \qquad (t \geq 0). \tag{3.9}$$

We refer to ([51], Proposition 1.3, p.16) for the proof of Theorem 3.1. Let us just note that, since $L < \infty$ a.s., $N_t \xrightarrow[t \to \infty]{} 0$ a.s. We now make some further remarks about Theorem 3.1:

• Note first that $\log(S_\infty)$ is exponentially distributed, since, from Lemma 2.1,

$$\log(S_\infty) \stackrel{(law)}{=} \log\left(\frac{1}{U}\right) \tag{3.10}$$

where $U$ is uniform on $[0, 1]$.

• Then, the additive decomposition of the supermartingale $\dfrac{N_t}{S_t}$ is given by:

$$\frac{N_t}{S_t} = 1 + \int_0^t \frac{dN_u}{S_u} - \log(S_t) = \mathbb{E}\left[\log(S_\infty)|\mathscr{F}_t\right] - \log(S_t). \qquad (3.11)$$

Indeed, from Itô's formula, we have:

$$\frac{N_t}{S_t} = 1 + \int_0^t \frac{dN_u}{S_u} - \int_0^t \frac{N_u dS_u}{S_u^2}$$

$$= 1 + \int_0^t \frac{dN_u}{S_u} - \int_0^t \frac{dS_u}{S_u}$$

$$\text{(since } dS_u \text{ only charges the set } \{u \geq 0; N_u = S_u\})$$

$$= 1 + \int_0^t \frac{dN_u}{S_u} - \log(S_t). \qquad (3.12)$$

Then:

$$0 \leq \frac{N_t}{S_t} + \log(S_t) = 1 + \int_0^t \frac{dN_u}{S_u} \leq 1 + \log(S_\infty)$$

since $S_0 = N_0 = 1$ and $S$ is an increasing process. Therefore, the martingale $\left(\int_0^t \frac{dN_u}{S_u}, t \geq 0\right)$ belongs to $H^1$, and we can pass to the limit when $t$ tends to $+\infty$ in (3.12). Since $\dfrac{N_t}{S_t} \xrightarrow[t\to\infty]{} 0$ a.s., this yields:

$$0 = 1 + \int_0^{+\infty} \frac{dN_u}{S_u} - \log(S_\infty),$$

thus:

$$\log(S_\infty) = 1 + \int_0^{+\infty} \frac{dN_u}{S_u}$$

and

$$\mathbb{E}\left[\log(S_\infty)|\mathscr{F}_t\right] = \mathbb{E}\left[1 + \int_0^{+\infty} \frac{dN_u}{S_u}\Big|\mathscr{F}_t\right] = 1 + \int_0^t \frac{dN_u}{S_u}.$$

• The martingale $\mathbb{E}\left[\log(S_\infty)|\mathscr{F}_t\right]$ belongs to $(BMO)$ (see [70], Exercise 3.16, p.75) since, from (3.11):

$$\mathbb{E}\left[\log(S_\infty) - \log(S_t)|\mathscr{F}_t\right] \leq 1. \qquad (3.13)$$

### 3.1.4 Application of the Representation Theorem 3.1 to the Supermartingale $\left(\mathbb{P}\left(\mathscr{G}_K > t|\mathscr{F}_t\right), t \geq 0\right)$, when $M_\infty = 0$

Let $(M_t, t \geq 0)$ be a positive, continuous local martingale such that $M_0 = a > 0$ and $M_\infty = 0$ a.s. We assume that $(M_t, t \geq 0)$ is defined on a filtered probability space

$\left( \Omega, (\mathscr{F}_t, t \geq 0), \mathscr{F}_\infty := \bigvee_{t \geq 0} \mathscr{F}_t, \mathbb{P} \right)$, and we apply Theorem 3.1 with $L := \mathscr{G}_K = \sup\{t \geq 0; M_t = K\}$. This leads to:

**Theorem 3.2.** *Let $M_0 = a \geq K$. Then:*

$$\mathbb{P}(\mathscr{G}_K > t | \mathscr{F}_t) = \left( \frac{M_t}{K} \wedge 1 \right) = \frac{N_t}{S_t} \tag{3.14}$$

*where $(N_t, t \geq 0)$ is the positive local martingale converging to 0 as $t \to \infty$ defined by:*

$$N_t := \left( \frac{M_t}{K} \wedge 1 \right) \exp\left( \frac{1}{2K} L_t^K \right), \tag{3.15}$$

$$S_t := \sup_{s \leq t} N_s = \exp\left( \frac{1}{2K} L_t^K \right) \tag{3.16}$$

*and $(L_t^K, t \geq 0)$ denotes the local time of $(M_t, t \geq 0)$ at level $K$.*

*Remark 3.1.* Note that it follows from the previous remarks that $\frac{1}{2K} L_\infty^K \ (= \log(S_\infty))$ is exponentially distributed.

*Proof.* From Tanaka's formula, Theorem 2.1 and Theorem 3.1, we have:

$$\mathbb{P}(\mathscr{G}_K > t | \mathscr{F}_t) = \frac{M_t}{K} \wedge 1 = 1 + \frac{1}{K} \int_0^t 1_{\{M_s \leq K\}} dM_s - \frac{1}{2K} L_t^K = \frac{N_t}{S_t}. \tag{3.17}$$

The comparison of (3.11) and (3.17) implies:

$$\int_0^t \frac{dN_u}{S_u} = \frac{1}{K} \int_0^t 1_{\{M_s \leq K\}} dM_s, \tag{3.18}$$

$$\text{and} \quad \frac{1}{2K} L_t^K = \log(S_t). \tag{3.19}$$

Hence

$$N_t = \left( \frac{M_t}{K} \wedge 1 \right) S_t$$

$$= \left( \frac{M_t}{K} \wedge 1 \right) \exp\left( \frac{1}{2K} L_t^K \right). \tag{3.20}$$

Note that, since $M_t \xrightarrow[t \to \infty]{} 0$ a.s., it follows from (3.20) that $N_t \xrightarrow[t \to \infty]{} 0$ a.s., since $L_t^K = L_\infty^K < \infty$ for $t$ large enough.

$\square$

*Remark 3.2.* On the other hand, if $K > a$:

$$\mathbb{P}(\mathscr{G}_K = 0) = \mathbb{P}\left(\sup_{t \geq 0} M_t < K\right) = 1 - \frac{a}{K}$$

from Lemma 2.1. Hence, for the stopping time $T \equiv 0$, $\mathbb{P}(\mathscr{G}_K = T = 0) > 0$ and hypothesis (A) is not fulfilled.

### 3.1.5 A Remark on Theorem 3.2

**a)** We now compare the results of Theorems 3.1 and 3.2. We remark that not every supermartingale of the form $\left(\dfrac{N_t}{S_t}, t \geq 0\right)$ can be written as $\left(\dfrac{M_t}{K} \wedge 1, t \geq 0\right)$.
Indeed, assuming:

$$\frac{M_t}{K} \wedge 1 = \frac{N_t}{S_t},$$

we deduce that:

$$d\langle N\rangle_t = \exp\left(\frac{L_t^K}{K}\right) 1_{\{M_t < K\}} \frac{d\langle M\rangle_t}{K^2}. \tag{3.21}$$

Now, in a Brownian setting, we have $d\langle N\rangle_t = n_t^2 dt$ and $d\langle M\rangle_t = m_t^2 dt$, for two $(\mathscr{F}_t, t \geq 0)$ predictable processes $(m_t, t \geq 0)$ and $(n_t, t \geq 0)$. Then (3.21) implies:

$$n_t^2 = \frac{1}{K^2} \exp\left(\frac{L_t^K}{K}\right) 1_{\{M_t < K\}} m_t^2 \qquad dt \cdot \mathbb{P} \text{ a.s.}$$

Consequently:

$$n_t^2 = 0 \quad \text{on } \{(t, \omega); M_t(\omega) > K\} \quad dt \cdot \mathbb{P} \text{ a.s.}$$

However, this cannot be satisfied if we start, for example, from a local martingale $N$ such that $n_t^2 > 0$ for all $t$. Note that the random set $\{t \geq 0; M_t > K\}$ is not empty. If it were, then the local time at level $K$ of $M$ would be equal to 0, which is in contradiction with $S_\infty \overset{(law)}{=} \dfrac{M_0}{U}$.

**b)** Let $(M_t, t \geq 0) \in \mathscr{M}_+^{0,c}$ as in Chapter 2. It is now natural to ask the following question: for which functions $h : \mathbb{R}^+ \rightarrow [0, 1]$ is it true that, for any $(M_t, t \geq 0)$ belonging to $\mathscr{M}_+^{0,c}$, $(h(M_t), t \geq 0)$ is an Azéma supermartingale, i.e. may be written:

$$h(M_t) = \frac{N_t}{S_t} = \widetilde{M_t} \wedge 1. \tag{3.22}$$

(We choose $K = 1$ in (3.14) without loss of generality.) We shall call such a function $h$ an Azéma function. Here is a partial answer to this question:

**Proposition 3.1.** *Assume that $h$ is an Azéma function such that:*

*i)* $\{x; h(x) < 1\} = [0, b[$ *for some positive real $b$.*
*ii)* $h''$, *in the Schwartz distribution sense, is a bounded measure.*

*Then:*

$$h(x) = \left(\frac{x}{b}\right) \wedge 1. \tag{3.23}$$

*Proof.* From (*ii*) and the Itô-Tanaka formula, for any $(M_t, t \geq 0) \in \mathcal{M}_+^{0,c}$, we have:

$$h(M_t) = h(M_0) + \int_0^t h'(M_s)dM_s + \frac{1}{2}\int_0^\infty h''(dx)L_t^x(M) \tag{3.24}$$

where $(L_t^x(M), t \geq 0, x \geq 0)$ is the family of local times of $M$. Since $(h(M_t), t \geq 0)$ is an Azéma supermartingale, its increasing process, in (3.24), is carried by the set $\{s \geq 0;\ h(M_s) = 1\}$ (from (3.11)). Therefore:

$$\int_0^t 1_{\{h(M_s)<1\}}\int_0^\infty h''(dx)d_s L_s^x(M) = 0. \tag{3.25}$$

Now, the LHS of (3.25) equals:

$$\int_0^\infty h''(dx)1_{\{h(x)<1\}}L_t^x(M) = \int_{[0,b[} h''(dx)L_t^x(M)$$

as a consequence of hypothesis (*i*). This is equivalent to $h''(dx) = 0$ on $[0, b[$, thus $h(x) = cx + d$ on $[0, b[$. Furthermore, $h(0) = 0$, since $\lim_{t\to\infty} h(M_t) = 0$ for any $M \in \mathcal{M}_+^{0,c}$. Thus, $h(x) = cx$ on $[0, b[$, and applying (*i*) once again yields the result, since $h \leq 1$ from (3.22).

$\square$

## 3.2 Study of the Pre $\mathcal{G}_K$- and Post $\mathcal{G}_K$-processes, when $M_\infty = 0$

Let $(M_t, t \geq 0) \in \mathcal{M}_+^{0,c}$, $K \geq 0$ such that $M_0 = a \geq K$, and $\mathcal{G}_K := \sup\{t \geq 0; M_t = K\}$. We shall now study the pre $\mathcal{G}_K$- and post $\mathcal{G}_K$-processes $(M_s, s \leq \mathcal{G}_K)$ and $(M_{\mathcal{G}_K+s}, s \geq 0)$.

### 3.2.1 Enlargement of Filtration Formulae

For that study, we shall consider the enlarged filtration $(\mathcal{F}_t^K, t \geq 0)$, i.e. the smallest filtration containing the initial filtration $(\mathcal{F}_t, t \geq 0)$ and which makes $\mathcal{G}_K$ a stopping time. We assume that hypotheses (C) and (A) are fulfilled (which implies, in particular, that $K \leq a$; see Remark 3.2). Let $(Z_t, t \geq 0)$ denote the Azéma supermartingale studied in Subsection 3.1.4:

$$Z_t = \mathbb{P}(\mathcal{G}_K > t | \mathcal{F}_t) = \frac{N_t}{S_t}. \tag{3.26}$$

Then, for every local martingale $(X_t, t \geq 0)$, there exists a $(\mathscr{F}_t^K, t \geq 0)$-local martingale $(\widetilde{X}_t, t \geq 0)$ such that:

$$X_t = \widetilde{X}_t + \int_0^{t \wedge \mathscr{G}_K} \frac{d\langle X, Z \rangle_s}{Z_s} - \int_{t \wedge \mathscr{G}_K}^t \frac{d\langle X, Z \rangle_s}{1 - Z_s}. \tag{3.27}$$

Now, since $Z_t = \dfrac{N_t}{S_t} = \left( \dfrac{M_t}{K} \right) \wedge 1$, we have:

$$X_t = \widetilde{X}_t + \int_0^{t \wedge \mathscr{G}_K} 1_{\{M_s < K\}} \frac{d\langle X, M \rangle_s}{M_s} - \int_{t \wedge \mathscr{G}_K}^t \frac{d\langle X, M \rangle_s}{K - M_s}. \tag{3.28}$$

It is of some interest to take $X_t = M_t$. Formula (3.28) then becomes:

$$M_t = \widetilde{M}_t + \int_0^{t \wedge \mathscr{G}_K} 1_{\{M_s < K\}} \frac{d\langle M \rangle_s}{M_s} - \int_{t \wedge \mathscr{G}_K}^t \frac{d\langle M \rangle_s}{K - M_s}. \tag{3.29}$$

### 3.2.2  Study of the Post $\mathscr{G}_K$-Process

We recall that $K \leq a = M_0$. Hence, $\mathscr{G}_K < \infty$ a.s.

**Proposition 3.2.** *Let us define* $(R_t = K - M_{\mathscr{G}_K + t}, t \geq 0)$. *Then, there exists a 3-dimensional Bessel process* $(\rho_u, u \geq 0)$, *starting at 0 and considered up to* $T_K(\rho) := \inf\{u \geq 0; \rho_u = K\}$ *such that:*

$$(R_t, t \geq 0) = \left( \rho_{\langle \widehat{M} \rangle_t}, t \geq 0 \right) \tag{3.30}$$

*where, with the notation of (3.29),* $\langle \widehat{M} \rangle_t = \langle \widetilde{M} \rangle_{\mathscr{G}_K + t} - \langle \widetilde{M} \rangle_{\mathscr{G}_K}$, *and* $\langle \widehat{M} \rangle_\infty = T_K(\rho)$.

*Proof.* From (3.29), we may write:

$$M_{\mathscr{G}_K + t} = K + \widehat{M}_t - \int_0^t \frac{d\langle M \rangle_{\mathscr{G}_K + s}}{K - M_{\mathscr{G}_K + s}}, \tag{3.31}$$

with $\widehat{M}_t = \widetilde{M}_{\mathscr{G}_K + t} - \widetilde{M}_{\mathscr{G}_K}$, i.e.

$$R_t = -\widehat{M}_t + \int_0^t \frac{d\langle M \rangle_{\mathscr{G}_K + s}}{R_s}. \tag{3.32}$$

Now, there exists a Brownian motion $(\beta_u, u \geq 0)$ such that $\widehat{M}_t = \beta_{\langle \widehat{M} \rangle_t}$. Hence, since $\langle \widehat{M} \rangle_t = \langle \widetilde{M} \rangle_{\mathscr{G}_K + t} - \langle \widetilde{M} \rangle_{\mathscr{G}_K}$,

$$R_t = -\beta_{\langle \widehat{M} \rangle_t} + \int_0^t \frac{d\langle \widehat{M} \rangle_s}{R_s}. \tag{3.33}$$

Therefore, by change of time, there exists a Bessel process $\rho$ of dimension 3 starting at 0 such that $(R_t = \rho_{\langle \widehat{M} \rangle_t})$. Since $M_t \xrightarrow[t \to \infty]{} 0$ a.s, $R_t \xrightarrow[t \to \infty]{} K$ a.s, and $\rho$ is considered up to $T_K(\rho) := \inf\{u \geq 0; \rho_u = K\} = \langle \widehat{M} \rangle_\infty$.

$\square$

### 3.2.3 Study of the Pre $\mathcal{G}_K$-Process

In Section 2.4, we have already described the law of the r.v. $\mathcal{G}_K$. We shall now describe the law of the pre $\mathcal{G}_K$-process $(M_s, s \leq \mathcal{G}_K)$ conditionally on $\mathcal{G}_K$. To proceed, we assume, as in Section 2.4, that:

i) for every $t > 0$, there exists a measurable function $(t,x) \mapsto m_t(x)$ such that $(m_t(x), x \geq 0)$ is the density of the law of $M_t$,
ii) there exists a predictable process $(\sigma_t, t \geq 0)$ such that

$$d\langle M \rangle_t = \sigma_t^2 dt. \tag{3.34}$$

We shall consider a measurable choice of the function
$(t,x) \mapsto \theta_t(x) := \mathbb{E}\left[\sigma_t^2 | M_t = x\right].$

However, note the difference with our hypothesis in Subsection 2.4.1: we no longer make continuity assumptions for these functions.

**Theorem 3.3.** Let $(\Phi_u, u \geq 0)$ denote a positive $(\mathcal{F}_u, u \geq 0)$ predictable process. Then:

i) We have:

$$\mathbb{E}\left[\Phi_{\mathcal{G}_K}\right] = \mathbb{E}\left[\Phi_0\left(1 - \frac{M_0}{K}\right)^+\right] + \frac{1}{2K}\int_0^\infty \mathbb{E}\left[\Phi_s \sigma_s^2 | M_s = K\right] m_s(K)ds \quad dK \text{ a.e.} \tag{3.35}$$

ii) As a consequence of (i), we recover:

$$\mathbb{P}(\mathcal{G}_K \in dt) = \left(1 - \frac{M_0}{K}\right)^+ \delta_0(dt) + \frac{1}{2K}\theta_t(K)m_t(K)1_{\{t>0\}}dt \quad dK \text{ a.e.} \tag{3.36}$$

iii) Furthermore:

$$\mathbb{E}\left[\Phi_{\mathcal{G}_K} | \mathcal{G}_K = t\right] = \frac{\mathbb{E}\left[\Phi_t \sigma_t^2 | M_t = K\right]}{\mathbb{E}\left[\sigma_t^2 | M_t = K\right]} \quad \mathbb{P}(\mathcal{G}_K \in dt) \text{ a.e.} \tag{3.37}$$

*Proof.* We apply the balayage formula (2.98) to the process $Y_t = (K - M_t)^+$, with $t = +\infty$, and we take expectations. We obtain:

$$\mathbb{E}\left[\Phi_{\mathcal{G}_K}(K - M_\infty)^+\right] = \mathbb{E}\left[\Phi_0(K - M_0)^+\right] + \frac{1}{2}\mathbb{E}\left[\int_0^\infty \Phi_s dL_s^K\right], \tag{3.38}$$

since $\mathscr{G}_Y(\infty) = \mathscr{G}_K$, and $dL_s^K$ only charges the set $\{s \geq 0; M_s = K\} = \{s \geq 0; \mathscr{G}_K(s) = s\}$. Now, since $M_\infty = 0$ a.s.:

$$\mathbb{E}\left[\Phi_{\mathscr{G}_K}\right] = \mathbb{E}\left[\Phi_0\left(1 - \frac{M_0}{K}\right)^+\right] + \frac{1}{2K}\mathbb{E}\left[\int_0^\infty \Phi_s dL_s^K\right]. \tag{3.39}$$

Hence, (3.35) will be proven if we show:

$$\mathbb{E}\left[\int_0^\infty \Phi_s dL_s^K\right] = \int_0^\infty \mathbb{E}\left[\Phi_s \sigma_s^2 | M_s = K\right] m_s(K) ds \qquad dK \text{ a.e.} \tag{3.40}$$

In order to prove (3.40), we use the density of occupation formula (2.45). Integrating $\Phi_s$ on both sides of (2.45) yields, using hypothesis (*ii*):

$$\int_0^\infty \Phi_s f(M_s) \sigma_s^2 ds = \int_0^\infty dK f(K) \left(\int_0^\infty \Phi_s dL_s^K\right). \tag{3.41}$$

Taking expectations on both sides, we obtain, with hypothesis (*i*):

$$\int_0^\infty f(K) dK \int_0^\infty \mathbb{E}\left[\Phi_s \sigma_s^2 | M_s = K\right] m_s(K) ds = \int_0^\infty f(K) dK \mathbb{E}\left[\int_0^\infty \Phi_s dL_s^K\right] \tag{3.42}$$

which is easily seen to imply (3.40). Then, replacing in (3.35) $\Phi_s$ by $\Phi_s g(s)$, for a generic Borel function $g : \mathbb{R}^+ \to \mathbb{R}^+$, we deduce (3.36) and (3.37).

□

*Example 3.2.3.* The particular case where $(M_t, t \geq 0)$ is Markovian, e.g. the Black-Scholes situation where $M_t = \mathscr{E}_t$, allows for some simplifications of the above formula. In this case, $\sigma_s = \sigma(s, M_s)$ with $\sigma(s, x)$ a deterministic function, and we obtain from (3.37):

$$\mathbb{E}\left[\Phi_{\mathscr{G}_K} | \mathscr{G}_K = s\right] = \mathbb{E}\left[\Phi_s | M_s = K\right],$$

i.e. conditionally on $\{\mathscr{G}_K = s\}$, the pre $\mathscr{G}_K$-process is the bridge (of $M$) on the time interval $[0, s]$, ending at $K$.

### 3.2.4 Some Predictable Compensators

The computations we have made in the previous section make it possible to derive, without further efforts, some explicit expressions for several predictable compensators. Let $(C_t, t \geq 0)$ an increasing process defined on a filtered probability space $(\Omega, (\mathscr{F}_t, t \geq 0), \mathbb{P})$ such that $C_0 = 0$ and which is not necessarily adapted. Then, there exists a unique predictable increasing process $(A_t, t \geq 0)$ with $A_0 = 0$ such that for all predictable, positive processes $(\Phi_t, t \geq 0)$:

$$\mathbb{E}\left[\int_0^\infty \Phi_s dC_s\right] = \mathbb{E}\left[\int_0^\infty \Phi_s dA_s\right]. \tag{3.43}$$

$(A_t, t \geq 0)$ is called the predictable compensator of the process $(C_t, t \geq 0)$. Let $(M_t, t \geq 0) \in \mathcal{M}_+^{0,c}$. For $K \geq 0$, let $(\Lambda_K(t), t \geq 0)$ be the increasing process defined by:

$$\left(\Lambda_K(t) = 1_{\{0 < \mathcal{G}_K \leq t\}}, t \geq 0\right). \tag{3.44}$$

Let us also define:

$$S_{[t,+\infty[} := \sup_{s \geq t} M_s, \quad S_\infty := \sup_{s \geq 0} M_s, \quad S_t' = S_\infty - S_{[t,+\infty[}. \tag{3.45}$$

**Proposition 3.3.**

i)  *The predictable compensator of $(\Lambda_K(t), t \geq 0)$ is the process*

$$\left(\frac{1}{2K} L_t^K, t \geq 0\right). \tag{3.46}$$

ii) *The predictable compensator of $(S_t', t \geq 0)$ is the process*

$$\left(\frac{1}{2}\int_0^t \frac{d\langle M\rangle_s}{M_s}, t \geq 0\right) \tag{3.47}$$

*with the convention* $\dfrac{1}{M_s} = 0$ *for* $s \geq T_0(M) := \inf\{t \geq 0; M_t = 0\}$.

*Proof.* We first prove Point $(i)$
It is an immediate consequence of (3.38). Indeed, with $\Phi_0 = 0$:

$$\mathbb{E}\left[\int_0^\infty \Phi_s d_s\left(1_{\{0 < \mathcal{G}_K \leq s\}}\right)\right] = \mathbb{E}\left[\Phi_{\mathcal{G}_K}\right] = \frac{1}{2K}\mathbb{E}\left[\int_0^\infty \Phi_s dL_s^K\right], \tag{3.48}$$

since $M_\infty = 0$.
We now prove Point $(ii)$
We integrate both sides of (3.48) with respect to $f(K)dK$, for $f$ Borel and positive. This yields:

$$\mathbb{E}\left[\int_0^\infty f(K)\Phi_{\mathcal{G}_K} dK\right] = \mathbb{E}\left[\frac{1}{2}\int_0^\infty f(K)\frac{dK}{K}\left(\int_0^\infty \Phi_s dL_s^K\right)\right]$$
$$= \mathbb{E}\left[\frac{1}{2}\int_0^\infty f(M_s)\Phi_s\frac{d\langle M\rangle_s}{M_s}\right]. \tag{3.49}$$

from the density of occupation formula. We note that $\{\mathcal{G}_K \leq t\} = \{S_{[t,+\infty[} < K\}$, i.e. the inverse of $K \mapsto \mathcal{G}_K$ is $t \mapsto S_{[t,+\infty[}$. As a consequence, we may express the LHS of (3.49) as:

$$\mathbb{E}\left[\int_0^\infty f(S_{[t,+\infty[})\Phi_t dS_t'\right]$$

which we compare to the RHS of (3.49). Taking $f(x) = 1$, we obtain:

$$\mathbb{E}\left[\int_0^\infty \Phi_t dS_t'\right] = \frac{1}{2}\left[\int_0^\infty \Phi_t \frac{d\langle M\rangle_t}{M_t}\right] \tag{3.50}$$

which is Point $(ii)$ of Proposition 3.3.

$\square$

*Remark 3.3.* Applying (3.50) with $\Phi_t = f(M_t)$, we have:

$$\mathbb{E}\left[\int_0^\infty f(M_t)dS_t'\right] = -\mathbb{E}\left[\int_0^\infty f(S_{[t,+\infty[})dS_{[t,+\infty[}\right]$$

$$\text{(since the support of } S_t' \text{ is the set } \{t; M_t = S_{[t,+\infty[}\}$$
$$\text{and } dS_t' = -dS_{[t,+\infty[})$$

$$= \mathbb{E}\left[\int_0^{S_\infty} f(x)dx\right]$$

$$\text{(after the change of variable } S_{[t,+\infty[} = x)$$

$$= \int_0^\infty f(x)\left(\frac{a}{x} \wedge 1\right)dx$$

$$\text{(since } S_\infty \stackrel{(law)}{=} \frac{a}{U}, \text{ with } U \text{ uniform on } [0,1]).$$

Hence, from (3.50):

$$\int_0^\infty f(x)\left(\frac{a}{x} \wedge 1\right)dx = \frac{1}{2}\mathbb{E}\left[\int_0^\infty f(M_t)\frac{d\langle M\rangle_t}{M_t}\right]. \tag{3.51}$$

We now see that, under the hypotheses $(i)$ and $(ii)$ of Subsection 2.4.1, the RHS of (3.51) equals:

$$\frac{1}{2}\mathbb{E}\left[\int_0^\infty f(M_t)\frac{d\langle M\rangle_t}{M_t}\right] = \int_0^\infty dt \int_0^\infty m_t(K)\theta_t(K)f(K)\frac{dK}{2K}. \tag{3.52}$$

Comparing (3.51) and (3.52), we obtain:

$$\frac{a}{K} \wedge 1 = \int_0^\infty m_t(K)\theta_t(K)\frac{dt}{2K}. \tag{3.53}$$

Formula (3.53) agrees with the expression (2.41) of the law of $\mathcal{G}_K$:

$$\mathbb{P}(\mathcal{G}_K > 0) = \int_0^\infty m_t(K)\theta_t(K)\frac{dt}{2K}$$

$$\left(= \left(\frac{a}{K} \wedge 1\right) \quad \text{from (3.14)}\right).$$

*Remark 3.4.* In Proposition 3.3, we computed the predictable compensators of $\left(1_{\{0 < \mathcal{G}_K \leq t\}}, t \geq 0\right)$ and of $\left(S_t' = S_\infty - S_{[t,+\infty[}, t \geq 0\right)$ when $(M_t, t \geq 0) \in \mathcal{M}_+^{0,c}$. If we remove the assumption $M_t \xrightarrow[t \to \infty]{} 0$ a.s., we can nevertheless make a similar computation, using this time identity (2.97) from Theorem 2.5:

$$\mathbb{E}\left[1_{\{\mathscr{G}_K \leq t\}} (K - M_\infty)^+ | \mathscr{F}_t\right] = (K - M_t)^+. \tag{3.54}$$

This yields:

**Proposition 3.4.** *We remove assumption $M_\infty = 0$ a.s. Then:*

i) *The predictable compensator of $\left(1_{\{0 < \mathscr{G}_K \leq t\}}(K - M_\infty)^+, t \geq 0\right)$ is the process* $\left(\frac{1}{2K} L_t^K, t \geq 0\right)$.

ii) *Let $\left(I_t := (S_\infty - M_\infty)^2 - \left(S_{[t,+\infty[} - M_\infty\right)^2, t \geq 0\right)$. Then, the predictable compensator of $(I_t, t \geq 0)$ is the process $(\langle M \rangle_t, t \geq 0)$.*

These results originate from Azéma-Yor [3].

### 3.2.5 Expression of the Azéma supermartingale $(\mathbb{P}(\mathscr{G}_K > t | \mathscr{F}_t), t \geq 0)$ when $M_\infty \neq 0$

In the previous section, we gave an expression for the Azéma supermartingale $(\mathbb{P}(\mathscr{G}_K > t | \mathscr{F}_t), t \geq 0)$, with $K \leq a = M_0$, when the local martingale $M$ converges towards 0 a.s. We shall now remove this assumption and give a general expression when $M_\infty \neq 0$ a.s. Let

$$\mathscr{F}_{\mathscr{G}_K^-} := \sigma\left\{H_{\mathscr{G}_K}; H \text{ predictable process}\right\}$$

and $\nu_K$ the law of $M_\infty$ conditionally on $\mathscr{F}_{\mathscr{G}_K^-}$. Thus, for every positive, Borel function:

$$\mathbb{E}\left[f(M_\infty) | \mathscr{F}_{\mathscr{G}_K^-}\right] = \int_0^\infty f(m) \nu_K(dm). \tag{3.55}$$

In particular, we have:

$$\mu_{\mathscr{G}_K} := \mathbb{E}\left[(K - M_\infty)^+ | \mathscr{F}_{\mathscr{G}_K^-}\right] = \int_0^\infty (K - m)^+ \nu_K(dm). \tag{3.56}$$

Then, there exists a predictable process $(\mu_u^{(K)}, u \geq 0)$ (the predictable projection of $\left(1_{\{\mathscr{G}_K \leq u\}} \mu_{\mathscr{G}_K}, u \geq 0\right)$, see [70], p.173) such that, for all predictable stopping times $T$:

$$\mathbb{E}\left[1_{\{\mathscr{G}_K \leq T\}} \mu_{\mathscr{G}_K} 1_{\{T < \infty\}} | \mathscr{F}_{T^-}\right] = \mu_T^{(K)} 1_{\{T < \infty\}}. \tag{3.57}$$

### 3.2.6 Computation of the Azéma Supermartingale

**Theorem 3.4.** *We assume $M_\infty \neq 0$ and $K \leq a = M_0$. Then, the supermartingale $Z_t = \mathbb{P}(\mathscr{G}_K > t | \mathscr{F}_t)$ is given by:*

$$Z_t = \mathbb{E}\left[\frac{(K - M_\infty)^+}{\mu_{\mathscr{G}_K}} \Big| \mathscr{F}_t\right] - \frac{(K - M_t)^+}{\mu_{\mathscr{G}_K(t)}} \tag{3.58}$$

$$= \frac{1}{2}\mathbb{E}\left[\int_0^\infty \frac{dL_s^K}{\mu_s} - \int_0^t \frac{dL_s^K}{\mu_s} \Big| \mathscr{F}_t\right]. \tag{3.59}$$

*Proof.* We first prove (3.58)
From (2.100), with $t = 0$, we have:

$$\mathbb{E}\left[\Phi_{\mathscr{G}_K}(K - M_\infty)^+\right] = \mathbb{E}\left[\Phi_0(K - M_0)^+\right] + \frac{1}{2}\mathbb{E}\left[\int_0^\infty \Phi_s dL_s^K\right] \tag{3.60}$$

for every positive predictable process $(\Phi_s, s \geq 0)$. But, from (3.56):

$$\mathbb{E}\left[\Phi_{\mathscr{G}_K}(K - M_\infty)^+\right] = \mathbb{E}\left[\Phi_{\mathscr{G}_K}\mathbb{E}\left[(K - M_\infty)^+ | \mathscr{F}_{\mathscr{G}_K^-}\right]\right] = \mathbb{E}\left[\Phi_{\mathscr{G}_K}\mu_{\mathscr{G}_K}\right]. \tag{3.61}$$

Thus,

$$\mathbb{E}\left[\Phi_{\mathscr{G}_K}\mu_{\mathscr{G}_K}\right] = \mathbb{E}\left[\Phi_0(K - M_0)^+\right] + \frac{1}{2}\mathbb{E}\left[\int_0^\infty \Phi_s dL_s^K\right]. \tag{3.62}$$

Replacing in (3.62) $\Phi_s$ by $\Phi_s/\mu_s$, we obtain:

$$\mathbb{E}\left[\Phi_{\mathscr{G}_K}\right] = \mathbb{E}\left[\frac{\Phi_0}{\mu_0}(K - M_0)^+\right] + \frac{1}{2}\mathbb{E}\left[\int_0^\infty \frac{\Phi_s}{\mu_s} dL_s^K\right]. \tag{3.63}$$

In particular, for $\Phi_u := 1_{[0,T]}(u)$, with $T$ a generic stopping time, we obtain:

$$\mathbb{P}(\mathscr{G}_K \leq T) = \mathbb{E}\left[\frac{(K - M_0)^+}{\mu_0}\right] + \frac{1}{2}\left[\int_0^T \frac{dL_s^K}{\mu_s}\right] \tag{3.64}$$

$$= \mathbb{E}\left[\frac{(K - M_T)^+}{\mu_{\mathscr{G}_K(T)}}\right] \tag{3.65}$$

from the balayage formula (2.98), with $\mathscr{G}_K(T) := \sup\{s \leq T; M_s = K\}$. Now, to a set $\Gamma_t \in \mathscr{F}_t$, we associate the stopping time:

$$T = \begin{cases} t & \text{on } \Gamma_t, \\ +\infty & \text{on } \Gamma_t^c. \end{cases} \tag{3.66}$$

Hence, (3.65) becomes:

$$\mathbb{E}\left[1_{\Gamma_t}1_{\{\mathscr{G}_K\leq t\}}\right]+\mathbb{E}\left[1_{\Gamma_t^c}\right]=\mathbb{E}\left[1_{\Gamma_t}\frac{(K-M_t)^+}{\mu_{\mathscr{G}_K(t)}}\right]+\mathbb{E}\left[1_{\Gamma_t^c}\frac{(K-M_\infty)^+}{\mu_{\mathscr{G}_K}}\right].\qquad(3.67)$$

Then, by simply writing $1_{\Gamma_t^c}=1-1_{\Gamma_t}$, formula (3.67) may be written equivalently as:

$$\mathbb{E}\left[1_{\Gamma_t}1_{\{\mathscr{G}_K>t\}}\right]=\mathbb{E}\left[1_{\Gamma_t}\left(\frac{(K-M_\infty)^+}{\mu_{\mathscr{G}_K}}-\frac{(K-M_t)^+}{\mu_{\mathscr{G}_K(t)}}\right)\right]\qquad(3.68)$$

which is (3.58).

Formula (3.59) is then obtained by developing $\left(\dfrac{(K-M_t)^+}{\mu_{\mathscr{G}_K(t)}},t\geq 0\right)$ with the balayage formula:

$$\frac{(K-M_t)^+}{\mu_{\mathscr{G}_K(t)}}=\frac{(K-M_0)^+}{\mu_0}+\int_0^t\frac{1_{\{M_s<K\}}dM_s}{\mu_{\mathscr{G}_K(s)}}+\frac{1}{2}\int_0^t\frac{dL_s^K}{\mu_s}\qquad(3.69)$$

and by using (3.68).

$\square$

## 3.3 A Wider Framework: the Skorokhod Submartingales

### 3.3.1 Introduction

Let $(M_t,t\geq 0)\in\mathscr{M}_+^{0,c}$. We wish to explain how our basic formula (2.6) which we now write as:

$$\mathbb{E}\left[F_t\left(1-\frac{M_t}{K}\right)^+\right]=\mathbb{E}\left[F_t1_{\{\mathscr{G}_K\leq t\}}\right]\qquad(3.70)$$

for every positive and $\mathscr{F}_t$-measurable variable $F_t$, is a particular case of the following representation problem for certain (Skorokhod) submartingales. Let us consider, on a filtered space $(\Omega,\mathscr{F},(\mathscr{F}_t))$:

a) a probability $\mathbb{P}$, and a positive process $(X_t,t\geq 0)$ which is $(\mathscr{F}_t)$-adapted and integrable;
b) a $\sigma$-finite measure $\mathbb{Q}$ on $(\Omega,\mathscr{F})$. $\mathbb{Q}$ may be finite, even a probability, but we are also interested in the more general case where $\mathbb{Q}$ is only $\sigma$-finite.
c) a positive $\mathscr{F}$-measurable r.v. $\mathscr{G}$ such that:

$$\forall\,\Gamma_t\in\mathscr{F}_t,\quad\mathbb{E}[\Gamma_tX_t]=\mathbb{Q}\left(\Gamma_t1_{\{\mathscr{G}\leq t\}}\right).\qquad(3.71)$$

Note that it follows immediately from (3.71) that $(X_t,t\geq 0)$ is a $(\mathbb{P},(\mathscr{F}_t))$-submartingale since, for $s<t$ and $\Gamma_s\in\mathscr{F}_s$:

$$\mathbb{E}[\Gamma_s(X_t-X_s)]=\mathbb{Q}\left[\Gamma_s1_{\{s\leq\mathscr{G}\leq t\}}\right]\geq 0.\qquad(3.72)$$

Conversely, we would like to find out which positive submartingales $X$, with respect to $(\Omega, (\mathscr{F}_t), \mathbb{P})$ may be represented in the form (3.71); that is, we seek a pair $(\mathbb{Q}, \mathscr{G})$ such that (3.71) is satisfied. So far, we have not solved this problem in its full generality, but we have three set-ups where the problem is solved. The next three subsections are devoted to the discussion of each of these cases.

### 3.3.2 Skorokhod Submartingales

The three cases are concerned with what we shall call Skorokhod submartingales, i.e.: $(X_t, t \geq 0)$ is a submartingale such that:

$$X_t = -\mathscr{M}_t + L_t, \quad t \geq 0 \tag{3.73}$$

with:

1. $X_t \geq 0$; $X_0 = 0$, $\hspace{4cm}$ (3.74)
2. $(L_t, t \geq 0)$ is increasing, and $dL_t$ is carried by the zeroes of $(X_t, t \geq 0)$, (3.75)
3. $(\mathscr{M}_t, t \geq 0)$ is a true martingale, not necessarily positive. $\hspace{1cm}$ (3.76)

As is well known (see Skorokhod's Lemma, [70] Chap. VI, p.239), this implies that:

$$L_t = S_t(\mathscr{M}) := \sup_{s \leq t} \mathscr{M}_s. \tag{3.77}$$

Note, therefore, that we can assume that $\mathscr{M}_0 = 0$ without loss of generality. We shall also assume that the submartingale $(X_t, t \geq 0)$ is continuous, hence, so are $(M_t, t \geq 0)$ and $(L_t, t \geq 0)$. The three main cases we will consider are:

$i)$ $\hspace{2cm}$ $X_t = (1 - Y_t)^+, \hspace{1cm} (t \geq 0),$ $\hspace{2.5cm}$ (3.78)

where $(Y_t, t \geq 0)$ is a positive continuous martingale, which converges to 0 as $t \to \infty$ and with $Y_0 = 1$.

$ii)$ $\hspace{2cm}$ $X_t = S_t(\mathscr{N}) - \mathscr{N}_t, \hspace{1cm} (t \geq 0),$ $\hspace{2cm}$ (3.79)

where $(\mathscr{N}_t, t \geq 0)$ is a positive continuous martingale, with $\mathscr{N}_0 = 1$, which converges to 0 as $t \to \infty$, and $S_t(\mathscr{N}) := \sup_{s \leq t} \mathscr{N}_s$.

$iii)$ $\hspace{2cm}$ $X_t = |B_t|, \hspace{1cm} (t \geq 0),$ $\hspace{3cm}$ (3.80)

where $(B_t, t \geq 0)$ is a standard Brownian motion.

### 3.3.2.1 Case 1

Denote:
$$\mathcal{G} := \sup\{t \geq 0; \, Y_t = 1\} = \sup\{t \geq 0; \, X_t = 0\}.$$

Then, we have shown in Theorem 2.1 that:

$$\mathbb{P}(\mathcal{G} \leq t | \mathcal{F}_t) = (1 - Y_t)^+. \tag{3.81}$$

Therefore, in this case, we may write:

$$\mathbb{E}[\Gamma_t X_t] = \mathbb{E}\left[\Gamma_t 1_{\{\mathcal{G} \leq t\}}\right] \tag{3.82}$$

for every $\Gamma_t \in \mathcal{F}_t$. Thus, $\mathbb{Q} = \mathbb{P}$ is convenient in this situation.

### 3.3.2.2 Case 2

Again, we introduce:

$$\mathcal{G} := \sup\{t \geq 0; \, \mathcal{N}_t = S_t(\mathcal{N})\} = \sup\{t \geq 0; \, X_t = 0\}. \tag{3.83}$$

We have:

$$\mathbb{P}(\mathcal{G} > t | \mathcal{F}_t) = \mathbb{P}\left(\sup_{s \geq t} \mathcal{N}_s > S_t(\mathcal{N}) | \mathcal{F}_t\right)$$

$$= \mathbb{P}\left(\frac{\mathcal{N}_t}{U} > S_t(\mathcal{N}) | \mathcal{F}_t\right)$$

(from Lemma 2.1, with $U$ uniform on $[0,1]$ and independent from $\mathcal{F}_t$)

$$= \mathbb{P}\left(U < \frac{\mathcal{N}_t}{S_t(\mathcal{N})} | \mathcal{F}_t\right)$$

$$= \frac{\mathcal{N}_t}{S_t(\mathcal{N})}, \tag{3.84}$$

and thus:

$$\mathbb{E}\left[\Gamma_t\left(1 - \frac{\mathcal{N}_t}{S_t(\mathcal{N})}\right)\right] = \mathbb{E}\left[\Gamma_t 1_{\{\mathcal{G} \leq t\}}\right]. \tag{3.85}$$

Since (3.85) is valid for every $\Gamma_t \in \mathcal{F}_t$ and every $t \geq 0$, we may replace $\Gamma_t$ by $\Gamma_t S_t(\mathcal{N})$ and rewrite (3.85) in the equivalent form:

$$\mathbb{E}[\Gamma_t (S_t(\mathcal{N}) - \mathcal{N}_t)] = \mathbb{E}\left[\Gamma_t S_t(\mathcal{N}) 1_{\{\mathcal{G} \leq t\}}\right]. \tag{3.86}$$

However, on $\{\mathcal{G} \leq t\}$, $S_t = S_\infty$ and (3.86) equals:

$$\mathbb{E}[\Gamma_t (S_t(\mathcal{N}) - \mathcal{N}_t)] = \mathbb{E}\left[\Gamma_t S_\infty(\mathcal{N}) 1_{\{\mathcal{G} \leq t\}}\right]. \tag{3.87}$$

Therefore, a solution of (3.71) is:

$$\mathbb{Q} = S_\infty(\mathcal{N}) \cdot \mathbb{P}. \qquad (3.88)$$

Note that, in this case, $\mathbb{Q}$ has infinite total mass, since:

$$\mathbb{P}\left(S_\infty(\mathcal{N}) \in dt\right) = \frac{dt}{t^2} 1_{\{t \geq 1\}},$$

i.e., from Lemma 2.1 again, $S_\infty(\mathcal{N}) = \dfrac{1}{U}$ with $U$ uniform on $[0,1]$.


### 3.3.2.3  Case 3

This study has been the subject of many considerations within the penalizations procedures of Brownian paths studied in [58] and [77]. In fact, on the canonical space $\mathscr{C}(\mathbb{R}^+, \mathbb{R})$, where we denote by $(X_t, t \geq 0)$ the coordinate process, and $\mathscr{F}_t := \sigma\{X_s, s \leq t\}$, if $W$ denotes the Wiener measure, a $\sigma$-finite measure $\mathscr{W}$ has been constructed in [58] and [77] such that:

$$\forall \Gamma_t \in \mathscr{F}_t, \quad W[\Gamma_t|X_t|] = \mathscr{W}[\Gamma_t 1_{\{\mathscr{G} \leq t\}}], \qquad (3.89)$$

where $\mathscr{G} = \sup\{s \geq 0; X_s = 0\}$ is finite a.s. under $\mathscr{W}$. Thus, now a solution to (3.71) is given by

$$\mathbb{Q} = \mathscr{W}.$$

We note that $W$ and $\mathscr{W}$ are mutually singular.


## 3.3.3  A Comparative Analysis of the Three Cases

We note that in case 1 and case 2, $X$ converges $\mathbb{P}$-a.s., and the solution to (3.71) may be written, in both cases:

$$\mathbb{E}[\Gamma_t X_t] = \mathbb{E}\left[\Gamma_t X_\infty 1_{\{\mathscr{G} \leq t\}}\right], \qquad (3.90)$$

where: $\mathscr{G} = \sup\{t \geq 0; X_t = 0\}$.

Is this the general case for continuous Skorokhod submartingales which converge a.s.? Here is a partial answer to this question:

1) We assume $\mathbb{E}[X_\infty] < \infty$:

   a) If the convergence of $X_t$ towards $X_\infty$ also holds in $L^1$, then $(\mathscr{M}_t, t \geq 0)$ is uniformly integrable and relation (3.90) holds.
   b) If the convergence of $X_t$ towards $X_\infty$ does not hold in $L^1$, then (3.90) is not satisfied.

2) Without the assumption $\mathbb{E}[X_\infty] < \infty$, relation (3.90) is known to hold in the following cases (this list is by no means exhaustive):

   a) when $(\mathcal{M}_t, t \geq 0)$ is a positive continuous local martingale converging towards 0 (in this set-up, $\mathbb{E}[X_\infty] = \infty$, this has been proven in the previous Subsubsection 3.3.2.2 Case 2),

   b) when $(\mathcal{M}_t, t \geq 0)$ is uniformly integrable,

   c) when $\mathcal{M}_t \xrightarrow[t \to \infty]{} \mathcal{M}_\infty$ with $|\mathcal{M}_\infty| < \infty$ a.s. and $\mathcal{M}^-$ belongs to the class $\mathscr{D}$ (see [40], p.24):

$$\left\{ \mathcal{M}_T^- ;\ T \text{ a.s finite stopping time} \right\} \text{ is uniformly integrable.} \qquad (3.91)$$

The proofs of the two last statements $(b)$ and $(c)$ are given in the next subsection.

### 3.3.4 Two Situations Where the Measure $\mathbb{Q}$ Exists

We end this Section by discussing two situations where we can prove the existence of the measure $\mathbb{Q}$.

#### 3.3.4.1 Case Where $(\mathcal{M}_t, t \geq 0)$ is Uniformly Integrable

In this case, there exists a r.v. $\mathcal{M}_\infty$ such that:

$$\mathcal{M}_t = \mathbb{E}\left[\mathcal{M}_\infty | \mathscr{F}_t\right]. \qquad (3.92)$$

As previously, let $X_t := -\mathcal{M}_t + S_t(\mathcal{M})$ and $\mathscr{G} := \sup\{t \geq 0;\ X_t = 0\}$. Let us define, for $t \geq 0$ the stopping time $T^{(t)}$ by:

$$T^{(t)} := \inf\{s \geq t;\ \mathcal{M}_s = S_t\}$$

where we denote, to simplify, $S_t$ for $S_t(\mathcal{M})$. We have, for every $t \geq 0$, $\Gamma_t \in \mathscr{F}_t$ and $u > t$:

$$\mathbb{E}\left[\Gamma_t \mathcal{M}_t\right] = \mathbb{E}\left[\Gamma_t \mathcal{M}_{u \wedge T^{(t)}}\right]$$

$$= \mathbb{E}\left[\Gamma_t \mathcal{M}_u \mathbf{1}_{\left\{\sup_{t \leq v \leq u} \mathcal{M}_v < S_t\right\}}\right] + \mathbb{E}\left[\Gamma_t S_t \mathbf{1}_{\left\{\sup_{t \leq v \leq u} \mathcal{M}_v \geq S_t\right\}}\right]. \qquad (3.93)$$

We let $u$ tend to $+\infty$ in (3.93). Since $(\mathcal{M}_t, t \geq 0)$ is uniformly integrable, $\mathcal{M}_t$ converges to $\mathcal{M}_\infty$ a.s. and in $L^1$ as $t \to \infty$. Then:

$$\mathbb{E}\left[\Gamma_t \mathcal{M}_t\right] = \mathbb{E}\left[\Gamma_t \mathcal{M}_\infty \mathbf{1}_{\{\mathscr{G} \leq t\}}\right] + \mathbb{E}\left[\Gamma_t S_t \mathbf{1}_{\{\mathscr{G} \geq t\}}\right]. \qquad (3.94)$$

This relation is valid for every $\Gamma_t \in \mathcal{F}_t$, and we can replace $\Gamma_t$ by $\Gamma_t 1_{\{S_t < a\}}$. Then:

$$\mathbb{E}\left[\Gamma_t 1_{\{S_t < a\}} \mathcal{M}_t\right] = \mathbb{E}\left[\Gamma_t 1_{\{S_t < a\}} \mathcal{M}_\infty 1_{\{\mathcal{G} \leq t\}}\right] + \mathbb{E}\left[\Gamma_t 1_{\{S_t < a\}} S_t 1_{\{\mathcal{G} \geq t\}}\right],$$

or equivalently:

$$\mathbb{E}\left[\Gamma_t 1_{\{S_t < a\}} (S_t - \mathcal{M}_t)\right] = \mathbb{E}\left[\Gamma_t 1_{\{S_t < a\}} (S_t - \mathcal{M}_\infty) 1_{\{\mathcal{G} \leq t\}}\right], \tag{3.95}$$

i.e., since $S_t = S_\infty$ on $\{\mathcal{G} \leq t\}$:

$$\mathbb{E}\left[\Gamma_t 1_{\{S_t < a\}} X_t\right] = \mathbb{E}\left[\Gamma_t 1_{\{S_t < a\}} X_\infty 1_{\{\mathcal{G} \leq t\}}\right].$$

Finally, letting $a$ tend to $+\infty$ and applying the monotone convergence theorem, we obtain:

$$\mathbb{E}\left[\Gamma_t X_t\right] = \mathbb{E}\left[\Gamma_t X_\infty 1_{\{\mathcal{G} \leq t\}}\right]. \tag{3.96}$$

Therefore:

$$\mathbb{Q} = X_\infty \cdot \mathbb{P}.$$

Note that in this set-up $X_\infty$ is not necessarily integrable.

*Remark 3.5.* However, let us assume for a moment that $\mathbb{E}[X_\infty] < \infty$.
**a)** If $X_t$ converges towards $X_\infty$ in $L^1$, the relation, $\forall t \geq 0$, $\mathbb{E}[X_t] = \mathbb{E}[L_t]$, yields, applying the monotone convergence theorem: $\mathbb{E}[X_\infty] = \mathbb{E}[L_\infty] < \infty$. Thus, we can write $\mathcal{M}_t = X_t - L_t$, and this martingale converges in $L^1$ (towards $X_\infty - L_\infty$). Therefore $(\mathcal{M}_t, t \geq 0)$ is uniformly integrable and the above applies.

Note that the hypothesis that $X_t$ converges in $L^1$ towards $X_\infty$ is satisfied for any uniformly bounded Skorokhod submartingale. Indeed, under this boundedness condition, i.e. $X_t \leq K$ for some $K > 0$, we get: $\mathcal{M}_t = L_t - X_t$, and $|\mathcal{M}_t| \leq K + L_\infty$. But $\mathbb{E}[L_t] = \mathbb{E}[X_t] \leq K$, and, as a consequence, $\mathbb{E}[L_\infty] \leq K$, so that $(\mathcal{M}_t, t \geq 0)$ is uniformly integrable, hence it converges in $L^1$, and finally, so does $X_t = L_t - \mathcal{M}_t$.

**b)** If $X_t$ does not converge in $L^1$ towards $X_\infty$, then relation (3.90) cannot hold. Indeed, otherwise, relation (3.90) would imply, for $K \geq 0$:

$$\mathbb{E}\left[X_t 1_{\{X_t \geq K\}}\right] = \mathbb{E}\left[X_\infty 1_{\{X_t \geq K\}} 1_{\{\mathcal{G} \leq t\}}\right]$$

$$\leq \mathbb{E}\left[X_\infty 1_{\left\{\sup_{t \geq 0} X_t \geq K\right\}}\right] \xrightarrow[K \to \infty]{} 0 \text{ uniformly in } t,$$

since $\mathbb{E}[X_\infty] < \infty$ and $\sup_{t \geq 0} X_t < \infty$ a.s. Thus, the family $(X_t, t \geq 0)$ would be uniformly integrable, and the convergence of $X_t$ towards $X_\infty$ would also hold in $L^1$. An example of such a Skorokhod submartingale is given, for $a > 0$, by

$$X_t = -B_{t \wedge T_a} + \sup_{s \leq t} B_{s \wedge T_a},$$

where $(B_t, t \geq 0)$ is a standard Brownian motion started from 0 and $T_a := \inf\{t \geq 0, B_t = a\}$. Note that in this case: $X_\infty = 0$.

### 3.3.4.2 Extension of the Previous Case

We now assume that:

$$\mathscr{M}_t \xrightarrow[t \to \infty]{} \mathscr{M}_\infty \quad \text{a.s.} \tag{3.97}$$

with $|\mathscr{M}_\infty| < \infty$ a.s. and that $\mathscr{M}^-$ belongs to the class $\mathscr{D}$.

Let $a > 0$, $T_a = \inf\{t \geq 0; \mathscr{M}_t = a\}$ and $\left(\mathscr{M}_t^{(a)} := \mathscr{M}_{t \wedge T_a}, t \geq 0\right)$. This martingale being uniformly integrable, we can apply the result of Subsubsection 3.3.4.1. This leads to, with obvious notations:

$$\mathbb{E}\left[\Gamma_t X_t^{(a)}\right] = \mathbb{E}\left[\Gamma_t X_\infty^{(a)} 1_{\{\mathscr{G}^{(a)} \leq t\}}\right]. \tag{3.98}$$

We let $a$ tend to $+\infty$ in (3.98).

• $X_t^{(a)}$ being an increasing function of $a$, applying the monotone convergence theorem, the LHS of (3.98) converges towards $\mathbb{E}[\Gamma_t X_t]$.

• On the other hand:

$$\begin{cases} \mathscr{G}^{(a)} = +\infty & \text{on } \{T_a < \infty\} \\ \mathscr{G}^{(a)} = \mathscr{G} & \text{on } \{T_a = \infty\} \end{cases} \tag{3.99}$$

Hence:

$$\begin{aligned}
\mathbb{E}\left[\Gamma_t X_\infty^{(a)} 1_{\{\mathscr{G}^{(a)} \leq t\}}\right] &= \mathbb{E}\left[\Gamma_t X_\infty^{(a)} 1_{\{\mathscr{G}^{(a)} \leq t\}} 1_{\{T_a < \infty\}}\right] + \mathbb{E}\left[\Gamma_t X_\infty^{(a)} 1_{\{\mathscr{G}^{(a)} \leq t\}} 1_{\{T_a = \infty\}}\right] \\
&= \mathbb{E}\left[\Gamma_t X_\infty^{(a)} 1_{\{\mathscr{G} \leq t\}} 1_{\{T_a = \infty\}}\right] \\
&\quad (\text{since } 1_{\{\mathscr{G}^{(a)} \leq t\}} = X_\infty^{(a)} = 0 \text{ on } \{T_a < \infty\}) \\
&\xrightarrow[a \to \infty]{} \mathbb{E}\left[\Gamma_t X_\infty 1_{\{\mathscr{G} \leq t\}}\right]
\end{aligned}$$

from the monotone convergence theorem, since $X_\infty^{(a)}$ and $1_{\{T_a = \infty\}}$ are increasing functions of $a$.

$\square$

**Problem 3.1 (Multiplicative decomposition of supermartingales, change of probability and enlargement of filtration, see [72]).**

**A.** Let $(\Omega, (\mathscr{F}_t, t \geq 0), \mathscr{F}_\infty = \bigvee_{t \geq 0} \mathscr{F}_t, \mathbb{P})$ a filtered probability space and $(M_t, t \geq 0)$ a strictly positive, continuous $(\mathscr{F}_t, t \geq 0)$ martingale such that $M_0 = 1$. Let us define:

$$\underline{M}_t := \inf_{0 \leq s \leq t} M_s.$$

**1)** Prove that $\left( Y_t := \dfrac{M_t}{\underline{M}_t} - 1, t \geq 0 \right)$ is a continuous, non-negative, local $\mathbb{P}$-martingale such that $\mathbb{P}(Y_0 = 0) = 1$.

**2)** Let $(l_t, t \geq 0)$ be the non-decreasing, continuous process, such that $l_0 = 0$ and $(Y_t - l_t, t \geq 0)$ is a continuous local martingale. Prove that:

i) The support of $dl_t$ is included in $\{t \geq 0; \ Y_t = 0\}$.

ii) $\underline{M}_t = e^{-l_t}$.

(<u>Hint</u>: From Itô's formula, $dY_t = \dfrac{d\underline{M}_t}{\underline{M}_t} - \dfrac{M_t}{\underline{M}_t^2} d\underline{M}_t$, hence $dl_t = -\dfrac{M_t}{\underline{M}_t^2} d\underline{M}_t = -\dfrac{1}{\underline{M}_t} d\underline{M}_t$.)

    **B.** We now assume that $M_\infty = 0$ $\mathbb{P}$-a.s.

**1)** Prove that: $M_\infty = 0$ $\mathbb{P}$-a.s. $\iff$ $\underline{M}_\infty = 0$ $\mathbb{P}$-a.s. $\iff$ $l_\infty = +\infty$ $\mathbb{P}$-a.s.

Let $\mathbb{Q}$ denote the probability induced on $\mathscr{F}_\infty$ by:

$$\mathbb{Q}_{|\mathscr{F}_t} = M_t \cdot \mathbb{P}_{|\mathscr{F}_t}.$$

**2)** Prove that $\mathbb{Q}(M_t = 0) = 0$ and that $\left( \dfrac{1}{M_t}, t \geq 0 \right)$ is a $\mathbb{Q}$-martingale such that

$$\frac{1}{M_t} \xrightarrow[t \to \infty]{} 0 \quad \mathbb{Q}\text{-a.s.}$$

**3)** Prove that $\underline{M}_\infty$ is a $\mathbb{Q}$-finite r.v. with uniform distribution on $[0, 1]$.

(<u>Hint</u>: Introduce $T_c(M) := \inf\{s \geq 0; \ M_s = c\}$ and remark that: $\mathbb{Q}(\underline{M}_t < c) = \mathbb{Q}(T_c(M) < t)$).

**4)** Let $g := \sup\{t \geq 0; \ M_t = \underline{M}_\infty\}$ $(= 0$ if this set is empty$)$, and denote $(Z_t := \mathbb{Q}(g > t | \mathscr{F}_t), t \geq 0)$ the $\mathbb{Q}$-Azéma supermartingale associated to $g$. Prove that:

$Z_t = \dfrac{\underline{M}_t}{M_t}$ $\mathbb{Q}$-a.s.

(<u>Hint</u>: Introduce $\sigma_t := \inf\{s > t; \ M_s \leq \underline{M}_t\}$ and compute, for $\Gamma_t \in \mathscr{F}_t$:

$\mathbb{Q}(\Gamma_t \cap \{g > t\}) = \mathbb{Q}(\Gamma_t \cap \{\sigma_t < \infty\}) = \lim\limits_{n \to \infty} \mathbb{Q}(\Gamma_t \cap \{\sigma_t < t + n\}))$.

**5)** Deduce from **4)** that $\mathbb{Q}(0 < g < \infty) = 1$.

**6)** Observe that, from Girsanov's Theorem:

$$\left( \widetilde{M}_t := M_t - \int_0^t \frac{d\langle M \rangle_u}{M_u}, t \geq 0 \right)$$

is the martingale part of the $\mathbb{Q}$-semimartingale $(M_t, t \geq 0)$.

**7)** Prove that:

$$Z_t = 1 - \int_0^t \frac{M_u}{M_u^2} d\widetilde{M}_u + \log(\underline{M}_t) \qquad (\mathbb{Q}\text{-a.s.}).$$

    **C.** We now summarize the previous results denoting by $(N_t, t \geq 0)$ the $\mathbb{Q}$-martingale defined by:

$$N_t := \frac{1}{M_t} \qquad (t \geq 0).$$

We obtained:

let $(N_t, t \geq 0)$ be a positive, continuous $\mathbb{Q}$-martingale such that $\mathbb{Q}(N_0 = 1) = 1$ and $\lim_{t \to \infty} N_t = 0$ $\mathbb{Q}$-a.s. Then:

1) $\sup_{t \geq 0} N_t \overset{(law)}{=} \dfrac{1}{U}$, where $U$ is uniformly distributed in $[0, 1]$.

2) Let $g := \sup\{t \geq 0; N_t = \sup_{u \geq 0} N_u\}$, then $\mathbb{Q}(0 < g < \infty) = 1$. Let $Z_t := \mathbb{Q}(g > t | \mathscr{F}_t)$.

   Then:

   i) $Z_t = \dfrac{N_t}{\overline{N}_t}$, with $\overline{N}_t := \sup_{u \leq t} N_u$.

   ii) $(Z_t, t \geq 0)$ is a positive $\mathbb{Q}$-supermartingale, with Doob-Meyer decomposition:
$Z_t = M'_t - \log(\overline{N}_t)$, where $(M'_t, t \geq 0)$ denotes a $\mathbb{Q}$-martingale.

**D.** We now suppose that $\mathbb{P}$ is the Wiener measure on the canonical space $\mathscr{C}(\mathbb{R}^+, \mathbb{R})$ (where $(X_t, t \geq 0)$ is the coordinate process) such that $\mathbb{P}(X_0 = 0) = 1$. We then have:

i) $M_t = 1 + \displaystyle\int_0^t m_s dX_s$ where $(m_s, s \geq 0)$ is a predictable process such that, for any $t \geq 0$, $\displaystyle\int_0^t m_s^2 ds < \infty$ a.s.

ii) $\left( \beta_t = X_t - \displaystyle\int_0^t \dfrac{m_u}{M_u} du, t \geq 0 \right)$ is, from Girsanov's Theorem, a $\mathbb{Q}$-Brownian motion.

**1)** Prove that there exists a $(\mathscr{G}_t, \mathbb{Q})$ Brownian motion $(\widetilde{\beta}_t, t \geq 0)$ such that:

$$\beta_t = \widetilde{\beta}_t + \int_0^{t \wedge g} \frac{d\langle Z, \beta \rangle_u}{Z_u} - \int_{t \wedge g}^t \frac{d\langle Z, \beta \rangle_u}{1 - Z_u} \qquad (t \geq 0),$$

i.e.

$$\beta_t = \widetilde{\beta}_t - \int_0^{t \wedge g} \frac{m_u}{M_u} du + \int_{t \wedge g}^t \frac{M_u m_u}{M_u (M_u - \underline{M}_u)} du \qquad (t \geq 0)$$

where $(\mathscr{G}_t, t \geq 0)$ is the smallest filtration containing $(\mathscr{F}_t, t \geq 0)$ and making $g$ a $(\mathscr{G}_t, t \geq 0)$ stopping time.

(Hint: Apply the enlargement formulae (3.27)).

**2)** Deduce from **1)** that:

$$X_t = \widetilde{\beta}_t + \int_{t \wedge g}^t \frac{m_u}{M_u - \underline{M}_u} du \qquad \text{(under } \mathbb{Q}).$$

**3)** Let $\varphi : \mathbb{R}^+ \to \mathbb{R}^+$ be a positive Borel function such that $\int_0^\infty \varphi(y) dy = 1$ and let us denote $\Phi(y) = \int_0^y \varphi(z) dz$. Let:

$$M_t^\varphi := \varphi(S_t)(S_t - X_t) + \int_{S_t}^\infty \varphi(z) dz \qquad \text{with } S_t := \sup_{s \leq t} X_s.$$

Prove that $(M_t^\varphi, t \geq 0)$ is a positive, continuous martingale such that $M_0^\varphi = 1$ and $\lim_{t \to \infty} M_t^\varphi = 0$.

(Hint: Apply the balayage formula (2.98)).

Show that:

$$M_t^\varphi = 1 - \int_0^t \varphi(S_u)dX_u \qquad (t \geq 0).$$

**4)** Let us denote by $\mathbb{Q}^\varphi$ the probability on $\mathscr{F}_\infty$ induced by: $\mathbb{Q}^\varphi_{|\mathscr{F}_t} = M_t^\varphi \cdot \mathbb{P}_{|\mathscr{F}_t}$. Prove that $\underline{M}_t^\varphi = 1 - \Phi(S_t)$ and $\underline{M}_\infty^\varphi = 1 - \Phi(S_\infty)$. Deduce from this formula and **C.1)** that, under $\mathbb{Q}^\varphi$, $S_\infty$ admits $\varphi$ as a density probability function.

**5)** Let $g := \sup\{u \geq 0; M_u^\varphi = \underline{M}_\infty^\varphi\}$ and $g' := \sup\{u \geq 0; X_u = S_\infty\}$. Prove that $g = g'$, $\mathbb{Q}$-a.s.

**6)** Prove that, under $\mathbb{Q}^\varphi$:

i)   the processes $(X_u, u \leq g)$ and $(X_g - X_{g+u}, u \geq 0)$ are independent.

ii)  $(X_g - X_{g+u}, u \geq 0)$ is distributed as a 3-dimensional Bessel process.

iii) Conditionally on $S_\infty = s$, $(X_u, u \leq g)$ is a Brownian motion started at 0 and stopped at its first hitting time of level $s$.

## 3.4 Notes and Comments

Theorem 3.1 which expresses the decomposition of the supermartingale $(\mathbb{P}(L > t|\mathscr{F}_t), t \geq 0)$ in the form $\mathbb{P}(L > t|\mathscr{F}_t) = \dfrac{N_t}{S_t}$, with $S_t := \sup_{s \leq t} N_s$ is found in A. Nikeghbali and M. Yor (see [60]). The reader may also refer to R. Mansuy and M. Yor ([51] p.13-16) for a presentation of this result. Formula (3.10), i.e. the fact that $\log(S_\infty)$ is an exponential r.v., is a general result due to J. Azéma (see [37]). The progressive enlargement formulae used in Subsection 3.2.1 are due to T. Jeulin ([37]) and may be also found in R. Mansuy and M. Yor ([51], p.12-15). The predictable compensators computations in Subsection 3.2.4, as well as the computation of the Azéma supermartingale given in Theorem 3.4, are taken from A. Bentata and M. Yor ([5]). The example of Case 3 in Section 3.3 relative to the Skorokhod submartingales originates from the works about Brownian penalizations, for which one may refer to the monograph by J. Najnudel, B. Roynette and M. Yor ([59]). A study of general Skorokhod submartingales and of this representation problem are found in Najnudel-Nikeghbali [57].

# Chapter 4
# An Interesting Family of Black-Scholes Perpetuities

**Abstract** We obtain the Laplace transform and integrability properties of the integral over $\mathbb{R}^+$ of the call quantity associated with the geometric Brownian motion with negative drift, thus adding a new element to the list of already studied Brownian perpetuities.

## 4.1 Introduction

Let $(X_t, t \geq 0)$ be a process. It is often of interest to consider the r.v.:

$$\Sigma_\mu^X := \int_0^\infty X_t d\mu(t) \tag{4.1}$$

where $\mu$ is a measure on $\mathbb{R}^+$ such that the integral converges a.s. Can we describe the law of $\Sigma_\mu^X$, compute its expectation, ...?

### 4.1.1 A First Example

As in Chapter 1, let $(B_t, t \geq 0)$ denote a 1-dimensional Brownian motion starting from 0; associated to $B$, one considers the geometric Brownian motion:

$$\left( \mathscr{E}_t := \exp\left( B_t - \frac{t}{2} \right), t \geq 0 \right)$$

which is a positive martingale converging a.s. to 0 as $t \to \infty$. We define:

$$\Sigma_\mu^\pm := \int_0^\infty (\mathscr{E}_t - 1)^\pm \, \mu(dt). \tag{4.2}$$

Let us compute its expectation:

C. Profeta et al., *Option Prices as Probabilities*, Springer Finance,
DOI 10.1007/978-3-642-10395-7_4, © Springer-Verlag Berlin Heidelberg 2010

$$\mathbb{E}\left[\Sigma_\mu^\pm\right] = \mathbb{E}\left[\int_0^\infty (\mathscr{E}_t - 1)^\pm \mu(dt)\right] = \int_0^\infty \mathbb{E}\left[(\mathscr{E}_t - 1)^\pm\right]\mu(dt)$$

$$= \int_0^\infty \mathbb{P}\left(4B_1^2 \le t\right)\mu(dt) \quad \text{(from (1.19))}$$

$$= \mathbb{E}\left[\int_{4B_1^2}^\infty \mu(dt)\right]$$

$$= \mathbb{E}\left[\overline{\mu}(4B_1^2)\right]$$

with $\overline{\mu}(x) = \mu[x, +\infty[$. In particular, if $\mu(dt) = e^{-\lambda t}1_{\{t \ge 0\}}dt$:

$$\mathbb{E}\left[\Sigma_\mu^\pm\right] = \frac{1}{\lambda}\mathbb{E}\left[e^{-4\lambda B_1^2}\right] = \frac{1}{\lambda\sqrt{1+8\lambda}}. \tag{4.3}$$

### 4.1.2 Other Perpetuities

Other perpetuities have already been studied, in particular by Dufresne [21], Salminen-Yor [80], [81], [82] and Yor [98]. They obtained, for example, for $a \ne 0$ and $\nu > 0$, the identity in law:

$$\int_0^\infty \exp\left(aB_t - \nu t\right)dt \stackrel{(law)}{=} \frac{2}{a^2\gamma_{2\nu/a^2}} \tag{4.4}$$

where $\gamma_b$ is a gamma r.v. of parameter $b$:

$$\mathbb{P}(\gamma_b \in dt) = \frac{1}{\Gamma(b)}e^{-t}t^{b-1}dt \qquad (t \ge 0).$$

In particular, with $a = 1$ and $\nu = 1/2$:

$$\int_0^\infty \exp\left(B_t - \frac{t}{2}\right)dt \stackrel{(law)}{=} \frac{2}{\mathbf{e}} \tag{4.5}$$

where $\mathbf{e}$ is a standard exponential r.v.

### 4.1.3 A Family of Perpetuities Associated to the Black-Scholes Formula

As was hinted at in the Comments on Chapter 1, in recent years, the following questions have been asked to the third author, in connection with European option pricing[1]:

---

[1] A small notational point: throughout this Chapter, the usual strike parameter $K$ shall be denoted as $k$, since in the sequel the McDonald function $K_\nu$ plays an essential role.

i)  to express as simply as possible the quantity:

$$\int_0^\infty \lambda\, e^{-\lambda t}\, \mathbb{E}\left[(\mathscr{E}_t - k)^+\right] dt \qquad (k \geq 0),$$

ii)  to find the law of $\int_0^s (\mathscr{E}_t - k)^+ dt$, for fixed $s$,

iii) to find the law of $\int_0^\infty \lambda\, e^{-\lambda t}(\mathscr{E}_t - k)^+ dt$.

From Theorem 1.2, we see that ($i$) may be solved quite explicitly, following the same computations as in Example 4.1.1. However, questions ($ii$) and ($iii$) are harder to solve explicitly, as, indeed, one may look for double Laplace transforms of either quantities, e.g.:

$$\int_0^\infty e^{-\mu t}\, \mathbb{E}\left[e^{-\lambda \int_0^t (\mathscr{E}_s - k)^+ ds}\right] dt.$$

In the present Chapter, we are studying thoroughly the law of $\int_0^\infty (\mathscr{E}_t - k)^+ dt$ which seems to be slightly less difficult than either question ($ii$) or ($iii$). However, the results we obtain are not particularly simple.

### 4.1.4 Notation

These motivations having been presented, we now concentrate exclusively on the law of $\int_0^\infty (\mathscr{E}_t - k)^+ dt$. It will be convenient to consider the two-parameter process:

$$\left(\mathscr{E}_t^{(x)} := x \exp\left(B_t - \frac{t}{2}\right) = \exp\left(\log x + B_t - \frac{t}{2}\right), t \geq 0, x \geq 0\right). \qquad (4.6)$$

$(\mathscr{E}_t^{(x)}, t \geq 0, x > 0)$ is a Markov process taking values in $\mathbb{R}^+$ which, most often, we shall denote as: $(\mathscr{E}_t, t \geq 0; \mathbb{P}_x, x > 0)$ since $\mathscr{E}_0^{(x)} = x$.

To simplify notation, we shall write $\mathbb{P}$ for $\mathbb{P}_1$ and $\mathscr{E}_t$ for $\mathscr{E}_t^{(1)}$.

For any $k > 0$, let us define:

$$\Sigma_k^{(x)} := \int_0^\infty (\mathscr{E}_t^{(x)} - k)^+ dt. \qquad (4.7)$$

Again, to simplify, we shall write $\Sigma_k$ for $\Sigma_k^{(1)}$. Since $\mathscr{E}_t^{(x)} \xrightarrow[t\to\infty]{} 0$ a.s., the integral which defines (4.7) is a.s. convergent, as $(\mathscr{E}_t^{(x)} - k)^+ = 0$ for $t \geq \mathscr{G}_k := \sup\{t \geq 0; \mathscr{E}_t^{(x)} = k\}$, and $\mathscr{G}_k < \infty$ a.s.

### 4.1.5 Reduction of the Study

We now explain about some reductions of the study of the laws of $\Sigma_k^{(x)}$, to closely related problems.

*i*) From Itô's formula, we deduce:

$$\mathscr{E}_t^{(x)} = x + \int_0^t \mathscr{E}_s^{(x)} dB_s = \beta_{A_t^{(x)}} \tag{4.8}$$

where $(\beta_u, u \geq 0)$ denotes the Dambis, Dubins, Schwarz Brownian motion associated with $(\mathscr{E}_t^{(x)}, t \geq 0)$ (and $\beta_0 = x$) and:

$$A_t^{(x)} = \langle \mathscr{E}^{(x)} \rangle_t = \int_0^t (\mathscr{E}_s^{(x)})^2 ds. \tag{4.9}$$

Hence:

$$\Sigma_k^{(x)} = \int_0^\infty (\beta_{A_s^{(x)}} - k)^+ ds = \int_0^{T_0(\beta)} \frac{(\beta_v - k)^+}{\beta_v^2} dv \tag{4.10}$$

(after the change of variable $A_s^{(x)} = v$ and where $T_0(\beta) = \inf\{u \geq 0; \beta_u = 0\}$)

$$= \int_k^\infty \frac{(y-k)}{y^2} L_{T_0(\beta)}^y dy \qquad \left(= \int_k^\infty dz \int_z^\infty \frac{dy}{y^2} L_{T_0(\beta)}^y\right) \tag{4.11}$$

from the density of occupation formula, and where $L_{T_0}^y$ denotes the local time at time $T_0$ and level $y$ of the Brownian motion $(\beta_u, u \geq 0)$.

*ii*) When $x = k$, the first Ray-Knight Theorem (see Appendix B.2.3) allows us to write, from (4.11):

$$\Sigma_k^{(k)} = \int_k^\infty \frac{(y-k)}{y^2} \lambda_{y-k} dy \tag{4.12}$$

where $(\lambda_z, z \geq 0)$, conditionally on $\lambda_0 = l$, is a 0-dimensional squared Bessel process starting at $l$, and where $\lambda_0$ is an exponential variable with parameter $\frac{1}{2k}$, i.e. with expectation $2k$.

### 4.1.6 Scaling Properties

From the elementary relations:

$$\left(xe^{B_t - \frac{t}{2}} - k\right)^+ = x\left(e^{B_t - \frac{t}{2}} - \frac{k}{x}\right)^+ = k\left(\frac{x}{k}(e^{B_t - \frac{t}{2}}) - 1\right)^+ \tag{4.13}$$

valid for every $x, k > 0$, we deduce that the law of $\Sigma_k$ under $\mathbb{P}_x$ is that of $x\Sigma_{\frac{k}{x}}$ under $\mathbb{P}$, or that of $k\Sigma_1$ under $\mathbb{P}_{\frac{x}{k}}$. In other words, for every Borel positive function $\varphi$, we have:

$$\mathbb{E}_x\left[\varphi(\Sigma_k)\right] = \mathbb{E}\left[\varphi(x\Sigma_{\frac{k}{x}})\right] = \mathbb{E}_{\frac{x}{k}}\left[\varphi(k\Sigma_1)\right]. \tag{4.14}$$

These relations allow us to reduce our study of the law of $\Sigma_k$ under $\mathbb{P}_x$ to that of $\Sigma_1$ under $\mathbb{P}_{\frac{x}{k}}$. We might as well limit ourselves to the study of $\Sigma_k$ under $\mathbb{P}$.

## 4.1.7 General Case of the Brownian Exponential Martingale of Index $\nu \neq 0$

Let $\nu \neq 0$ and:

$$\mathscr{E}_t^{(x,\nu)} := \exp\left(\nu\left(\log x + B_t\right) - \frac{\nu^2 t}{2}\right) = x^\nu \exp\left(\nu B_t - \frac{\nu^2 t}{2}\right)$$

and define, for $k > 0$:

$$\Sigma_k^{(x,\nu)} := \int_0^\infty (\mathscr{E}_s^{(x,\nu)} - k)^+ ds. \tag{4.15}$$

Since, by scaling, $(\mathscr{E}_t^{(x,\nu)}, t \geq 0) \overset{(\text{law})}{=} (x^\nu \mathscr{E}_{\nu^2 t}, t \geq 0)$, we have:

$$\Sigma_k^{(x,\nu)} \overset{(\text{law})}{=} \int_0^\infty (x^\nu \mathscr{E}_{\nu^2 s} - k)^+ ds = \frac{x^\nu}{\nu^2} \int_0^\infty \left(\mathscr{E}_u - \frac{k}{x^\nu}\right)^+ du$$

$$\overset{(\text{law})}{=} \frac{x^\nu}{\nu^2} \Sigma_{\frac{k}{x^\nu}}. \tag{4.16}$$

Thus, the study of the law of $\Sigma_k^{(x,\nu)}$ may be reduced very simply to that of $\Sigma_{\frac{k}{x^\nu}}$. This is the reason why we have chosen to limit ourselves to $\nu = 1$.

## 4.1.8 Statement of the Main Results

Here are the main results of this Chapter.

**Theorem 4.1.** *Let $\alpha \geq 0$. Then, for every $x > 0$, $\mathbb{E}_x\left[(\Sigma_1)^\alpha\right] < \infty$ if and only if $\alpha < 1$.*

**Theorem 4.2.** *Let $\alpha < 0$ and $x > 0$. Then:*

i)  *For every $x > 1$, $\mathbb{E}_x\left[(\Sigma_1)^\alpha\right] < \infty$.*
ii) *For every $x < 1$, $\mathbb{P}_x(\Sigma_1 = 0) = 1 - x$ ; hence $\mathbb{E}_x\left[(\Sigma_1)^\alpha\right] = +\infty$.*
iii) *For $x = 1$, $\mathbb{E}_1\left[(\Sigma_1)^\alpha\right] < \infty$ if and only if $\alpha > -\frac{1}{3}$.*

**Theorem 4.3 (Laplace transform of $\Sigma_1$).**

*For every $\theta \geq 0$:*

$$\mathbb{E}_x\left[e^{-\frac{\theta}{2}\Sigma_1}\right] = \begin{cases} \dfrac{\sqrt{x}\,K_\gamma(\sqrt{4\theta x})}{\frac{1}{2}K_\gamma(\sqrt{4\theta}) - \sqrt{\theta}\,K_\gamma'(\sqrt{4\theta})} & \text{if } x \geq 1 \\[4mm] 1 + x\dfrac{\frac{1}{2}K_\gamma(\sqrt{4\theta}) + \sqrt{\theta}K_\gamma'(\sqrt{4\theta})}{\frac{1}{2}K_\gamma(\sqrt{4\theta}) - \sqrt{\theta}K_\gamma'(\sqrt{4\theta})} & \text{if } 0 < x \leq 1 \end{cases} \tag{4.17}$$

*where $K_\gamma$ denotes the McDonald function with index $\gamma$ $\left(\text{see Appendix B.1}\right)$ and where $\gamma = \sqrt{1 - 4\theta}$ if $4\theta \leq 1$ and $\gamma = i\sqrt{4\theta - 1}$ if $4\theta \geq 1$.*

We shall also prove, as a consequence of Theorem 4.3:

**Theorem 4.4.** *Let $(\lambda_x, x \geq 0)$ denote a squared Bessel process with dimension 0 started at $l$ and denote its law by $\mathbb{Q}_l^{(0)}$. Then:*
- *If $4\theta \geq 1$:*

$$\mathbb{Q}_l^{(0)}\left[\exp\left(-\frac{\theta}{2}\int_1^\infty \frac{(x-1)}{x^2}\lambda_{x-1}dx\right)\right] = \mathbb{Q}_l^{(0)}\left[\exp\left(-\frac{\theta}{2}\int_0^\infty \frac{x}{(x+1)^2}\lambda_x dx\right)\right]$$
$$= \exp\left(\frac{l}{2}\left(\frac{1}{2} + \frac{\sqrt{\theta}K_{i\nu}'(\sqrt{4\theta})}{K_{i\nu}(\sqrt{4\theta})}\right)\right) \tag{4.18}$$

*with $\nu = \sqrt{4\theta - 1}$.*

- *If $4\theta \leq 1$:*

$$\mathbb{Q}_l^{(0)}\left[\exp\left(-\frac{\theta}{2}\int_1^\infty \frac{x-1}{x^2}\lambda_{x-1}dx\right)\right] = \exp\left(\frac{l}{2}\left(\frac{1}{2} + \frac{\sqrt{\theta}\,K_\nu'(\sqrt{4\theta})}{K_\nu(\sqrt{4\theta})}\right)\right)$$

*with $\nu = \sqrt{1 - 4\theta}$.*

In Section 4.3, we shall study the asymptotic behavior, as $\theta \to \infty$, of $\mathbb{E}\left[e^{-\frac{\theta}{2}\Sigma_1}\right]$ and we shall obtain:

$$\mathbb{E}\left[e^{-\frac{\theta}{2}\Sigma_1}\right] \underset{\theta \to \infty}{\sim} \frac{C}{\theta^{\frac{1}{3}}} \qquad (C > 0). \tag{4.19}$$

Finally, in a short Section 4.4, we shall indicate how Theorems 4.1, 4.2 and 4.3 extend when we replace $\Sigma_k$ by $\Sigma_k^{(\rho)}$, with:

$$\Sigma_k^{(\rho)} := \int_0^\infty \left(e^{\rho(B_t - \frac{t}{2})} - k\right)^+ dt \qquad (\rho, k > 0)$$
$$\overset{\text{(law)}}{=} \frac{1}{\rho^2}\int_0^\infty \left(e^{(B_u - \frac{u}{2\rho})} - k\right)^+ du \qquad \text{(by scaling)} \tag{4.20}$$

which, from (4.20), corresponds to consider an extension of our previous perpetuities relative to $\left(B_t - \frac{1}{2}, t \geq 0\right)$ to Brownian motion with drift $-\frac{1}{2\rho}$, i.e. $\left(B_t - \frac{t}{2\rho}, t \geq 0\right)$.

## 4.2 Proofs of Theorems 4.1, 4.2, 4.3 and 4.4

### 4.2.1 A First Proof of Theorem 4.1

*i)* We first prove that for $x = 1$, $\mathbb{E}_1\left[\Sigma_1\right] = \infty$

Indeed, from (4.11):

$$\mathbb{E}_1[\Sigma_1] = \int_1^\infty \frac{(y-1)}{y^2}\mathbb{E}_1[L_{T_0}^y]dy$$

$$= \int_1^\infty \frac{(y-1)}{y^2}\mathbb{E}_1[L_{T_0}^1]dy$$

$$= 2\int_1^\infty \frac{(y-1)}{y^2}dy = +\infty \qquad (4.21)$$

(since $(L_{T_0}^y, y \geq 0)$ is a martingale and $L_{T_0}^1$ is an exponential variable with parameter $1/2$, hence for $y \geq 1$, $\mathbb{E}_1\left[L_{T_0}^y\right] = \mathbb{E}_1\left[L_{T_0}^1\right] = 2$).

*ii)* We now prove that, for every $x > 0$, $\mathbb{E}_x[\Sigma_1] = \infty$

• For $x \geq 1$, this is clear since $\Sigma_k^{(x)}$ is an increasing function of $x$ (and a decreasing function of $k$). Hence:

$$\mathbb{E}_x[\Sigma_1] \geq \mathbb{E}_1[\Sigma_1] = +\infty \qquad \text{(from (4.21))}.$$

• For $x < 1$, from the Markov property:

$$\Sigma_1^{(x)} \overset{\text{(law)}}{=} 1_{\left\{T_{\log(\frac{1}{x})}^{(-\frac{1}{2})} < \infty\right\}} \cdot \Sigma_1 \qquad (4.22)$$

with $T_{\log(\frac{1}{x})}^{(-\frac{1}{2})} := \inf\left\{t \geq 0; B_t - \frac{t}{2} = \log\left(\frac{1}{x}\right)\right\}$, and in (4.22), $\Sigma_1$ and $T_{\log(\frac{1}{x})}^{(-\frac{1}{2})}$ are assumed to be independent. Hence:

$$\mathbb{E}\left[\Sigma_1^{(x)}\right] = \mathbb{P}\left(T_{\log(\frac{1}{x})}^{(-\frac{1}{2})} < \infty\right)\mathbb{E}[\Sigma_1] = +\infty$$

from (4.21) and since $\mathbb{P}\left(T_{\log(\frac{1}{x})}^{(-\frac{1}{2})} < \infty\right) = x$ (see (1.10)).

*iii)* We now prove that, for $x > 0$ and $0 \leq \alpha < 1$, $\mathbb{E}_x\left[(\Sigma_1)^\alpha\right] < \infty$
We have:

$$\mathbb{E}_x\left[(\Sigma_1)^\alpha\right] = x^\alpha \mathbb{E}\left[(\Sigma_{\frac{1}{x}})^\alpha\right] \quad \text{(from (4.14))}$$

$$\leq x^\alpha \mathbb{E}\left[(\Sigma_0)^\alpha\right]$$

(since $\Sigma_k$ is a decreasing function of $k$)

$$\leq x^\alpha \int_0^\infty \frac{2}{y^\alpha} e^{-y} dy < \infty \quad \text{(from (4.5))}.$$

## 4.2.2 Second Proof of Theorem 4.1

It hinges upon:

**Lemma 4.1.** *Let $G_a^{(\nu)}$ and $T_a^{(\nu)}$ defined by (1.7) and (1.6).*
*i) If $\nu$ and $a$ have the same sign:*

- $\mathbb{E}\left[e^{\mu G_a^{(\nu)}}\right] < \infty$ *if and only if* $\mu < \dfrac{\nu^2}{2}$,  $\qquad$ (4.23)

- *for every real $\alpha$,* $\mathbb{E}\left[(G_a^{(\nu)})^\alpha\right] < \infty$. $\qquad$ (4.24)

*ii) If $\nu$ and $a$ have opposite signs:*

- $\mathbb{E}\left[e^{\mu G_a^{(\nu)}}\right] < \infty$ *if and only if* $\mu < \dfrac{\nu^2}{2}$, $\qquad$ (4.25)

- *for every real $\alpha > 0$,* $\mathbb{E}\left[(G_a^{(\nu)})^\alpha\right] < \infty$. $\qquad$ (4.26)

*Hence, for $\alpha < 0$,* $\mathbb{E}\left[(G_a^{(\nu)})^\alpha\right] = +\infty$.

*Proof.* The proof of this Lemma is obvious, and hinges on the well-known formulae
(1.8)–(1.14).

$\square$

We now give a second proof of Theorem 4.1
- We first show that, for $0 \leq \alpha < 1$, $\mathbb{E}_x\left[(\Sigma_1)^\alpha\right] < \infty$. From the relation:

$$\Sigma_1^{(x)} = x \int_0^\infty \left(\mathscr{E}_t - \frac{1}{x}\right)^+ dt = x \int_0^{G_{-\log x}^{\left(-\frac{1}{2}\right)}} \left(\mathscr{E}_t - \frac{1}{x}\right)^+ dt$$

$$\leq x \left(\sup_{t \geq 0} \mathscr{E}_t\right) \cdot G_{-\log x}^{\left(-\frac{1}{2}\right)}$$

we deduce that, for $0 \le \alpha < 1$, $\alpha p < 1$ and $\dfrac{1}{p} + \dfrac{1}{q} = 1$ $(p, q > 1)$:

$$\mathbb{E}_x \left[ (\Sigma_1)^\alpha \right] \le x^\alpha \mathbb{E} \left[ \left( \sup_{t \ge 0} \mathcal{E}_t \right)^\alpha \left( G^{(-\frac{1}{2})}_{-\log x} \right)^\alpha \right]$$

$$\le x^\alpha \mathbb{E} \left[ \left( \sup_{t \ge 0} \mathcal{E}_t \right)^{\alpha p} \right]^{\frac{1}{p}} \mathbb{E} \left[ \left( G^{(-\frac{1}{2})}_{(-\log x)} \right)^{\alpha q} \right]^{\frac{1}{q}}$$

$$\le C x^\alpha \mathbb{E} \left[ \left( \sup_{t \ge 0} \mathcal{E}_t \right)^{\alpha p} \right]^{\frac{1}{p}} \qquad \text{(from Lemma 4.1)}.$$

As $(\mathcal{E}_t, t \ge 0)$ is a positive martingale, starting from 1, and converging a.s. to 0 as $t \to \infty$, we get, applying Doob's maximal identity (2.8):

$$\sup_{s \ge 0} \mathcal{E}_s \overset{\text{(law)}}{=} \frac{1}{U} \tag{4.27}$$

with $U$ uniform on $[0, 1]$. Thus:

$$\mathbb{E} \left[ \left( \sup_{t \ge 0} \mathcal{E}_t \right)^{\alpha p} \right] = \int_0^1 \frac{1}{u^{\alpha p}} du < \infty \tag{4.28}$$

since $\alpha p < 1$.

• We then show that $\mathbb{E}_x[\Sigma_1] = \infty$. First of all, it follows from (1.21) with $K = 1$ that:

$$\mathbb{E}_1[\Sigma_1] = \int_0^\infty \mathbb{P} \left( G^{(\frac{1}{2})}_0 \le t \right) dt = \mathbb{E} \left[ \int_{G^{(\frac{1}{2})}_0}^\infty dt \right] = +\infty$$

since $G^{(\frac{1}{2})}_0 < +\infty$ a.s.

Likewise:

$$\mathbb{E}_x[\Sigma_1] = x \mathbb{E}_1 \left[ \Sigma_{\frac{1}{x}} \right] \qquad \text{(from (4.14))}$$

$$= x \mathbb{E} \left[ \int_{G^{(\frac{1}{2})}_{\log \frac{1}{x}}}^\infty dt \right] = \infty$$

from (1.21) and since $G^{(\frac{1}{2})}_{\log \frac{1}{x}} < \infty$ a.s.

### 4.2.3 Proof of Theorem 4.2

$i$) We already prove Point $(i)$

Let $x > 1$ and $x'$ such that $1 < x' < x$. It is then obvious that:

$$\Sigma_1^{(x)} \geq (x' - 1) T^{(-\frac{1}{2})}_{\log(\frac{x'}{x})} \tag{4.29}$$

$\left(\text{with } T_a^{(-\frac{1}{2})} := \inf\{t \geq 0; B_t - \frac{t}{2} = a\}, \text{ since, if } t < T^{(-\frac{1}{2})}_{\log \frac{x'}{x}}, \text{ then } e^{\log x + B_t - \frac{t}{2}} - 1 \geq x' - 1\right).$ Hence, with $\gamma > 0$:

$$\begin{aligned}
\mathbb{E}_x\left[\frac{1}{(\Sigma_1)^\gamma}\right] &= \int_0^\infty \mathbb{P}_x\left(\frac{1}{(\Sigma_1)^\gamma} \geq t\right) dt = \int_0^\infty \mathbb{P}_x(\Sigma_1 < v) \frac{\gamma}{v^{1+\gamma}} dv \\
&\leq \int_0^\infty \frac{\gamma}{v^{1+\gamma}} \mathbb{P}\left(T^{(-\frac{1}{2})}_{\log(\frac{x'}{x})} \leq \frac{v}{x'-1}\right) dv \qquad \text{(from (4.29))} \\
&= \int_0^\infty \frac{\gamma}{v^{1+\gamma}} \mathbb{P}\left(G^{(\log(\frac{x}{x'}))}_{\frac{1}{2}} \geq \frac{x'-1}{v}\right) dv \qquad \text{(from (1.8) and (1.9))} \\
&= \int_0^\infty \frac{\gamma}{(x'-1)^\gamma} u^{\gamma-1} \mathbb{P}\left(G^{(\log(\frac{x}{x'}))}_{\frac{1}{2}} \geq u\right) du \\
&= \frac{1}{(x'-1)^\gamma} \mathbb{E}\left[\left(G^{(\log \frac{x}{x'})}_{\frac{1}{2}}\right)^\gamma\right] < \infty \tag{4.30}
\end{aligned}$$

$\left(\text{from Point } (i) \text{ of Lemma 4.1}\right).$

$ii$) We now prove Point $(ii)$

It is clear that, for $x < 1$:

$$\{\Sigma_1^{(x)} = 0\} = \left\{T^{(-\frac{1}{2})}_{\log \frac{1}{x}} = \infty\right\}$$

Thus, $\mathbb{P}\left(\Sigma_1^{(x)} = 0\right) = \mathbb{P}\left(T^{(-\frac{1}{2})}_{\log \frac{1}{x}} = \infty\right) = \mathbb{P}\left(G^{(\log x)}_{\frac{1}{2}} = 0\right) = 1 - x > 0$ from (1.9) and (1.10).

$iii$) We now prove Point $(iii)$

For this purpose, we write, for $\gamma > 0$:

$$\mathbb{E}_1\left[\frac{1}{(\Sigma_1)^\gamma}\right] = \frac{1}{\Gamma(\gamma)} \int_0^\infty \mathbb{E}\left[e^{-\theta \Sigma_1}\right] \theta^{\gamma-1} d\theta \tag{4.31}$$

and we show, in the next Section 4.3, that:

$$\mathbb{E}_1\left[e^{-\theta \Sigma_1}\right] \underset{\theta \to \infty}{\sim} \frac{C}{\theta^{\frac{1}{3}}} \qquad (C > 0).$$

Thus, $\mathbb{E}\left[\dfrac{1}{(\Sigma_1)^\gamma}\right] < \infty$ if and only if $\displaystyle\int_1^\infty \theta^{\gamma-1-\frac{1}{3}}\,d\theta < \infty$, that is, if and only if $\gamma < \dfrac{1}{3}$.

### 4.2.4 Proof of Theorem 4.3

#### a) A useful Lemma

**Lemma 4.2.** *Let $\theta \geq 0$ and $K_\gamma$ the McDonald function with index $\gamma$, such that $\gamma = \sqrt{1-4\theta}$ if $\theta \leq \dfrac{1}{4}$ and $\gamma = i\sqrt{4\theta-1}$ if $\theta \geq \dfrac{1}{4}$.*

*1) Define the function $\varphi_\theta : \mathbb{R}^+ \to \mathbb{R}$ by:*

$$\varphi_\theta(y) = \sqrt{y}\,K_\gamma(\sqrt{4\theta y}) \qquad (y \geq 0). \tag{4.32}$$

*Then:*

*i)  $\varphi_\theta$ is a real valued function which satisfies:*

$$\varphi_\theta''(y) + \left(-\frac{\theta}{y} + \frac{\theta}{y^2}\right)\varphi_\theta(y) = 0. \tag{4.33}$$

*ii)  $\varphi_\theta$, restricted to $[1,+\infty[$ is positive, convex, bounded and decreasing.*

*2) We define the function $\widetilde{\varphi}_\theta : \mathbb{R}^+ \to \mathbb{R}^+$ by:*

$$\widetilde{\varphi}_\theta(y) = \begin{cases} \varphi_\theta(y) & \text{if } y \geq 1, \\ \big(\varphi_\theta(1) - \varphi_\theta'(1)\big) + y\varphi_\theta'(1) & \text{if } 0 \leq y \leq 1. \end{cases} \tag{4.34}$$

*Then $\widetilde{\varphi}_\theta$ is a bounded, positive, convex, decreasing function which satisfies:*

$$\widetilde{\varphi}_\theta''(y) + \left(-\frac{\theta}{y} + \frac{\theta}{y^2}\right)1_{\{y\geq1\}} \cdot \widetilde{\varphi}_\theta(y) = 0. \tag{4.35}$$

#### b) Proof of Lemma 4.2

*i)* Relation (4.33) (as well as relation (4.35)) follows from a direct computation, using the equation $K_\gamma''(z) + \dfrac{1}{z}K_\gamma'(z) - \left(1 + \dfrac{\gamma^2}{z^2}\right)K_\gamma(z) = 0$ $\big($see Appendix B.1$\big)$ and the fact that $\gamma^2 = 1 - 4\theta$ $\big($see Petiau [46, F], p. 306, formula (8), which needs to be corrected by replacing $a$ by $-a$, or Kamke [32, F], p. 440$\big)$.

We distinguish two cases:

**Case 1**: $4\theta \leq 1$, $\gamma = \sqrt{1-4\theta}$. The function $K_\gamma$, hence also $\varphi_\theta$, is positive. Furthermore $\varphi_\theta$ is bounded on $\mathbb{R}^+$ since:

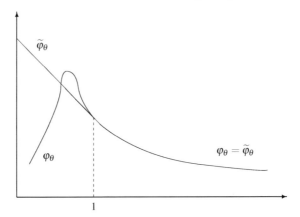

**Fig. 3** Graphs of $\varphi_\theta$ and $\widetilde{\varphi}_\theta$

$$\varphi_\theta(y) \underset{y\to 0}{\sim} Cy^{\frac{1-\gamma}{2}}, \quad \varphi_\theta(y) \underset{y\to\infty}{\sim} C'y^{\frac{1}{4}}e^{-\sqrt{4\theta y}} \qquad \left(\text{see Appendix B.1}\right).$$

On the other hand, from (4.33), the function $\varphi_\theta$ is convex on the interval $[1,+\infty[$. As it is convex, positive, and bounded, it is decreasing. Lemma 4.2 is thus proven in this case.

**Case 2**: $4\theta \geq 1$, $\gamma = i\sqrt{4\theta - 1}$ and here $\varphi_\theta(y) = \sqrt{y}\,K_{i\nu}\left(\sqrt{4\theta y}\right)$ with $\nu = \sqrt{4\theta - 1}$. From the integral representation formula $\left(\text{see Appendix B.1}\right)$:

$$K_{i\nu}(y) = \int_0^\infty e^{-y\cosh u}\cos(\nu u)du \qquad (y \geq 0) \tag{4.36}$$

we deduce that $K_{i\nu}(y)$ is real valued, hence so is $\varphi_\theta(y)$.

On the other hand, $\varphi_\theta$ is bounded on $[0,+\infty[$. Indeed, for $y \geq 0$, from (4.36):

$$\left|\sqrt{4\theta}\,\varphi_\theta\left(\frac{y^2}{4\theta}\right)\right| = y\left|K_{i\nu}(y)\right| \leq y\int_0^\infty e^{-y\frac{e^u}{2}}du = y\int_{\frac{y}{2}}^\infty e^{-z}\frac{dz}{z}.$$

Thus:

$$\left|\sqrt{4\theta}\,\varphi_\theta\left(\frac{y^2}{4\theta}\right)\right| \leq 2e^{-\frac{y}{2}}.$$

Hence:

$$\left|\sqrt{4\theta}\,\varphi_\theta\left(\frac{y^2}{4\theta}\right)\right| \underset{y\to\infty}{\longrightarrow} 0 \quad \text{and} \quad \left|\sqrt{4\theta}\,\varphi_\theta\left(\frac{y^2}{4\theta}\right)\right| \leq y\left(C+|\log(y)|\right) \underset{y\to 0}{\longrightarrow} 0.$$

Moreover, it is clear, from (4.36), that $K_{i\nu}(y) > 0$ hence $\varphi_\theta(y) > 0$, for $y$ large enough.

*ii*) We now show that $K_{i\nu}(y)$ is decreasing in $y$ on $[\nu, +\infty[$. If not, there would exist a point $y_0 > \nu$ such that:

$$K_{i\nu}(y_0) > 0, \quad K'_{i\nu}(y_0) = 0, \quad K''_{i\nu}(y_0) \leq 0.$$

However,

$$K''_{i\nu}(y_0) + \frac{1}{y_0} K'_{i\nu}(y_0) = \left(1 - \frac{\nu^2}{y_0^2}\right) K_{i\nu}(y_0).$$

Thus $K''_{i\nu}(y_0) > 0$, which is absurd. Since $K_{i\nu}$ is decreasing on $[\nu, +\infty[$, and positive near $+\infty$, $K_{i\nu}$ is positive on $[\nu, +\infty[$. Thus, $\varphi_\theta$ is positive on $[1, +\infty[$ (since, if $y \geq 1, \sqrt{4\theta y} \geq \sqrt{4\theta} \geq \sqrt{4\theta - 1} = \nu$). From the relation (4.33), we then deduce that $\varphi_\theta$ is convex on $[1, +\infty[$. Since it is bounded, convex and positive, it is decreasing. Lemma 4.2 is thus proven.

$\square$

### c) End of the proof of Theorem 4.3

Let $\left( M_t^\theta := \widetilde{\varphi_\theta}(B_t) \exp\left(-\frac{\theta}{2} \int_0^t \frac{\widetilde{\varphi_\theta}''}{\widetilde{\varphi_\theta}}(B_s) ds\right), t \geq 0 \right)$. Then $(M_t^\theta, t \geq 0)$ is a local martingale. It is equal, from (4.35) to:

$$M_t^\theta = \widetilde{\varphi_\theta}(B_t) \exp\left(-\frac{\theta}{2} \int_0^t \frac{(B_s - 1)^+}{B_s^2} ds\right) \tag{4.37}$$

and, from Lemma 4.2, for every $x \geq 0$, $(M_{t \wedge T_0}^\theta, t \geq 0)$ is bounded. Thus, from Doob's optional stopping Theorem:

$$\widetilde{\varphi_\theta}(x) = \mathbb{E}_x[M_0^\theta] = \mathbb{E}_x\left[\widetilde{\varphi_\theta}(B_{T_0}) \exp\left(-\frac{\theta}{2} \int_0^{T_0(B)} \frac{(B_s - 1)^+}{B_s^2} ds\right)\right]$$

$$= \widetilde{\varphi_\theta}(0) \mathbb{E}_x\left[\exp\left(-\frac{\theta}{2} \Sigma_1\right)\right] \quad \text{(from (4.10))}.$$

Thus:

$$\mathbb{E}_x\left[e^{-\frac{\theta}{2} \Sigma_1}\right] = \frac{\widetilde{\varphi_\theta}(x)}{\widetilde{\varphi_\theta}(0)}.$$

This is precisely Theorem 4.3, owing to formula (4.34) which yields $\widetilde{\varphi_\theta}$ explicitly.

$\square$

*Remark 4.1.* Theorem 4.3 allows to recover formula (4.5): $\Sigma_0 \overset{\text{(law)}}{=} \frac{2}{\mathbf{e}}$. Indeed, on the one hand, from (B.9):

$$\mathbb{E}_1\left[\exp\left(-\frac{\theta}{2} \Sigma_0\right)\right] = \mathbb{E}\left[\exp\left(-\frac{\theta}{2} \frac{2}{\mathbf{e}}\right)\right] = \int_0^\infty e^{-\frac{\theta}{z} - z} dz = 2\sqrt{\theta} K_1(\sqrt{4\theta})$$

and, on the other hand:

$$\mathbb{E}_1\left[e^{-\frac{\theta}{2}\Sigma_0}\right] = \lim_{\varepsilon\downarrow 0}\mathbb{E}_1\left[e^{-\frac{\theta}{2}\Sigma_\varepsilon}\right] = \lim_{\varepsilon\downarrow 0}\mathbb{E}\left[\exp\left(-\frac{\theta}{2}\int_0^\infty(e^{B_t-\frac{t}{2}}-\varepsilon)^+dt\right)\right]$$

$$= \lim_{\varepsilon\downarrow 0}\mathbb{E}\left[\exp\left(-\frac{\theta\varepsilon}{2}\int_0^\infty\left(\frac{1}{\varepsilon}e^{B_t-\frac{t}{2}}-1\right)^+dt\right)\right] = \lim_{\varepsilon\downarrow 0}\mathbb{E}_{1/\varepsilon}\left[e^{-\frac{\theta\varepsilon}{2}\Sigma_1}\right]$$

$$= \lim_{\varepsilon\downarrow 0}\frac{\frac{1}{\sqrt{\varepsilon}}K_{\sqrt{1-4\theta\varepsilon}}(\sqrt{4\theta})}{\frac{1}{2}K_{\sqrt{1-4\theta\varepsilon}}(\sqrt{4\theta\varepsilon}) - \frac{1}{2}\sqrt{4\theta\varepsilon}\,K'_{\sqrt{1-4\theta\varepsilon}}(\sqrt{4\theta\varepsilon})}$$

(from (4.17), replacing $\theta$ by $\theta\varepsilon$ and $x$ by $1/\varepsilon$)

$$= \lim_{\varepsilon\downarrow 0}\frac{\frac{1}{\sqrt{\varepsilon}}K_1(\sqrt{4\theta}) + O(1)}{\frac{1}{2}K_{\sqrt{1-4\theta\varepsilon}}(\sqrt{4\theta\varepsilon})\left[1-\sqrt{1-4\theta\varepsilon}\right] + \sqrt{\theta\varepsilon}\,K_{\sqrt{1-4\theta\varepsilon}+1}(\sqrt{4\theta\varepsilon})}$$

$$\left(\text{since } zK'_\nu(z) = \nu K_\nu(z) - zK_{\nu+1}(z); \text{ see (B.4)}\right)$$

$$= \lim_{\varepsilon\downarrow 0}\frac{\frac{1}{\sqrt{\varepsilon}}K_1(\sqrt{4\theta}) + O(1)}{O(1) + \sqrt{\theta\varepsilon}\left(\frac{1}{2\theta\varepsilon} + O(1)\right)} = 2\sqrt{\theta}\,K_1(\sqrt{4\theta}).$$

### 4.2.5 Proof of Theorem 4.4

From Theorem 4.3, we know that, for $\theta \geq \frac{1}{4}$:

$$\mathbb{E}_1\left[e^{-\frac{\theta}{2}\Sigma_1}\right] = \frac{2K_{i\nu}(\sqrt{4\theta})}{K_{i\nu}(\sqrt{4\theta}) - \sqrt{4\theta}\,K'_{i\nu}(\sqrt{4\theta})}$$

$$= \frac{1}{\frac{1}{2} - \sqrt{\theta}\frac{K'_{i\nu}(\sqrt{4\theta})}{K_{i\nu}(\sqrt{4\theta})}}$$

$$= \frac{1}{2}\int_0^\infty\exp\left(-\frac{l}{2} + \frac{l}{2}\left(\frac{1}{2} + \sqrt{\theta}\frac{K'_{i\nu}(\sqrt{4\theta})}{K_{i\nu}(\sqrt{4\theta})}\right)\right)dl. \quad (4.38)$$

On the other hand, from (4.12):

$$\mathbb{E}_1\left[e^{-\frac{\theta}{2}\Sigma_1}\right] = \mathbb{E}\left[\exp\left(-\frac{\theta}{2}\int_1^\infty\frac{(y-1)}{y^2}\lambda_{y-1}\,dy\right)\right]$$

$$= \frac{1}{2}\int_0^\infty e^{-\frac{l}{2}}dl\,\mathbb{E}\left[\exp\left(-\frac{\theta}{2}\int_1^\infty\frac{(y-1)}{y^2}\lambda_{y-1}dy\right)\Big|\lambda_0 = l\right]$$

$$= \frac{1}{2}\int_0^\infty e^{-\frac{l}{2}}dl\,\mathbb{Q}_l^{(0)}\left[\exp\left(-\frac{\theta}{2}\int_0^\infty\frac{y}{(1+y)^2}\lambda_y\,dy\right)\right] \quad (4.39)$$

where $\mathbb{Q}_l^{(0)}$ denotes the expectation relative to a squared Bessel process of dimension 0, starting from $l$. The comparison of (4.39) and (4.38) implies Theorem 4.4 in the case $\theta \geq \dfrac{1}{4}$, since the Laplace transform is one-to-one. The proof of Theorem 4.4 in the case $\theta \leq \dfrac{1}{4}$ is the same.

$\square$

*Remark 4.2.* Let $\mu$ denote a positive $\sigma$-finite measure on $\mathbb{R}^+$. Let $\Phi$ denote the unique decreasing, positive solution on $\mathbb{R}^+$ of the Sturm-Liouville equation $\Phi'' = \mu \Phi$ (and such that $\Phi(0) = 1$). It is well known (see [70], Chap. IX, p. 444) that:

$$Q_l^{(0)} \left[ \exp \left( -\frac{1}{2} \int_0^\infty \lambda_y \mu(dy) \right) \right] = \exp \left( \frac{l}{2} \Phi'(0^+) \right). \tag{4.40}$$

(Observe that, as $(\lambda_y, y \geq 0)$ has compact support a.s., $\int_0^\infty \lambda_y \mu(dy) < \infty$ if $\mu$ is $\sigma$-finite on $\mathbb{R}^+$). Theorem 4.4 may be recovered easily by applying formula (4.40) with $\mu(dy) = \dfrac{y}{1+y^2} dy$.

On the other hand, if $\mu(dy) = a(y)dy$, with

$$c_1 \, 1_{[0,\gamma_1]}(y) \leq a(y) \leq c_2 \, 1_{[0,\gamma_2]}(y) \qquad (0 < c_1 \leq c_2, \, 0 < \gamma_1 \leq \gamma_2),$$

we deduce from [70], Chap. XI, Corollary 1.8, that:

$$\exp \left( -\frac{l}{2} c_2' \sqrt{\theta} \right) \leq Q_l^{(0)} \left[ \exp \left( -\frac{\theta}{2} \int_0^\infty \lambda_y \mu(dy) \right) \right] \leq \exp \left( -\frac{l}{2} c_1' \sqrt{\theta} \right) \tag{4.41}$$

for $\theta$ large enough, whereas, as we shall show in Section 4.3:

$$Q_l^{(0)} \left[ \exp \left( -\frac{\theta}{2} \int_0^\infty \frac{y}{1+y^2} \lambda_y \, dy \right) \right] \underset{\theta \to \infty}{\sim} C \exp \left( -\frac{l}{2} \theta^{\frac{1}{3}} \right). \tag{4.42}$$

# 4.3 Asymptotic Behavior of $\mathbb{E}_1 \left[ \exp \left( -\frac{\theta}{2} \Sigma_1 \right) \right]$ as $\theta \to \infty$

We shall now end the proof of Theorem 4.2 by showing:

**Theorem 4.5.** *There is the equivalence result:*

$$\mathbb{E}_1 \left[ e^{-\frac{\theta}{2} \Sigma_1} \right] \underset{\theta \to \infty}{\sim} \frac{C}{\theta^{\frac{1}{3}}}. \tag{4.43}$$

**Proof of Theorem 4.5**

We recall that, from Theorem 4.3:

$$\mathbb{E}_1 \left[ e^{-\frac{\theta}{2} \Sigma_1} \right] = \frac{2 K_{i\nu}(\sqrt{4\theta})}{K_{i\nu}(\sqrt{4\theta}) - \sqrt{4\theta} \, K'_{i\nu}(\sqrt{4\theta})} \tag{4.44}$$

with $\nu = \sqrt{4\theta - 1}$ $\left(\theta \geq \frac{1}{4}\right)$. We shall successively find an equivalent of the numerator and of the denominator of (4.44), the difficulty arising from the fact that, in $K_{i\nu}(\sqrt{4\theta})$ $\left(\text{and } K'_{i\nu}(\sqrt{4\theta})\right)$ the argument $\sqrt{4\theta}$ and index $i\nu = i\sqrt{4\theta - 1}$ tend both to infinity as $\theta \to \infty$. To overcome this difficulty, we shall use some results about Bessel functions found in Watson ([90], p. 245-248), which we now recall.

i) Let $H_{i\nu}^{(1)}$ the first Hankel function (see [46], p. 120); it is related to $K_{i\nu}$ via the formula:

$$K_{i\nu}(z) = \frac{i\pi}{2} e^{-\frac{\nu\pi}{2}} H_{i\nu}^{(1)}(ze^{\frac{i\pi}{2}}). \tag{4.45}$$

We define $\varepsilon$ by the formula: $i\nu = iz(1 - \varepsilon)$ and assume that, as $z \to \infty$, $\varepsilon$ remains bounded (with, of course, $\nu$ depending on $z$). Then, there is the second order asymptotic expansion:

$$H_\nu^{(1)}(z) \underset{+\infty}{=} -\frac{2}{3\pi}\left\{ e^{\frac{2}{3}\pi i}\left(\sin\frac{\pi}{3}\right)\frac{\Gamma\left(\frac{1}{3}\right)}{\left(\frac{1}{6}z\right)^{\frac{1}{3}}} + e^{\frac{4}{3}\pi i}(\varepsilon z)\left(\sin\frac{2\pi}{3}\right)\frac{\Gamma\left(\frac{2}{3}\right)}{\left(\frac{1}{6}z\right)^{\frac{2}{3}}} + o\left(\frac{1}{z^{\frac{2}{3}}}\right)\right\}. \tag{4.46}$$

ii) Let us study the numerator of (4.44):

$$\begin{aligned}
N &= 2K_{i\nu}(\sqrt{4\theta}) = 2\frac{i\pi}{2}e^{-\frac{\nu\pi}{2}}H_{i\nu}^{(1)}(e^{\frac{i\pi}{2}}\sqrt{4\theta}) \\
&= (i\pi)e^{-\frac{\nu\pi}{2}}\left(-\frac{2}{3\pi}\right)\left\{ C_1 e^{\frac{2}{3}\pi i}\frac{1}{(e^{\frac{i\pi}{2}}\sqrt{4\theta})^{\frac{1}{3}}} + o\left(\frac{1}{\theta^{\frac{1}{6}}}\right)\right\} \tag{4.47}
\end{aligned}$$

$$\left(\text{with } C_1 = \left(\sin\frac{\pi}{3}\right)\Gamma\left(\frac{1}{3}\right)6^{\frac{1}{3}}\right)$$

$$\underset{\theta\to\infty}{\sim} \frac{2^{\frac{2}{3}}}{3}C_1 e^{-\frac{\nu\pi}{2}}\frac{1}{\theta^{\frac{1}{6}}}. \tag{4.48}$$

Here, we have used the first order expansion (4.46) with: $\exp\left(\frac{2}{3}i\pi\right)\exp\left(-\frac{i\pi}{6}\right) = i$ and the fact that:

$$\begin{aligned}
\varepsilon z &= z - \nu \\
&= i\sqrt{4\theta} - i\sqrt{4\theta - 1} = i\sqrt{4\theta}\left(1 - \sqrt{1 - \frac{1}{4\theta}}\right) \underset{\theta\to\infty}{\longrightarrow} 0.
\end{aligned}$$

iii) We now study the denominator of (4.44)

$$\begin{aligned}
D &= K_{i\nu}(\sqrt{4\theta}) - \sqrt{4\theta}\,K'_{i\nu}(\sqrt{4\theta}) \\
&= K_{i\nu}(\sqrt{4\theta}) + \sqrt{4\theta}\,K_{i\nu-1}(\sqrt{4\theta}) + i\sqrt{4\theta - 1}\,K_{i\nu}(\sqrt{4\theta}) \tag{4.49}
\end{aligned}$$

$\left(\text{after using } \dfrac{d}{dz}\left(z^\mu K_\mu(z)\right) = -z^\mu K_{\mu-1}(z); \text{ see (B.4)}\right)$. Since we have already studied the asymptotic behavior of $K_{i\nu}(\sqrt{4\theta})$, it remains to study that of

$$\Delta(\theta) := \sqrt{4\theta}\, K_{i\nu-1}(\sqrt{4\theta}) + i\sqrt{4\theta - 1}\, K_{i\nu}(\sqrt{4\theta}). \tag{4.50}$$

Now, since $\sqrt{4\theta}\, K_{i\nu-1}(\sqrt{4\theta}) + i\sqrt{4\theta - 1}\, K_{i\nu}(\sqrt{4\theta})$ is real $\left(\text{from (4.36) and (4.44)}\right)$ and since $i\sqrt{4\theta - 1}\, K_{i\nu}(\sqrt{4\theta})$ is purely imaginary, the development of $\Delta(\theta)$ as $\theta$ tends to $\infty$ is that of the real part of $\sqrt{4\theta}\, K_{i\nu-1}(\sqrt{4\theta})$. Then, we obtain:

$$
\begin{aligned}
\sqrt{4\theta}\, K_{i\nu-1}(\sqrt{4\theta}) &= \sqrt{4\theta}\, \frac{i\pi}{2}\, e^{\frac{i\pi}{2}(i\nu-1)} H^{(1)}_{i\nu-1}\left(e^{\frac{i\pi}{2}}\sqrt{4\theta}\right) \\
&= \sqrt{4\theta}\, \frac{\pi}{2}\, e^{-\frac{\nu\pi}{2}} H^{(1)}_{i\nu-1}\left(e^{\frac{i\pi}{2}}\sqrt{4\theta}\right)
\end{aligned} \tag{4.51}
$$

from (4.45), by replacing $i\nu$ by $i\nu - 1$. This time, we use the second order development (4.46) with here $z = e^{\frac{i\pi}{2}}\sqrt{4\theta}$ and:

$$i\nu - 1 = z(1-\varepsilon) = i\sqrt{4\theta}(1-\varepsilon), \quad \text{i.e.:} \quad \varepsilon z = \varepsilon e^{\frac{i\pi}{2}}\sqrt{4\theta}$$

$$\text{or} \quad \varepsilon z = \varepsilon e^{\frac{i\pi}{2}}\sqrt{4\theta} = 1 + \frac{1}{4\sqrt{\theta}} + o\left(\frac{1}{\sqrt{\theta}}\right) \quad \left(\nu = \sqrt{4\theta - 1}\right) \tag{4.52}$$

We obtain:

$$
\sqrt{4\theta}\, K_{i\nu-1}(\sqrt{4\theta}) = \sqrt{4\theta}\, \frac{\pi}{2}\, e^{-\frac{\nu\pi}{2}}\left(-\frac{2}{3\pi}\right)
$$

$$
\times \left\{ C_1 e^{\frac{2i\pi}{3}} \frac{1}{\left(e^{\frac{i\pi}{2}}\sqrt{4\theta}\right)^{\frac{1}{3}}} + C_2 e^{\frac{4i\pi}{3}}\left(1 + \frac{i}{4\sqrt{\theta}}\right) \cdot \frac{1}{\left(e^{\frac{i\pi}{2}}\sqrt{4\theta}\right)^{\frac{2}{3}}} + o\left(\frac{1}{\theta^{\frac{1}{3}}}\right) \right\} \tag{4.53}
$$

$$
\left(\text{with } C_1 = \left(\sin\frac{\pi}{3}\right)\Gamma\left(\frac{1}{3}\right) 6^{\frac{1}{3}} \text{ and } C_2 = \left(\sin\frac{2\pi}{3}\right)\Gamma\left(\frac{2}{3}\right) 6^{\frac{2}{3}} \quad (C_1, C_2 > 0)\right)
$$

$$
= -\frac{1}{3}\sqrt{4\theta}\, e^{-\frac{\nu\pi}{2}}\left\{ C_1 i \frac{1}{(\sqrt{4\theta})^{\frac{1}{3}}} - C_2\left(1 + \frac{i}{4\sqrt{\theta}}\right)\frac{1}{(\sqrt{4\theta})^{\frac{2}{3}}} + o\left(\frac{1}{\theta^{\frac{1}{3}}}\right) \right\}. \tag{4.54}
$$

We now consider the real part of (4.54) and we obtain, from (4.50):

$$
\begin{aligned}
\Delta(\theta) &= \mathscr{R}e\left(\sqrt{4\theta}\, K_{i\nu-1}(\sqrt{4\theta}) + i\sqrt{4\theta - 1}\, K_{i\nu}(\sqrt{4\theta})\right) \\
&= \frac{1}{3}\sqrt{4\theta}\, e^{-\frac{\nu\pi}{2}} C_2\left(\frac{1}{(\sqrt{4\theta})^{\frac{2}{3}}} + o\left(\frac{1}{\theta^{\frac{1}{3}}}\right)\right) \\
&\underset{\theta\to\infty}{\sim} \frac{2^{\frac{1}{3}} C_2}{3}\, e^{-\frac{\nu\pi}{2}}\, \theta^{\frac{1}{6}}.
\end{aligned} \tag{4.55}
$$

*iv*) We then gather (4.44), (4.48) and (4.55), to obtain:

$$\mathbb{E}_1\left[e^{-\frac{\theta}{2}\Sigma_1}\right] \underset{\theta\to\infty}{\sim} \frac{2\,2^{\frac{2}{3}}C_1\,e^{-\frac{\nu\pi}{2}}\frac{1}{\theta^{\frac{1}{6}}}}{2^{\frac{2}{3}}C_1\,e^{-\frac{\nu\pi}{2}}\frac{1}{\theta^{\frac{1}{6}}}+2^{\frac{1}{3}}C_2\,e^{-\frac{\nu\pi}{2}}\theta^{\frac{1}{6}}}$$

$$\underset{\theta\to\infty}{\sim} \frac{2^{\frac{4}{3}}C_1}{C_2}\frac{1}{\theta^{\frac{1}{3}}} = \frac{2}{3^{\frac{1}{3}}}\frac{\Gamma(\frac{1}{3})}{\Gamma(\frac{2}{3})}\frac{1}{\theta^{\frac{1}{3}}}\tag{4.56}$$

from the explicit formulae for $C_1$ and $C_2$. This is Theorem 4.5.

$\square$

*Remark 4.3.* Using Theorem 4.5, we obtain, for $0 < x \leq 1$:

$$\mathbb{E}_x\left[e^{-\frac{\theta}{2}\Sigma_1}\right] = 1 + x\frac{\frac{1}{2}K_{i\sqrt{4\theta-1}}(\sqrt{4\theta}) + \sqrt{\theta}K'_{i\sqrt{4\theta-1}}(\sqrt{4\theta})}{\frac{1}{2}K_{i\sqrt{4\theta-1}}(\sqrt{4\theta}) - \sqrt{\theta}K'_{i\sqrt{4\theta-1}}(\sqrt{4\theta})}.$$

It now follows easily from (4.48) and (4.55) that:

$$\mathbb{E}_x\left[e^{-\frac{\theta}{2}\Sigma_1}\right] \underset{\theta\to\infty}{\longrightarrow} 1 - x.\tag{4.57}$$

Now, on the other hand:

$$\mathbb{E}_x\left[e^{-\frac{\theta}{2}\Sigma_1}\right] \underset{\theta\to\infty}{\longrightarrow} \mathbb{P}_x(\Sigma_1 = 0).\tag{4.58}$$

We recall (and recover here) that $\mathbb{P}_x(\Sigma_1 = 0) = 1 - x$. This is Point (*ii*) of Theorem 4.2.

## 4.4 Extending the Preceding Results to the Variables $\Sigma_k^{(\rho,x)}$

Let

$$\Sigma_k^{(\rho,x)} := \int_0^\infty \left(e^{\rho(\log x + B_t - \frac{t}{2})} - k\right)^+ dt \qquad (\rho, x, k > 0).\tag{4.59}$$

In the preceding Sections, we have studied the case $\rho = 1$.

The analogue of Theorem 4.5 may be stated as:

**Theorem 4.6.** *The Laplace transform of $\Sigma_1^{(\rho)}$ under $\mathbb{P}_x$ is given by:*

$$\mathbb{E}_x\left[e^{-\frac{\theta}{2}\Sigma_1^{(\rho)}}\right] = \begin{cases} \dfrac{\sqrt{x}\,K_\gamma\left(\frac{\sqrt{4\theta}}{\rho}x^{\frac{\rho}{2}}\right)}{\frac{1}{2}K_\gamma\left(\frac{\sqrt{4\theta}}{\rho}\right)-\sqrt{\theta}\,K_\gamma'\left(\frac{\sqrt{4\theta}}{\rho}\right)} & \text{if } x \geq 1, \\[4mm] 1+x\dfrac{\frac{1}{2}K_\gamma\left(\frac{\sqrt{4\theta}}{\rho}\right)+\sqrt{\theta}\,K_\gamma'\left(\frac{\sqrt{4\theta}}{\rho}\right)}{\frac{1}{2}K_\gamma\left(\frac{\sqrt{4\theta}}{\rho}\right)-\sqrt{\theta}\,K_\gamma'\left(\frac{\sqrt{4\theta}}{\rho}\right)} & \text{if } 0 < x \leq 1, \end{cases} \tag{4.60}$$

with $\gamma = \dfrac{\sqrt{1-4\theta}}{\rho}$ if $\theta \leq \dfrac{1}{4}$ and $\gamma = i\dfrac{\sqrt{4\theta-1}}{\rho}$ if $\theta > \dfrac{1}{4}$.

The proof of this Theorem 4.6 is quite similar to that of Theorem 4.5. It hinges upon the following Lemma 4.3

**Lemma 4.3.**

i) Let, for $\theta \geq 0$ and $\rho > 0$, the function $\varphi_\theta^{(\rho)} : \mathbb{R}^+ \to \mathbb{R}$ defined by:

$$\varphi_\theta^{(\rho)}(y) = \sqrt{y}\,K_\gamma\left(\frac{\sqrt{4\theta}}{\rho}y^{\frac{\rho}{2}}\right) \qquad (y \geq 0). \tag{4.61}$$

Then, it satisfies:

$$\left(\varphi_\theta^{(\rho)}\right)''(y) + \theta\left(-y^{\rho-2} + \frac{1}{y^2}\right)\varphi_\theta^{(\rho)}(y) = 0 \tag{4.62}$$

$\big($see [46, F], p. 306, with $\rho - 2 = m$, formula (8), after taking care of replacing a by $-a\big)$.

ii) Let $\widetilde{\varphi}_\theta^{(\rho)} : \mathbb{R}^+ \to \mathbb{R}$ defined by:

$$\widetilde{\varphi}_\theta^{(\rho)}(y) = \begin{cases} \varphi_\theta^{(\rho)}(y) & \text{if } y \geq 1, \\ \varphi_\theta^{(\rho)}(1) - \left(\varphi_\theta^{(\rho)}\right)'(1) + y\left(\varphi_\theta^{(\rho)}\right)'(1) & \text{if } y \leq 1. \end{cases} \tag{4.63}$$

Then $\widetilde{\varphi}_\theta^{(\rho)}$ is positive, decreasing, convex and satisfies:

$$\left(\widetilde{\varphi}_\theta^{(\rho)}\right)''(y) + \theta\left(-y^{\rho-2} + \frac{1}{y^2}\right)1_{\{y \geq 1\}}\widetilde{\varphi}_\theta^{(\rho)}(y) = 0. \tag{4.64}$$

*Remark 4.4.* As a check, we note that formula (4.60) allows to recover the identity (4.4):

$$\int_0^\infty e^{\rho B_t - \frac{\rho t}{2}}\,dt \overset{(\text{law})}{=} \frac{2}{\rho^2\gamma_{1/\rho}}.$$

where $\gamma_b$ is a gamma r.v. of parameter $b$. Indeed, on one hand:

$$\mathbb{E}\left[\exp\left(-\frac{\theta}{2}\frac{2}{\rho^2}\gamma_{1/\rho}\right)\right] = \frac{1}{\Gamma(1/\rho)}\int_0^\infty e^{-\frac{\theta}{\rho^2 z}-z}z^{\frac{1}{\rho}-1}dz$$

$$= \frac{2}{\Gamma(1/\rho)}K_{\frac{1}{\rho}}\left(\frac{\sqrt{4\theta}}{\rho}\right)\left(\frac{\theta}{\rho^2}\right)^{\frac{1}{2\rho}}.$$

On the other hand:

$$\mathbb{E}_1\left[e^{-\frac{\theta}{2}\Sigma_0^{(\rho)}}\right] = \lim_{\varepsilon\downarrow 0}\mathbb{E}_1\left[\exp\left(-\frac{\theta}{2}\int_0^\infty\left(e^{\rho B_t-\frac{\rho t}{2}}-\varepsilon\right)^+ dt\right)\right]$$

$$= \lim_{\varepsilon\downarrow 0}\mathbb{E}_1\left[\exp\left(-\frac{\theta\varepsilon}{2}\int_{0^\cdot}^\infty\left(e^{\rho\left(\frac{\log\frac{1}{\varepsilon}}{\rho}+B_t\right)-\frac{\rho t}{2}}-1\right)^+ dt\right)\right]$$

$$= \lim_{\varepsilon\downarrow 0}\frac{\left(\frac{1}{\varepsilon}\right)^{\frac{1}{2\rho}}K_{\frac{\sqrt{1-4\theta\varepsilon}}{\rho}}\left(\frac{\sqrt{4\theta\varepsilon}}{\rho}\left(\frac{1}{\varepsilon}\right)^{\frac{1}{\rho}}\frac{\rho}{2}\right)}{\frac{1}{2}K_{\frac{\sqrt{1-4\theta\varepsilon}}{\rho}}\left(\frac{\sqrt{4\theta\varepsilon}}{\rho}\right)-\sqrt{\theta\varepsilon}K'_{\frac{\sqrt{1-4\theta\varepsilon}}{\rho}}\left(\frac{\sqrt{4\theta\varepsilon}}{\rho}\right)}$$

$$\left(\text{from (4.60), replacing } \theta \text{ by } \theta\varepsilon \text{ and } x \text{ by } \left(\frac{1}{\varepsilon}\right)^{\frac{1}{\rho}}\right)$$

$$\underset{\varepsilon\to 0}{\sim}\frac{\left(\frac{1}{\varepsilon}\right)^{\frac{1}{2\rho}}K_{\frac{1}{\rho}}\left(\frac{\sqrt{4\theta}}{\rho}\right)}{\frac{1}{2}K_{\frac{\sqrt{1-4\theta\varepsilon}}{\rho}}\left(\frac{\sqrt{4\theta\varepsilon}}{\rho}\right)-\frac{\rho}{2}\left(\frac{\sqrt{4\theta\varepsilon}}{\rho}\right)K'_{\frac{\sqrt{1-4\theta\varepsilon}}{\rho}}\left(\frac{\sqrt{4\theta\varepsilon}}{\rho}\right)}$$

$$\underset{\varepsilon\to 0}{\sim}\frac{\left(\frac{1}{\varepsilon}\right)^{\frac{1}{2\rho}}K_{\frac{1}{\rho}}\left(\frac{\sqrt{4\theta}}{\rho}\right)}{\frac{\rho}{2}\frac{\sqrt{4\theta\varepsilon}}{\rho}K_{\frac{1}{\rho}+1}\left(\frac{\sqrt{4\theta\varepsilon}}{\rho}\right)}\qquad\left(\text{since } zK'_\nu(z)=\nu K_\nu(z)-zK_{\nu+1}(z)\right)$$

$$\underset{\varepsilon\to 0}{\sim}\frac{\left(\frac{1}{\varepsilon}\right)^{\frac{1}{2\rho}}K_{\frac{1}{\rho}}\left(\frac{\sqrt{4\theta}}{\rho}\right)}{\frac{\rho}{2}\Gamma\left(\frac{1}{\rho}+1\right)\left(\frac{\sqrt{\theta\varepsilon}}{\rho}\right)^{-\frac{1}{\rho}}}\qquad\left(\text{since } K_\nu(z)\underset{z\to 0}{\sim}\frac{1}{2}\Gamma(\nu)\left(\frac{z}{2}\right)^{-\nu}\right)$$

$$\underset{\varepsilon\to 0}{\sim}\frac{\left(\frac{1}{\varepsilon}\right)^{\frac{1}{2\rho}}K_{\frac{1}{\rho}}\left(\frac{\sqrt{4\theta}}{\rho}\right)}{\frac{1}{2}\Gamma\left(\frac{1}{\rho}\right)}\frac{\varepsilon^{\frac{1}{2\rho}}\theta^{\frac{1}{2\rho}}}{\rho^{\frac{1}{\rho}}}$$

$$\underset{\varepsilon\downarrow 0}{\longrightarrow}\left(\frac{\theta}{\rho^2}\right)^{\frac{1}{2\rho}}\frac{2}{\Gamma\left(\frac{1}{\rho}\right)}K_{\frac{1}{\rho}}\left(\frac{\sqrt{4\theta}}{\rho}\right)=\mathbb{E}\left[e^{-\frac{\theta}{2}\Sigma_0^{(\rho)}}\right].$$

*Remark 4.5.* Taking up again the arguments of the proof of Theorem 4.5, it is not difficult to see that:

$$\mathbb{E}_1\left[e^{-\frac{\theta}{2}\Sigma_1^{(\rho)}}\right]\underset{\theta\to\infty}{\sim}\frac{C(\rho)}{\theta^{\frac{1}{3}}}$$

where $C(\rho)$ is a strictly positive constant, depending on $\rho$. We then deduce that, for $\alpha < 0$:

$$\mathbb{E}_1\left[(\Sigma_1^{(\rho)})^\alpha\right] < \infty \text{ if and only if } \alpha > -\frac{1}{3}.$$

On the other hand, it is not difficult to see that:

- If $x > 1$, for all $\alpha < 0$, $\mathbb{E}_x\left[(\Sigma_1^{(\rho)})^\alpha\right] < \infty$.

- If $x < 1$, for all $\alpha < 0$, $\mathbb{E}_x\left[(\Sigma_1^{(\rho)})^\alpha\right] = +\infty$.

Concerning the positive moments of $\Sigma_1^{(\rho)}$:

- If $0 < \alpha < \frac{1}{\rho}$, then

$$\mathbb{E}_x\left[(\Sigma_1^{(\rho)})^\alpha\right] < \infty. \tag{4.65}$$

Indeed (4.65), for $x = 1$, follows from:

$$\mathbb{E}_1\left[(\Sigma_1^{(\rho)})^\alpha\right] \le \mathbb{E}_1\left[(\Sigma_0^{(\rho)})^\alpha\right] = \frac{1}{\Gamma(\frac{1}{\rho})}\int_0^\infty \left(\frac{2}{\rho^2 z}\right)^\alpha e^{-z}z^{\frac{1}{\rho}-1}dz < \infty \qquad \text{if } \alpha < \frac{1}{\rho}$$

since $\Sigma_0 \overset{\text{(law)}}{=} \dfrac{2}{\rho^2\gamma_{\frac{1}{\rho}}}$ from (4.4).

The fact that $\mathbb{E}_x\left[(\Sigma_1^{(\rho)})^\alpha\right] < \infty$ for every $x > 0$, and every $\alpha < \dfrac{1}{\rho}$ may be obtained by using arguments close to those used in the proof of Theorem 4.1.

We believe (but have not been able to settle) that, for all $\rho > 0$:

$$\mathbb{E}_x\left[(\Sigma_1^{(\rho)})^{\frac{1}{\rho}}\right] = +\infty. \tag{4.66}$$

We now show (4.66), when $\rho > 1$

It then suffices, by using the arguments of the proof of Theorem 4.1, to see that (4.66) is true when $x = 1$.

- We first show that:

$$1 - \mathbb{E}_1\left[e^{-\frac{\theta}{2}\Sigma_1^{(\rho)}}\right] \underset{\theta\to 0}{\sim} C\theta^{\frac{1}{\rho}}. \tag{4.67}$$

Indeed, from Theorem 4.6, we have:

$$\mathbb{E}_1\left[\exp\left(-\frac{\theta}{2}\Sigma_1^{(\rho)}\right)\right] = 1 + \frac{\frac{1}{2}K_\gamma\left(\frac{\sqrt{4\theta}}{\rho}\right) + \sqrt{\theta}\,K'_\gamma\left(\frac{\sqrt{4\theta}}{\rho}\right)}{\frac{1}{2}K_\gamma\left(\frac{\sqrt{4\theta}}{\rho}\right) - \sqrt{\theta}\,K'_\gamma\left(\frac{\sqrt{4\theta}}{\rho}\right)}$$

with $\gamma = \dfrac{\sqrt{1-4\theta}}{\rho}$ (and $\theta \le \dfrac{1}{4}$). Thus:

$$1 - \mathbb{E}_1\left[e^{-\frac{\theta}{2}\Sigma_1^{(\rho)}}\right] = -\frac{\frac{1}{2}K_\gamma\left(\frac{\sqrt{4\theta}}{\rho}\right) + \frac{\rho}{2}\left\{\frac{\sqrt{4\theta}}{\rho}K_\gamma'\left(\frac{\sqrt{4\theta}}{\rho}\right)\right\}}{\frac{1}{2}K_\gamma\left(\frac{\sqrt{4\theta}}{\rho}\right) - \frac{\rho}{2}\left\{\frac{\sqrt{4\theta}}{\rho}K_\gamma'\left(\frac{\sqrt{4\theta}}{\rho}\right)\right\}}$$

$$= -\frac{\frac{1}{2}K_{\frac{\sqrt{1-4\theta}}{\rho}}\left(\frac{\sqrt{4\theta}}{\rho}\right)\left(1 - \frac{\sqrt{1-4\theta}}{\rho}\right) - \frac{\sqrt{\theta}}{\rho}K_{\frac{\sqrt{1-4\theta}}{\rho}-1}\left(\frac{\sqrt{4\theta}}{\rho}\right)}{\frac{1}{2}K_{\frac{\sqrt{1-4\theta}}{\rho}}\left(\frac{\sqrt{4\theta}}{\rho}\right)\left(1 + \frac{\sqrt{1-4\theta}}{\rho}\right) + \frac{\sqrt{\theta}}{\rho}K_{\frac{\sqrt{1-4\theta}}{\rho}-1}\left(\frac{\sqrt{4\theta}}{\rho}\right)}$$

(after using $zK_\mu'(z) = -\mu K_\mu(z) - zK_{\mu-1}(z)$).

Since $\rho > 1$, we replace $K_{\frac{\sqrt{1-4\theta}}{\rho}-1}$ by $K_{1-\frac{\sqrt{1-4\theta}}{\rho}}$ (recall that $K_\mu = K_{-\mu}$) and we deduce, from:

$$K_{1-\frac{\sqrt{1-4\theta}}{\rho}} \xrightarrow[\theta\to 0]{} K_{1-\frac{1}{\rho}}, \quad \text{with} \quad 1 - \frac{1}{\rho} > 0,$$

that:

$$1 - \mathbb{E}_1\left[e^{-\frac{\theta}{2}\Sigma_1^{(\rho)}}\right] \underset{\theta\to 0}{\sim} \frac{a\,\theta^{1-\frac{1}{2\rho}} + b\,\theta^{\frac{1}{2\rho}}}{a'\,\theta^{-\frac{1}{2\rho}} + b'\,\theta^{\frac{1}{2\rho}}} \qquad \left(\text{since, for } \mu > 0,\ K_\mu(z) \underset{z\to 0}{\sim} C_\mu z^{-\mu}\right)$$

$$\underset{\theta\to 0}{\sim} \frac{b}{a'}\,\theta^{\frac{1}{\rho}} \qquad \left(\text{since } \rho \geq 1 \text{ implies } 1 - \frac{1}{2\rho} \geq \frac{1}{2\rho}\right).$$

- From (4.67) we deduce:

$$\int_0^\infty e^{-\frac{\theta t}{2}}\mathbb{P}\left(\Sigma_1^{(\rho)} \geq t\right)dt = \frac{2}{\theta}\left(1 - \mathbb{E}\left[e^{-\frac{\theta}{2}\Sigma_1^{(\rho)}}\right]\right)$$

$$\underset{\theta\to 0}{\sim} C'\theta^{\frac{1}{\rho}-1}.$$

Hence, from the Tauberian Theorem:

$$\mathbb{P}\left(\Sigma_1^{(\rho)} \geq t\right) \underset{t\to\infty}{\sim} C''\frac{1}{t^{\frac{1}{\rho}}} \tag{4.68}$$

and

$$\mathbb{E}\left[(\Sigma_1^{(\rho)})^{\frac{1}{\rho}}\right] = \int_0^\infty \mathbb{P}\left((\Sigma_1^{(\rho)})^{\frac{1}{\rho}} \geq t\right)dt$$

$$= \int_0^\infty \mathbb{P}\left(\Sigma_1^{(\rho)} \geq t^\rho\right)dt = +\infty \qquad \text{(from (4.68))}.$$

We note that, for $\rho < 1$, the preceding argument cannot be applied since:

$$\int_1^\infty \exp\left(-\frac{\theta t}{2}\right)\frac{dt}{t^{\frac{1}{\rho}}} \xrightarrow[\theta\to 0]{} \int_1^\infty \frac{dt}{t^{\frac{1}{\rho}}} < \infty.$$

**Problem 4.1 (Explicit computations of the laws of some Brownian perpetuities).**
This problem takes up M. Yor's proof of Dufresne's identity, as presented in [98], p.14-22.

**A. Getoor's result on Bessel processes last passage times ([25])**

Let $(\widehat{R}_t, t \geq 0)$ be a Bessel process of dimension $\widehat{\delta} = 2(1+\nu)$ started at 0, with $\nu > 0$.

**1) i)** Prove that $\left( M_t := \dfrac{1}{\widehat{R}_t^{\widehat{\delta}-2}}, t \geq 1 \right)$ is a positive, continuous local martingale such that $\lim\limits_{t \to \infty} M_t = 0$ a.s.

**ii)** Show that: $\inf\limits_{t \geq 1} \widehat{R}_t \overset{(law)}{=} \widehat{R}_1 \cdot U^{\frac{1}{\widehat{\delta}-2}}$, with $U$ uniform on $[0,1]$ and independent of $\widehat{R}_1$.

**2)** Let us define: $\mathscr{G}_1^{(\widehat{R})} := \sup\{u \geq 0; \ \widehat{R}_u = 1\}$ (=0 if this set is empty). Note that $\mathscr{G}_1^{(\widehat{R})} > 0$ a.s. Prove that:

$$\mathscr{G}_1^{(\widehat{R})} \overset{(law)}{=} \dfrac{1}{\left( \inf\limits_{u \geq 1} \widehat{R}_u \right)^2}.$$

(Hint: Use the scaling property of $\widehat{R}$ and $\{\mathscr{G}_1^{(\widehat{R})} \geq t\} = \{\inf\limits_{u \geq t} \widehat{R}_u \leq 1\}$).

**3)** Deduce from **1)**(*ii*) and **2)** that:

$$\dfrac{1}{\mathscr{G}_1^{(\widehat{R})}} \overset{(law)}{=} \widehat{R}_1^2 U^{1/\nu}.$$

Conclude that:

$$\mathscr{G}_1^{(\widehat{R})} \overset{(law)}{=} \dfrac{1}{2\gamma_\nu} \tag{1}$$

where $\gamma_\nu$ is a gamma r.v. with parameter $\nu$.

(Hint: $\widehat{R}_1^2 \overset{(law)}{=} 2\gamma_{\frac{\widehat{\delta}}{2}}$, $U^{1/\nu} \overset{(law)}{=} \beta(\nu,1)$ and, from the "beta-gamma algebra",

$\gamma_a \overset{(law)}{=} \gamma_{a+b} \cdot \beta(a,b))$.
*Comment: The result* (1) *is obtained in:*
*R. K. Getoor [25]. The Brownian escape process. Ann. Probab., 7(5):864-867, 1979.*
*Pitman-Yor [65] show that there is a general formula for last passage times of tran-sient diffusions, see Theorem 2.4 of the present monograph.*

**B. Dufresne's result about the geometric Brownian motion perpetuity (see [21])**
Let $(B_t, t \geq 0)$ be a standard Brownian motion started at 0.
**1)** Prove that:

$$e^{B_t - \nu t} = 1 + \int_0^t e^{B_s - \nu s} dB_s + \left( \dfrac{1}{2} - \nu \right) \int_0^t e^{B_s - \nu s} ds.$$

**2)** We now set:

$$A_t = \int_0^t e^{2(B_s - \nu s)} ds$$

and define the process $(\rho_u, u \leq A_\infty)$ via the implicit relation:

$$\exp(B_t - \nu t) = \rho_{A_t}. \tag{2}$$

Prove that:

$$\rho_u = 1 + \beta_u + \left(\frac{1}{2} - \nu\right) \int_0^u \frac{ds}{\rho_s}, \qquad u < A_\infty$$

where $(\beta_u, u < A_\infty)$ is a standard Brownian motion. Observe that $(\rho_u, u < A_\infty)$ is a Bessel process of dimension $\delta = 2(1 - \nu)$ starting at 1. Note that, as $u \to A_\infty$, $\rho_u \to 0$. It is therefore possible to extend $(\rho_u, u < A_\infty)$ by continuity until the time $A_\infty$. Prove that:

$$A_\infty = T_0(\rho) := \inf\{u \geq 0; \ \rho_u = 0\}.$$

**3)** Prove that $A_\infty \overset{(law)}{=} \mathscr{G}_1^{(\hat{R})}$.
(<u>Hint</u>: Use a time reversal argument)
**4)** Prove that:

$$\int_0^\infty e^{2(B_s - \nu s)} ds \overset{(law)}{=} \frac{1}{2\gamma_\nu}$$

and more generally:

$$\int_0^\infty ds \exp(aB_s - bs) \overset{(law)}{=} \frac{2}{a^2 \gamma_{\frac{2b}{a^2}}}.$$

<u>Comment</u>: Here, we have given a proof of Dufresne's result, see:
D. Dufresne [21]. The distribution of a perpetuity, with applications to risk theory and pension funding. Scand. Actuar. J., (1-2):39-79, 1990,
which may be reduced to Getoor's result on last passage times for Bessel processes.

**C.** We now recall <u>Lamperti's relation</u> (see [70], p.452)

$$\exp(B_t + \mu t) = R^{(\mu)} \left( \int_0^t ds \exp(2(B_s + \mu s)) \right)$$

where $\mu \geq 0$ and $(R_t^{(\mu)}, t \geq 0)$ is a Bessel process of index $\mu$ (i.e. dimension $2(1 + \mu)$) started at 1. Compare Lamperti's (implicit) relation (for $\mu \geq 0$) with (2) (for $-\nu < 0$).
**1)** Deduce from Lamperti's relation that:

$$R_t^{(\mu)} = \exp(B_u + \mu u)\big|_{u = H_t} \qquad \text{with } H_t = \int_0^t \frac{ds}{\left(R_s^{(\mu)}\right)^2}.$$

**2)** Prove that, for every $\alpha > 0$:

$$\int_0^\infty \frac{ds}{\left(R_s^{(\mu)}\right)^{\alpha+2}} = \int_0^\infty e^{-\alpha(B_u + \mu u)} du$$

and deduce then that:

$$\int_0^\infty \frac{ds}{\left(R_s^{(\mu)}\right)^{\alpha+2}} \overset{(law)}{=} \frac{2}{\alpha^2 \gamma_{\frac{2\mu}{\alpha}}} \qquad (\mu > 0,\ \alpha > 0).$$

## 4.5 Notes and Comments

This Chapter is taken from [76]. For studies of other perpetuities related to Brownian motion with drift, we refer the reader to Salminen-Yor [80], [81] and [82].

# Chapter 5
# Study of Last Passage Times up to a Finite Horizon

**Abstract** In Chapter 1, we have expressed the European put and call quantities in terms of the last passage time $\mathscr{G}_K^{(\mathscr{E})}$. However, since $\mathscr{G}_K^{(\mathscr{E})}$ is not a stopping time, formulae (1.20) and (1.21) are not very convenient for simulation purposes. To counter this drawback, we introduce in Section 5.1 of the present Chapter the $\mathscr{F}_t$-measurable random time:

$$\mathscr{G}_K^{(\mathscr{E})}(t) = \sup\{s \leq t;\ \mathscr{E}_s = K\}$$

and write the analogues of formulae (1.20) and (1.21) for these times $\mathscr{G}_K^{(\mathscr{E})}(t)$. This will lead us to the interesting notion of past-future martingales, which we shall study in details in Section 5.2.

## 5.1 Study of Last Passage Times up to a Finite Horizon for the Brownian Motion with Drift

### 5.1.1 Introduction and Notation

Let $\left(\Omega, (B_t, \mathscr{F}_t)_{t\geq 0}, \mathscr{F}_\infty = \bigvee_{t\geq 0} \mathscr{F}_t, \mathbb{P}\right)$ denote a Brownian motion starting at 0 and $(\mathscr{F}_t := \sigma(B_s, s \leq t), t \geq 0)$ its natural filtration. We define the past-future (or the two parameters) filtration:

$$\mathscr{F}_{s,t} = \sigma(B_u, u \leq s;\ B_h, h \geq t), \qquad 0 \leq s \leq t < \infty. \tag{5.1}$$

For every $\nu \in \mathbb{R}$, we denote $(B_t^{(\nu)}, t \geq 0) = (B_t + \nu t, t \geq 0)$ the Brownian motion with drift $\nu$. Let us define, for $a, \nu \in \mathbb{R}$ and $t > 0$:

$$G_a^{(\nu)}(t) := \sup\{u \leq t; B_u^{(\nu)} = a\}, \tag{5.2}$$

$$(= 0 \text{ if the set } \{u \leq t; B_u^{(\nu)} = a\} \text{ is empty}).$$

C. Profeta et al., *Option Prices as Probabilities*, Springer Finance,
DOI 10.1007/978-3-642-10395-7_5, © Springer-Verlag Berlin Heidelberg 2010

$G_a^{(\nu)}(t)$ is the last passage time of $B^{(\nu)}$ at level $a$ before time $t$. Of course, we have $G_a^{(\nu)}(t) \xrightarrow[t \to \infty]{} G_a^{(\nu)}$ a.s. with:

$$G_a^{(\nu)} := \sup\{u \geq 0; B_u^{(\nu)} = a\}, \tag{5.3}$$

$$(= 0 \text{ if the set } \{u \geq 0; B_u^{(\nu)} = a\} \text{ is empty}).$$

In Chapter 1, we described the link that exists between the last passage times of the drifted Brownian motion and some option prices defined with the help of the geometric Brownian motion $\left( \mathscr{E}_t^{(\nu)} := \exp\left( \nu B_t - \dfrac{\nu^2}{2} t \right), t \geq 0 \right)$. In particular, we showed that:

$$\mathbb{P}\left( G_{\log(K)}^{(-\nu)} \leq s | \mathscr{F}_s \right) = \left( 1 - \frac{\mathscr{E}_s^{(2\nu)}}{K^{2\nu}} \right)^+ \tag{5.4}$$

(see Theorem 1.3). However, this formula is not very convenient for simulation purposes, since the event $\{G_{\log(K)}^{(-\nu)} \leq s\}$ depends on the whole Brownian trajectory after time $s$. That is why, to overcome this drawback, we shall consider in this Chapter times such as $G_{\log(K)}^{(-\nu)}(t)$ instead of $G_{\log(K)}^{(-\nu)}$, so that the event $\{G_{\log(K)}^{(-\nu)}(t) \leq s\}$ only depends on the Brownian trajectory before time $t$. That is what we shall call working up to horizon $t$.

### 5.1.2  Statement of our Main Result

**Theorem 5.1.** *For every* $K \geq 0$, $\nu \in \mathbb{R}$, $s \leq t$:

i) $\mathbb{P}\left( G_{\log(K)}^{(\nu)}(t) \leq s | \mathscr{F}_{s,t} \right)$

$$= \left( 1 - \exp\left\{ -\frac{2}{t-s} \left( B_s^{(\nu)} - \log(K) \right) \left( B_t^{(\nu)} - \log(K) \right) \right\} \right)^+ \tag{5.5}$$

ii) $K^{-2\nu} \mathbb{P}\left( G_{\log(K)}^{(\nu)}(t) \leq s | B_s^{(\nu)} = \log(x) \right)$

$$= 1_{\{x<K\}} \left\{ K^{-2\nu} \mathbb{E}\left[ \mathscr{E}_u^{(-2\nu)} 1_{\{B_u^{(\nu)}>\log(x/K)\}} \right] - x^{-2\nu} \mathbb{P}\left( B_u^{(\nu)} < \log\left( \frac{x}{K} \right) \right) \right\} \tag{5.6}$$

$$+ 1_{\{x>K\}} \left\{ K^{-2\nu} \mathbb{E}\left[ \mathscr{E}_u^{(-2\nu)} 1_{\{B_u^{(\nu)}<\log(x/K)\}} \right] - x^{-2\nu} \mathbb{P}\left( B_u^{(\nu)} > \log\left( \frac{x}{K} \right) \right) \right\}.$$

$(s \leq t, u = t - s, x \geq 0)$

*iii) Yuri's Formula (this terminology is borrowed from [1]) holds:*

$$\mathbb{P}\left(G^{(\nu)}_{\log(K)}(t) \leq s\right) = \mathbb{E}\left[\left(1 - K^{2\nu}\mathcal{E}_s^{(2\nu)}\right)^+\right]$$

$$+ \mathbb{E}\left[\text{sgn}\left(1 - K^{2\nu}\mathcal{E}_s^{(2\nu)}\right)\text{sgn}\left(1 - K^{2\nu}\mathcal{E}_t^{(2\nu)}\right)\left(1 \wedge K^{2\nu}\mathcal{E}_t^{(2\nu)}\right)\right]. \quad (5.7)$$

*Remark 5.1.*
**a)** We have stated Theorem 5.1 by "decreasing order of conditioning": Point $(i)$ gives the conditional expectation of the event $\{G^{(\nu)}_{\log(K)}(t) \leq s\}$ with respect to $\mathscr{F}_{s,t}$, Point $(ii)$ gives its conditional expectation with respect to $\mathscr{F}_s$, and Point $(iii)$ its expectation without conditioning.
**b)** By comparing formulae (5.4) and (5.5), we see that Theorem 5.1 provides a kind of Black-Scholes formula up to horizon $t$.
**c)** Theorem 1.3, Point $(i)$, or formula (5.4), is a consequence of equation (5.5) on letting $t$ tend to infinity and observing that:

$$G^{(\nu)}_{\log(K)}(t) \xrightarrow[t \to \infty]{} G^{(\nu)}_{\log(K)} \text{ a.s.} \quad \text{and} \quad \frac{B^{(\nu)}_t}{t} \xrightarrow[t \to \infty]{} \nu \text{ a.s.}$$

**d)** An alternative form of equation (5.6) that we shall use is as follows. For $\nu > 0$:

$$K^{-2\nu}\mathbb{P}\left(G^{(\nu)}_{\log(K)}(t) \leq s \big| B^{(\nu)}_s = \log(x)\right)$$

$$= \left(K^{-2\nu} - x^{-2\nu}\right)^+ + \left(1_{\{x<K\}} - 1_{\{x>K\}}\right) \quad (5.8)$$

$$\times \left\{K^{-2\nu}\mathbb{E}\left[\mathcal{E}_u^{(-2\nu)}1_{\{B^{(\nu)}_u > \log(x/K)\}}\right] - x^{-2\nu}\mathbb{P}\left(B^{(\nu)}_u < \log\left(\frac{x}{K}\right)\right)\right\},$$

while, for $\nu < 0$:

$$K^{-2\nu}\mathbb{P}\left(G^{(\nu)}_{\log(K)}(t) \leq s \big| B^{(\nu)}_s = \log(x)\right)$$

$$= \left(K^{-2\nu} - x^{-2\nu}\right)^+ + \left(1_{\{x<K\}} - 1_{\{x>K\}}\right) \quad (5.9)$$

$$\times \left\{x^{-2\nu}\mathbb{P}\left(B^{(\nu)}_u > \log\left(\frac{x}{K}\right)\right) - K^{-2\nu}\mathbb{E}\left[\mathcal{E}_u^{(-2\nu)}1_{\{B^{(\nu)}_u < \log(x/K)\}}\right]\right\}.$$

Ones goes from (5.6) to (5.8) and (5.9) on observing that:

$$\mathbb{P}\left(B^{(\nu)}_u > \log\left(\frac{x}{K}\right)\right) = 1 - \mathbb{P}\left(B^{(\nu)}_u < \log\left(\frac{x}{K}\right)\right),$$

and

$$\mathbb{E}\left[\mathcal{E}_u^{(-2\nu)}1_{\{B^{(\nu)}_u < \log(x/K)\}}\right] = 1 - \mathbb{E}\left[\mathcal{E}_u^{(-2\nu)}1_{\{B^{(\nu)}_u > \log(x/K)\}}\right].$$

*Proof.* We first prove Point $(i)$

We have, from the definition of $G_l^{(\nu)}(t)$ for $l \in \mathbb{R}$:

$$\left\{ G_l^{(\nu)}(t) < s \right\} = A_{s,t}^+ \cup A_{s,t}^-$$

where

$$A_{s,t}^+ = \{\forall u \in ]s,t[, \ B_u^{(\nu)} > l\},$$
$$A_{s,t}^- = \{\forall u \in ]s,t[, \ B_u^{(\nu)} < l\}.$$

Therefore:

$$\mathbb{P}\left( G_l^{(\nu)}(t) < s | \mathscr{F}_{s,t} \right) = \mathbb{P}\left( A_{s,t}^+ | \mathscr{F}_{s,t} \right) + \mathbb{P}\left( A_{s,t}^- | \mathscr{F}_{s,t} \right) \tag{5.10}$$

$$= 1_{\{B_s^{(\nu)} > l\}} \mathbb{P}\left( \inf_{s \leq u \leq t} (B_u^{(\nu)} - B_s^{(\nu)}) > l - B_s^{(\nu)} | B_t^{(\nu)} - B_s^{(\nu)} \right)$$

$$+ 1_{\{B_s^{(\nu)} < l\}} \mathbb{P}\left( \sup_{s \leq u \leq t} (B_u^{(\nu)} - B_s^{(\nu)}) < l - B_s^{(\nu)} | B_t^{(\nu)} - B_s^{(\nu)} \right) \tag{5.11}$$

Thus, we need to compute for $l$ and $\lambda = l - B_s^{(\nu)}$ (with $\lambda(\lambda - m) > 0$):

$$1_{\{B_s^{(\nu)} > l\}} \mathbb{P}\left( \inf_{s \leq u \leq t} (B_u^{(\nu)} - B_s^{(\nu)}) > l - B_s^{(\nu)} | B_t^{(\nu)} - B_s^{(\nu)} = m \right)$$

$$= 1_{\{B_s^{(\nu)} > l\}} \mathbb{P}\left( T_\lambda^{(\nu)} > t - s | B_{t-s}^{(\nu)} = m \right), \quad (5.12)$$

and

$$1_{\{B_s^{(\nu)} < l\}} \mathbb{P}\left( \sup_{s \leq u \leq t} (B_u^{(\nu)} - B_s^{(\nu)}) < l - B_s^{(\nu)} | B_t^{(\nu)} - B_s^{(\nu)} = m \right)$$

$$= 1_{\{B_s^{(\nu)} < l\}} \mathbb{P}\left( T_\lambda^{(\nu)} > t - s | B_{t-s}^{(\nu)} = m \right). \quad (5.13)$$

We note that the quantities in equations (5.12) and (5.13) depend only on $t - s$ and not the pair $(s,t)$. Furthermore, the law of the bridge from $a$ to $b$ over the time $[0, u]$ of a Brownian motion with drift $\nu$ is independent of $\nu$. Hence, we need to compute:

$$\mathbb{P}(T_\lambda > t - s | B_{t-s} = m) = 1 - \mathbb{P}(T_\lambda < t - s | B_{t-s} = m),$$

and it is a well-known fact that:

$$P_{0 \to m}^{(u)}(T_\lambda < u) = \exp\left( -\frac{2(\lambda(\lambda - m))^+}{u} \right) \tag{5.14}$$

where $P_{0 \to m}^{(u)}$ denotes the law of a Brownian bridge of length $u$ starting at $0$ and ending at level m. Gathering (5.11), (5.12), (5.13) and (5.14) with $\lambda = l - B_s^{(\nu)}$ and $m = B_t^{(\nu)} - B_s^{(\nu)}$, we obtain (5.5).

*Remark 5.2.* (5.14) can be obtained by an application of the symmetry principle:

$$
\begin{aligned}
P_{0 \to m}^{(u)}(T_\lambda > u) &= 1 - P_{0 \to m}^{(u)}(T_\lambda < u) \\
&= \frac{\frac{1}{\sqrt{2\pi u}} \exp\left(-\frac{m^2}{2u}\right) - \frac{1}{\sqrt{2\pi u}} \exp\left(-\frac{(2\lambda - m)^2}{2u}\right)}{\frac{1}{\sqrt{2\pi u}} \exp\left(-\frac{m^2}{2u}\right)} \\
&= 1 - \exp\left(-\frac{(2\lambda - m)^2}{2u} + \frac{m^2}{2u}\right) \\
&= 1 - \exp\left(-\frac{2(\lambda(\lambda - m))^+}{u}\right).
\end{aligned}
$$

In fact, since formula (5.14) plays a key role in our proof of (5.5), we feel it is relevant to give some references where it also appears: Guasoni [27] (formula (10)-(11) p.85), Pagès [61] (Proposition 7.3) in his discussion of the Brownian bridge method for simulating the continuous Euler scheme, Pitman [64] while deriving the law of the local time at 0 for a Brownian bridge, Yor [97] while discussing Seshadri's identities ([83], p.11).

We now prove Point (*ii*) of Theorem 5.1
We start by writing:

$$
B_t^{(\nu)} \stackrel{(law)}{=} \widetilde{B}_s^{(\nu)} + B_{t-s}^{(\nu)} = \log(x) + B_u^{(\nu)} \quad \text{with } u = t - s \tag{5.15}
$$

where $\widetilde{B}^{(\nu)}$ is a Brownian motion with drift $\nu$ independent of $\mathscr{F}_u$. We may then write with $C = \log(x/K)$, from (5.5):

$$
\begin{aligned}
\mathbb{P}\left(G_{\log(K)}^{(\nu)}(t) < s \mid \mathscr{F}_s\right) &= \mathbb{E}\left[\left(1 - \exp\left\{-\frac{2}{u}\left(\log\frac{x}{K}\right)\left(\log\frac{x}{K} + B_u^{(\nu)}\right)\right\}\right)^+\right] \\
&= 1_{\{x<K\}} \mathbb{P}\left(B_u^{(\nu)} + C < 0\right) + 1_{\{x>K\}} \mathbb{P}\left(B_u^{(\nu)} + C > 0\right) \\
&\quad - 1_{\{x<K\}} \mathbb{E}\left[1_{\{B_u^{(\nu)}+C<0\}} \exp\left(-\frac{2C}{u}(C + B_u^{(\nu)})\right)\right] \\
&\quad - 1_{\{x>K\}} \mathbb{E}\left[1_{\{B_u^{(\nu)}+C>0\}} \exp\left(-\frac{2C}{u}(C + B_u^{(\nu)})\right)\right] \\
&:= \alpha + \beta - \gamma - \delta. \tag{5.16}
\end{aligned}
$$

We now make the computations for $\alpha, \beta, \gamma$ and $\delta$:

$$\alpha = 1_{\{x<K\}} \mathbb{P}\left(B_u^{(\nu)} < \log \frac{K}{x}\right)$$

$$= 1_{\{x<K\}} \mathbb{E}\left[1_{\{B_u^{(\nu)}<\log \frac{K}{x}\}} \frac{\mathscr{E}_u^{(-2\nu)}}{\mathscr{E}_u^{(-2\nu)}}\right]$$

$$= 1_{\{x<K\}} \mathbb{E}\left[1_{\{B_u-\nu u<\log \frac{K}{x}\}} e^{2\nu B_u - 2\nu^2 u}\right] \quad \text{(from the Cameron-Martin formula (1.5))}$$

$$= 1_{\{x<K\}} \mathbb{E}\left[1_{\{-B_u-\nu u<\log \frac{K}{x}\}} e^{-2\nu B_u - 2\nu^2 u}\right] \quad \text{(since } B_u \stackrel{(law)}{=} -B_u\text{)}$$

$$= 1_{\{x<K\}} \mathbb{E}\left[1_{\{B_u^{(\nu)}>\log \frac{x}{K}\}} \mathscr{E}_u^{(-2\nu)}\right]. \tag{5.17}$$

Similarly, one shows that:

$$\beta = 1_{\{x>K\}} \mathbb{P}\left(B_u^{(\nu)} > \log \frac{K}{x}\right) = 1_{\{x>K\}} \mathbb{E}\left[1_{\{B_u^{(\nu)}<\log \frac{x}{K}\}} \mathscr{E}_u^{(-2\nu)}\right]. \tag{5.18}$$

For $\gamma$, we have:

$$\gamma = 1_{\{x<K\}} \mathbb{E}\left[1_{\{B_u^{(\nu)}+C<0\}} \exp\left(-\frac{2C}{u}(C+B_u^{(\nu)})\right)\right]$$

$$= 1_{\{x<K\}} \mathbb{E}\left[1_{\{B_u^{(\nu)}+C<0\}} \exp\left(-\frac{2C}{u}(C+B_u^{(\nu)})\right) \frac{\mathscr{E}_u^{(-2\nu)}}{\mathscr{E}_u^{(-2\nu)}}\right]$$

$$= 1_{\{x<K\}} \mathbb{E}\left[1_{\{B_u^{(-\nu)}+C<0\}} \exp\left(-\frac{2C}{u}(C+B_u^{(-\nu)})+2\nu B_u - 2\nu^2 u\right)\right] \tag{5.19}$$

(from the Cameron-Martin formula (1.5))

$$= 1_{\{x<K\}} \exp\left(-\frac{2C^2}{u}+2C\nu - 2\nu^2 u\right) \mathbb{E}\left[1_{\{B_u<-C+\nu u\}} \exp\left(-\frac{B_u}{u}(2C-2\nu u)\right)\right]$$

$$= 1_{\{x<K\}} \exp\left(-\frac{2C^2}{u}+2C\nu - 2\nu^2 u\right) \int_{-\infty}^{-C+\nu u} \frac{dy}{\sqrt{2\pi u}} e^{-\frac{y^2}{2u}-\frac{y}{u}(2C-2\nu u)}$$

$$= 1_{\{x<K\}} \exp\left(-\frac{2C^2}{u}+2C\nu - 2\nu^2 u+\frac{(2C-2\nu u)^2}{2u}\right)$$

$$\times \int_{-\infty}^{-C+\nu u} \frac{dy}{\sqrt{2\pi u}} e^{-\frac{1}{2u}(y+2C-2\nu u)^2} \tag{5.20}$$

$$= 1_{\{x<K\}} \exp\left(-2C\nu\right) \int_{-\infty}^{C-\nu u} \frac{dz}{\sqrt{2\pi u}} e^{-\frac{z^2}{2u}}$$

(after the change of variable $y+2C-2\nu u = z$)

$$= 1_{\{x<K\}} K^{2\nu} x^{-2\nu} \mathbb{P}\left(B_u^{(\nu)} < \log \frac{x}{K}\right) \quad \left(\text{since } C = \log \frac{x}{K}\right). \tag{5.21}$$

A similar computation yields to:

$$\delta = 1_{\{x>K\}}\mathbb{E}\left[1_{\{B_u^{(\nu)}+C>0\}}\exp\left(-\frac{2C}{u}(C+B_u^{(\nu)})\right)\right]$$

$$= 1_{\{x>K\}}K^{2\nu}x^{-2\nu}\mathbb{P}\left(B_u^{(\nu)} > \log\frac{x}{K}\right). \tag{5.22}$$

Plugging the results (5.17)–(5.22) back into equation (5.16) yields then Point $(ii)$ of Theorem 5.1.

<u>We now prove Yuri's formula (5.7)</u>

We work in the case $\nu < 0$. The result for $\nu > 0$ may be deduced on utilizing the identity in law (1.8):

$$G_{\log(K)}^{(\nu)}(t) \overset{(law)}{=} G_{-\log(K)}^{(-\nu)}(t).$$

We shall begin with the equivalent form (5.9). In the following, $\log(x)$ shall denote the r.v. $\widetilde{B}_s + \nu s$ and $x^{-2\nu}$ the r.v. $\widetilde{\mathcal{E}}_s^{(-2\nu)} = \exp\left(-2\nu\widetilde{B}_s - 2\nu^2 s\right)$. These r.v. are independent of $B_u$ for $u = t - s$, see (5.15). We write:

$$K^{-2\nu}\mathbb{P}\left(G_{\log(K)}^{(\nu)}(t) < s\right)$$

$$=\widetilde{\mathbb{E}}\left[(K^{-2\nu} - x^{-2\nu})^+\right] - K^{-2\nu}\widetilde{\mathbb{E}}\left[1_{\{x<K\}}\mathbb{E}\left[\mathcal{E}_u^{(-2\nu)}1_{\{B_u^{(\nu)}<\log(x/K)\}}\right]\right]$$

$$+\widetilde{\mathbb{E}}\left[1_{\{x<K\}}x^{-2\nu}\mathbb{P}\left(B_u^{(\nu)} > \log\frac{x}{K}\right)\right] + K^{-2\nu}\widetilde{\mathbb{E}}\left[1_{\{x>K\}}\mathbb{E}\left[\mathcal{E}_u^{(-2\nu)}1_{\{B_u^{(\nu)}<\log(x/K)\}}\right]\right]$$

$$-\widetilde{\mathbb{E}}\left[1_{\{x>K\}}x^{-2\nu}\mathbb{P}\left(B_u^{(\nu)} > \log\frac{x}{K}\right)\right]$$

$$:=a - b + c + d - e.$$

On the other hand, Yuri's formula (5.7) asserts that:

$$K^{-2\nu}\mathbb{P}\left(G_{\log(K)}^{(\nu)}(t) < s\right)$$

$$=\widetilde{\mathbb{E}}\left[(K^{-2\nu} - x^{-2\nu})^+\right] - K^{-2\nu}\mathbb{P}\left(\mathcal{E}_s^{(-2\nu)} < K^{-2\nu}, \mathcal{E}_t^{(-2\nu)} > K^{-2\nu}\right)$$

$$+\mathbb{E}\left[1_{\{\mathcal{E}_s^{(-2\nu)}<K^{-2\nu}\}}1_{\{\mathcal{E}_t^{(-2\nu)}<K^{-2\nu}\}}\mathcal{E}_t^{(-2\nu)}\right]$$

$$+K^{-2\nu}\mathbb{P}\left(\mathcal{E}_s^{(-2\nu)} > K^{-2\nu}, \mathcal{E}_t^{(-2\nu)} > K^{-2\nu}\right)$$

$$-\mathbb{E}\left[1_{\{\mathcal{E}_s^{(-2\nu)}>K^{-2\nu}\}}1_{\{\mathcal{E}_t^{(-2\nu)}<K^{-2\nu}\}}\mathcal{E}_t^{(-2\nu)}\right]$$

$$:=a' - b' + c' + d' - e'.$$

The equality of $a$ and $a'$ is clear. We now examine the other terms, still making use of the Cameron-Martin formula (1.5).

- Analysis of the term $b$:

$$b = K^{-2\nu}\widetilde{\mathbb{E}}\left[1_{\{x<K\}}\mathbb{E}\left[\mathscr{E}_u^{(-2\nu)}1_{\{B_u^{(\nu)}<\log(x/K)\}}\right]\right]$$

$$= K^{-2\nu}\widetilde{\mathbb{E}}\left[1_{\{\widetilde{\mathscr{E}}_s^{(-2\nu)}<K^{-2\nu}\}}\mathbb{E}\left[\mathscr{E}_u^{(-2\nu)}1_{\{\exp(-2\nu B_u-2\nu^2 u)<K^{2\nu}\exp(-2\nu\widetilde{B}_s-2\nu^2 s)\}}\right]\right]$$

$$= K^{-2\nu}\widetilde{\mathbb{E}}\left[1_{\{\widetilde{\mathscr{E}}_s^{(-2\nu)}<K^{-2\nu}\}}\mathbb{P}\left(\exp(-2\nu B_u+2\nu^2 u)<K^{2\nu}\exp(-2\nu\widetilde{B}_s-2\nu^2 s)\right)\right]$$

$$= K^{-2\nu}\widetilde{\mathbb{E}}\left[1_{\{\widetilde{\mathscr{E}}_s^{(-2\nu)}<K^{-2\nu}\}}\mathbb{P}\left(\exp(2\nu(B_u-\widetilde{B}_s)-2\nu^2 t)>K^{-2\nu}\right)\right]\qquad (t=u+s)$$

$$= K^{-2\nu}\mathbb{P}\left(\mathscr{E}_s^{(-2\nu)}<K^{-2\nu},\mathscr{E}_t^{(-2\nu)}>K^{-2\nu}\right)$$

(since $B_u \overset{(law)}{=} -B_u$ and replacing $\widetilde{B}_s$ by $B_s$)

$$= b'.$$

- The equality of $d$ and $d'$ follows from a similar analysis.

- Analysis of the term $c$:

$$c = \widetilde{\mathbb{E}}\left[1_{\{x<K\}}x^{-2\nu}\mathbb{P}\left(B_u^{(\nu)}>\log\frac{x}{K}\right)\right]$$

$$= \widetilde{\mathbb{E}}\left[1_{\{\widetilde{\mathscr{E}}_s^{(-2\nu)}<K^{-2\nu}\}}\widetilde{\mathscr{E}}_s^{(-2\nu)}\mathbb{E}\left[\frac{\mathscr{E}_u^{(-2\nu)}}{\mathscr{E}_u^{(-2\nu)}}1_{\{\mathscr{E}_u^{(-2\nu)}>K^{2\nu}\widetilde{\mathscr{E}}_s^{(-2\nu)}\}}\right]\right]$$

$$= \widetilde{\mathbb{E}}\left[1_{\{\widetilde{\mathscr{E}}_s^{(-2\nu)}<K^{-2\nu}\}}\widetilde{\mathscr{E}}_s^{(-2\nu)}\mathbb{E}\left[\exp\left(2\nu B_u-2\nu^2 u\right)1_{\{\exp\left(-2\nu B_u+2\nu^2 u\right)>K^{2\nu}\widetilde{\mathscr{E}}_s^{(-2\nu)}\}}\right]\right]$$

$$= \widetilde{\mathbb{E}}\left[1_{\{\widetilde{\mathscr{E}}_s^{(-2\nu)}<K^{-2\nu}\}}\mathbb{E}\left[\exp\left(-2\nu(\widetilde{B}_s-B_u)-2\nu^2 t\right)1_{\{\exp\left(-2\nu(\widetilde{B}_s-B_u)-2\nu^2 t\right)<K^{-2\nu}\}}\right]\right]$$

$$= \mathbb{E}\left[1_{\{\mathscr{E}_s^{(-2\nu)}<K^{-2\nu}\}}\exp\left(-2\nu B_t-2\nu^2 t\right)1_{\{\mathscr{E}_t^{(-2\nu)}<K^{-2\nu}\}}\right]$$

(on replacing $B_u$ by $-B_u$ and $\widetilde{B}_s$ by $B_s$)

$$= \mathbb{E}\left[1_{\{\mathscr{E}_s^{(-2\nu)}<K^{-2\nu}\}}1_{\{\mathscr{E}_t^{(-2\nu)}<K^{-2\nu}\}}\mathscr{E}_t^{(-2\nu)}\right]$$

$$= c'.$$

- The equality of $e$ and $e'$ is established similarly.

This completes the proof of Theorem 5.1.

$\square$

*Remark 5.3.* In Chapter 1, we have chosen to prove Theorem 1.3 using the results of Theorem 1.2 together with the scaling properties of $T_a^{(\nu)}$, $G_a^{(\nu)}$ and $\mathscr{E}_t^{(\nu)}$. In fact, it is possible to ignore Theorem 1.2 completely, and to prove Theorem 1.3 as a corollary of Theorem 5.1. Indeed, points (*i*) and (*ii*) of Theorem 1.3 follow from

Theorem 5.1 by letting $t$ tend to $+\infty$ in (5.5), (5.6) and (5.7), and taking expectation. In completing this agenda, we show that Theorem 5.1 implies Point $(iii)$, relations (1.55) and (1.56) of Theorem 1.3. We shall only give the proof for $\nu > 0$ and $K \geq 1$ since the other cases may be obtained in the same manner. First, it is clear that for $x > 0$:

$$\mathbb{P}\left(G^{(\nu)}_{\log(K)}(t) < s \,\middle|\, B_s + \nu s = \log(x)\right) = \mathbb{P}\left(T^{(\nu)}_{\log(K/x)} > t - s\right), \tag{5.23}$$

and we get from (5.6) and (5.23) with $K = 1$, $0 < x < 1$ and $u = t - s$:

$$\begin{aligned}
\mathbb{P}\left(T^{(\nu)}_{\log(1/x)} \leq u\right) &= 1 - \mathbb{P}\left(T^{(\nu)}_{\log(1/x)} > u\right) \\
&= 1 - \mathbb{E}\left[\mathscr{E}^{(-2\nu)}_u \mathbf{1}_{\{B^{(\nu)}_u > \log(x)\}}\right] + x^{-2\nu}\mathbb{P}\left(B^{(\nu)}_u < \log(x)\right) \\
&= \mathbb{E}\left[\mathscr{E}^{(-2\nu)}_u \mathbf{1}_{\{B^{(\nu)}_u < \log(x)\}}\right] + x^{-2\nu}\mathbb{P}\left(B^{(\nu)}_u < \log(x)\right) \\
&= \mathbb{E}\left[e^{-2\nu B_u - 2\nu^2 u}\mathbf{1}_{\{B_u + \nu u < \log(x)\}}\right] + x^{-2\nu}\mathbb{P}\left(B^{(\nu)}_u < \log(x)\right) \\
&= \mathbb{E}\left[e^{2\nu B_u - 2\nu^2 u}\mathbf{1}_{\{B_u - \nu u > -\log(x)\}}\right] + x^{-2\nu}\mathbb{P}\left(B^{(\nu)}_u < \log(x)\right)
\end{aligned} \tag{5.24}$$

after changing $B_u$ to $-B_u$. Then, replacing in (5.24) $1/x$ by $K$, with $K \geq 1$, we obtain:

$$\mathbb{P}\left(T^{(\nu)}_{\log(K)} \leq u\right) = \mathbb{E}\left[\mathscr{E}^{(2\nu)}_u \mathbf{1}_{\{B^{(-\nu)}_u > \log(K)\}}\right] + K^{2\nu}\mathbb{P}\left(B^{(-\nu)}_u > \log(K)\right) \tag{5.25}$$

or equivalently:

$$\mathbb{P}\left(T^{(\nu)}_{\log(K)} \leq u\right) = \mathbb{E}\left[\mathscr{E}^{(2\nu)}_u \mathbf{1}_{\{\mathscr{E}^{(2\nu)}_u > K^{2\nu}\}}\right] + K^{2\nu}\mathbb{P}\left(\mathscr{E}^{(2\nu)}_u > K^{2\nu}\right). \tag{5.26}$$

We remark that we have proven (5.26) for $K \geq 1$ and $\nu > 0$. If instead of assuming $\nu > 0$, we had assumed $\nu < 0$, then, (5.26) would have become:

$$\mathbb{P}\left(T^{(\nu)}_{\log(K)} \leq u\right) = \mathbb{E}\left[\mathscr{E}^{(2\nu)}_u \mathbf{1}_{\{\mathscr{E}^{(2\nu)}_u < K^{2\nu}\}}\right] + K^{2\nu}\mathbb{P}\left(\mathscr{E}^{(2\nu)}_u < K^{2\nu}\right). \tag{5.27}$$

## 5.1.3 An Explicit Expression for the Law of $G^{(\nu)}_x(t)$

We recall that $G^{(\nu)}_x(t) = \sup\left\{s \leq t; B^{(\nu)}_s = x\right\}$. In the case $\nu = 0$, the following result has already been obtained ([96]):

$$\mathbb{P}\left(G^{(0)}_x(t) \in ds\right) = \frac{ds}{\sqrt{s(t-s)}}\exp\left(-\frac{x^2}{2s}\right) \quad (0 < s < t). \tag{5.28}$$

Note that this is a subprobability, since $\frac{ds}{\pi\sqrt{s(t-s)}}$ is the arcsine distribution on $[0,t]$.
Indeed,

$$\mathbb{P}\left(G_x^{(0)}(t)=0\right)=\mathbb{P}(T_x\geq t)=\mathbb{P}\left(|N|\leq\frac{x}{\sqrt{t}}\right)>0$$

with $N$ a standard Gaussian r.v.

We first prove (5.28)

We have:

$$\left\{0<G_x^{(0)}(t)\leq s\right\}=\{T_x\leq s\}\cap\left\{T_x+\widehat{G}_0^{(0)}(t-T_x)\leq s\right\}\qquad(5.29)$$

where $\widehat{G}_0^{(0)}(t-T_x):=\sup\{u\leq t;\ \widehat{B}_u=0\}$, with

$$\widehat{B}_u=B_{u+T_x}-B_{T_x}.\qquad(5.30)$$

Now, it is a well-known fact that, (conditionally on $T_x$) $\widehat{G}_0^{(0)}(t-T_x)$ follows an arcsine law (see [70], Exercise 3.20, p.112). Therefore, from (1.11):

$$\mathbb{P}\left(0<G_x^{(0)}(t)\leq s\right)=\int_0^s\frac{x}{\sqrt{2\pi a^3}}e^{-\frac{x^2}{2a}}da\int_0^{s-a}\frac{dy}{\pi\sqrt{y(t-a-y)}},\qquad(5.31)$$

and, by derivation:

$$\mathbb{P}\left(G_x^{(0)}(t)\in ds\right)=\left(\frac{1}{\pi}\int_0^s\frac{x}{\sqrt{2\pi a^3}}e^{-\frac{x^2}{2a}}\frac{da}{\sqrt{(s-a)(t-s)}}\right)ds$$

$$=\left(\frac{1}{\pi\sqrt{t-s}}\frac{1}{s}\int_0^\infty\frac{x}{\sqrt{2\pi b}}e^{-\frac{x^2}{2s}(b+1)}db\right)ds$$

(after the change of variable $b=\dfrac{s}{a}-1$)

$$=\left(\frac{e^{-\frac{x^2}{2s}}}{\pi\sqrt{s(t-s)}}\frac{1}{\sqrt{\pi}}\int_0^\infty\frac{e^{-c}}{\sqrt{c}}dc\right)ds$$

(after the change of variable $\dfrac{x^2}{2s}b=c$)

$$=\left(\frac{1}{\pi\sqrt{s(t-s)}}e^{-\frac{x^2}{2s}}\right)ds\qquad\text{(since }\Gamma(1/2)=\sqrt{\pi}\text{)}.$$

We are now interested in the general case $\nu\neq0$. We shall compute, for $0\leq s\leq t$:

$$\gamma_{x,t}^{(\nu)}(s) := \mathbb{P}\left(0 < G_x^{(\nu)}(t) \leq s\right) \tag{5.32}$$

$$= \mathbb{E}\left[1_{\{0 < G_x^{(0)}(t) \leq s\}} \exp\left(\nu B_t - \frac{\nu^2}{2}t\right)\right] \quad \text{(from (1.5))}$$

$$= \mathbb{E}\left[1_{\{T_x \leq s\}} 1_{\{T_x + \widehat{G}_0^{(0)}(t-T_x) \leq s\}} \exp\left(\nu(x + \widehat{B}_{t-T_x}) - \frac{\nu^2}{2}t\right)\right]$$

(from (5.29) and (5.30)) $\tag{5.33}$

$$= e^{\nu x - \frac{\nu^2}{2}t} \mathbb{E}\left[1_{\{T_x + \widehat{G}_0^{(0)}(t-T_x) \leq s\}} \exp\left(\nu\sqrt{(t-T_x) - \widehat{G}_0^{(0)}(t-T_x)}\,\widehat{m}_1\right)\right]$$

where we have used the "meander factorization" (see [10]):

$$\widehat{B}_{t-T_x} = \sqrt{(t-T_x) - \widehat{G}_0^{(0)}(t-T_x)}\,\widehat{m}_1$$

with $\widehat{m}_1 = \varepsilon m_1$ independent of $(T_x, \widehat{G}_0^{(0)}(t-T_x))$, $\varepsilon$ a symmetric Bernoulli r.v. and $m_1 \overset{(law)}{=} \sqrt{2\mathfrak{e}}$, where $\mathfrak{e}$ is a standard exponential r.v. We introduce the function:

$$\Phi(\lambda) := \mathbb{E}\left[\exp\left(\lambda \widehat{m}_1\right)\right] = \mathbb{E}\left[\cosh(\lambda m_1)\right] = \int_0^\infty e^{-t}\cosh(\lambda\sqrt{2t})dt.$$

An integration by parts shows that:

$$\Phi(\lambda) = 1 + e^{\frac{\lambda^2}{2}}|\lambda|\sqrt{2\pi}\mathbb{P}(|N| \leq |\lambda|)$$

with $N$ a standard Gaussian r.v. Consequently, we obtain:

$$\gamma_{x,t}^{(\nu)}(s) = e^{\nu x - \frac{\nu^2 t}{2}} \mathbb{E}\left[1_{\{0 < \widehat{G}_x^{(0)}(t) \leq s\}} \Phi\left(\nu\sqrt{t - \widehat{G}_x^{(0)}(t)}\right)\right] \tag{5.34}$$

which implies, from (5.28):

$$\mathbb{P}\left(G_x^{(\nu)}(t) \in ds\right) = e^{\nu x - \frac{\nu^2}{2}t} \frac{ds}{\pi\sqrt{s(t-s)}} \exp\left(-\frac{x^2}{2s}\right)\Phi(\nu\sqrt{t-s}). \tag{5.35}$$

**Exercise 5.1 (General computation of $\mathscr{G}_x(t) := \sup\{s \leq t; X_s = x\}$ for a linear diffusion).**
Let $(X_t, t \geq 0)$ be a regular linear diffusion taking values in $\mathbb{R}$ and started at 0. We denote by $s$ its scale function, $m(dx) = \rho(x)dx$ its speed measure, and $q(t,x,y)$ its transition density function with respect to $m$. Let us introduce the resolvent kernel of $(X_t, t \geq 0)$:

$$u_\lambda(x,y) = \int_0^\infty e^{-\lambda u}q(u,x,y)du.$$

**1)** We first assume that $\mathbb{P}(X_t \xrightarrow[t\to\infty]{} +\infty) = 1$. Prove that, for $x > 0$, $u_0(x,x) = -s(x)$ and rewrite Theorem 2.4 in terms of $u_\lambda$. In fact, it can be proven that this new expression is valid for all transient (regular) diffusions, see [12].

**2)** Let $\tau$ be an exponential r.v. of parameter $\alpha$, independent from $(X_t, t \geq 0)$. We define the diffusion $(\widehat{X}_t, t \geq 0)$ by:

$$\widehat{X}_t = \begin{cases} X_t & \text{if } t < \tau \\ \partial & \text{if } t \geq \tau \end{cases}$$

where $\partial$ is a cemetery point. Apply the result of Question **1)** to $(\widehat{X}_t, t \geq 0)$ to prove that:

$$\mathbb{P}_0\left(\mathscr{G}_x(\tau) \in dt\right) = \left(1 - \mathbb{E}\left[e^{-\alpha T_x}\right]\right)\delta_0(dt) + \frac{e^{-\alpha t} q(t,0,x)}{u_\alpha(x,x)}dt$$

where $T_x := \inf\{t \geq 0;\ X_t = x\}$.

**3)** We assume that $(X_t, t \geq 0)$ is a recurrent diffusion. Recall the following formula (see [42]):

$$\frac{1}{u_\lambda(x,x)} = \int_0^\infty (1 - e^{-\lambda u})n^x(u)du$$

where $n^x$ is the density of the Lévy measure $\nu^x$ of the subordinator $\tau_l^{(x)} := \inf\{t \geq 0;\ L_t^x > l\}$. Prove that $\dfrac{1}{\lambda u_\lambda(x,x)}$ is the Laplace transform of $u \mapsto \nu^x[u,+\infty[$.

(Hint: Compute $\int_a^\infty (1 - e^{-\lambda u})n^x(u)du$ with an integration by parts, and then let $a \to 0$.)

**4)** Deduce from **2)** and **3)** that, for $u \leq t$:

$$\mathbb{P}_0\left(\mathscr{G}_x(t) \in du\right) = \mathbb{P}_0(T_x > t)\delta_0(du) + q(u,0,x)\nu^x[t-u,+\infty[du.$$

**5)** In the case of Brownian motion, recover formula (5.28).

**6)** We now assume that $(X_t := B_t^{(\nu)}, t \geq 0)$ is a Brownian motion with drift $\nu$. Its resolvent kernel is given by:

$$u_\lambda^{(\nu)}(x,x) = \frac{1}{2\sqrt{2\lambda + \nu^2}}e^{-2\nu x}.$$

Prove then that:

$$\frac{\sqrt{\lambda + a}}{\lambda} = \int_0^\infty e^{-\lambda t}\left(\frac{e^{-at}}{\sqrt{\pi t}} + \sqrt{a}\,\mathrm{Erf}(\sqrt{at})\right)dt$$

where $\mathrm{Erf}(u) = \dfrac{2}{\pi}\int_0^u e^{-x^2}dx$, and recover formula (5.35).

## 5.2 Past-Future (Sub)-Martingales

### 5.2.1 Definitions

Let $(B_t, t \geq 0)$ be a Brownian motion started from 0. We have seen in the previous section the importance of the "past-future filtration" $(\mathscr{F}_{s,t}, 0 \leq s \leq t < \infty)$ defined by:

$$\mathscr{F}_{s,t} = \sigma(B_u, u \leq s; B_h, h \geq t), \qquad 0 \leq s \leq t < \infty. \tag{5.36}$$

Note that, for $s = t$, $\mathscr{F}_{s,s} = \mathscr{F}_\infty = \bigvee_{t \geq 0} \mathscr{F}_t$, with $\mathscr{F}_t := \sigma(B_u, u \leq t)$, and that, if $[s,t] \subset [s',t']$, $\mathscr{F}_{s',t'} \subset \mathscr{F}_{s,t}$. It is therefore quite natural to introduce the following notions:

**Definition 5.1.**

*i)* A $\mathbb{R}^+$-valued $\mathscr{F}_{s,t}$-adapted process $(\Delta_{s,t}, s \leq t)$ is called a $(\mathscr{F}_{s,t}, s \leq t)$ positive submartingale (or a past-future submartingale) if, for every $s, s', t, t'$ such that $[s,t] \subset [s',t']$, we have:

$$\mathbb{E}\left[\Delta_{s,t} | \mathscr{F}_{s',t'}\right] \geq \Delta_{s',t'}. \tag{5.37}$$

*ii)* A $\mathbb{R}^+$-valued $\mathscr{F}_{s,t}$-adapted process $(\Delta_{s,t}, s \leq t)$ is called a $(\mathscr{F}_{s,t}, s \leq t)$ positive martingale (or a past-future martingale) if, for every $s, s', t, t'$ such that $[s,t] \subset [s',t']$, we have:

$$\mathbb{E}\left[\Delta_{s,t} | \mathscr{F}_{s',t'}\right] = \Delta_{s',t'}. \tag{5.38}$$

Here are some particular examples of such submartingales: let $\Gamma$ be a Borel set in $\mathbb{R}$ and define

$$\Delta_{s,t}^\Gamma := \mathbb{P}\left(\forall u \in [s,t], B_u \in \Gamma | \mathscr{F}_{s,t}\right) \tag{5.39}$$

and the sets:

$$A_{s,t}^\Gamma := \{\forall u \in [s,t], B_u \in \Gamma\}. \tag{5.40}$$

They satisfy: $A_{s',t'}^\Gamma \subset A_{s,t}^\Gamma$ if $[s,t] \subset [s',t']$ which implies that $(\Delta_{s,t}^\Gamma, s \leq t)$ is a past-future submartingale. Moreover, from the Markov property of Brownian motion, there exists a function $f_\Gamma(s,t;x,y)$ (with $s \leq t, x, y \in \mathbb{R}$) such that:

$$\Delta_{s,t}^\Gamma = f_\Gamma(s,t;B_s,B_t). \tag{5.41}$$

This leads us to present the following definition:

**Definition 5.2.** A function $f : \mathbb{R}^+ \times \mathbb{R}^+ \times \mathbb{R} \times \mathbb{R} \to \mathbb{R}^+$ is a past-future subharmonic function (PFS-function) if $(f(s,t;B_s,B_t), s \leq t)$ is a past-future submartingale. A function $f : \mathbb{R}^+ \times \mathbb{R}^+ \times \mathbb{R} \times \mathbb{R} \to \mathbb{R}^+$ is a past-future harmonic function (PFH-function) if $(f(s,t;B_s,B_t), s \leq t)$ is a past-future martingale.

Let us go back, for a moment, to Theorem 5.1 and its proof. We have:

$$\{G_l^{(\nu)}(t) < s\} = A_{s,t}^{(\nu),\Gamma_l} = \{\forall u \in ]s,t[; B_u^{(\nu)} \in \Gamma_l\}$$

with $\Gamma_l = ]-\infty, l[\cup]l, +\infty[$. Hence, we deduce from Theorem 5.1 that:

$$\mathbb{P}\left(G_l^{(\nu)}(t) < s \,|\, \mathscr{F}_{s,t}\right) = \left(1 - \exp\left\{-\frac{2}{t-s}\left(B_s^{(\nu)} - \log(l)\right)\left(B_t^{(\nu)} - \log(l)\right)\right\}\right)^+$$

and, from (5.41), that:

$$h^{(l,\nu)}(s,t;x,y) = \left(1 - \exp\left\{-\frac{2}{t-s}\left(x+\nu s - \log(l)\right)\left(y+\nu t - \log(l)\right)\right\}\right)^+ \quad (5.42)$$

is a PFS-function. Besides, (5.42) leads us to believe that, for $l, \nu \in \mathbb{R}$, the function $f^{(l,\nu)}$ defined by:

$$f^{(l,\nu)}(s,t;x,y) = \exp\left\{-\frac{2}{t-s}\left(x+\nu s - \log(l)\right)\left(y+\nu t - \log(l)\right)\right\} \quad (5.43)$$

is a PFH-function. We shall prove this result in Subsection 5.2.3 below.

## 5.2.2 Properties and Characterization of PFH-Functions

Thanks to basic properties of Brownian motion, i.e.:

-   scaling: $\left(aB_{t/a^2}, t \geq 0\right) \overset{(law)}{=} (B_t, t \geq 0)$,
-   time inversion: $\left(\widehat{B}_t = tB_{1/t}, t \geq 0\right) \overset{(law)}{=} (B_t, t \geq 0)$ (where $\widehat{B}_0 = 0$ by continuity),
-   bridge property: conditionally on $B_t + \nu t = a$, the law of $(B_s + \nu s, s \leq t)$ does not depend on $\nu$,

the following Proposition is easily obtained:

**Proposition 5.1.** Let $f \, (= f(s,t;x,y))$ be a PFH-function. Then:

i) For any $a > 0$:

$$f^{(a)}(s,t;x,y) := f(a^2 s, a^2 t; ax, ay) \text{ is also a PFH-function,}$$

ii) $\widehat{f}$ defined by:

$$\widehat{f}(s,t;x,y) := f\left(\frac{1}{t}, \frac{1}{s}; \frac{y}{t}, \frac{x}{s}\right) \quad (5.44)$$

is also a PFH-function.

iii) For any $\nu, l \in \mathbb{R}$:

$$f^{(l,\nu)}(s,t;x,y) = f(s,t;x+\nu s+l, y+\nu t+l) \quad (5.45)$$

is also a PFH-function.

The proof of this Proposition is left to the reader.

Here is now a characterization of PFH-functions:

**Theorem 5.2.** *A regular function $h(s,t;x,y)$ $(s \leq t; x,y \in \mathbb{R})$ is a PFH-function if and only if:*

$$\frac{\partial h}{\partial s}(s,t;x,y) + \frac{y-x}{t-s}\frac{\partial h}{\partial x}(s,t;x,y) + \frac{1}{2}\frac{\partial^2 h}{\partial x^2}(s,t;x,y) = 0, \qquad (5.46)$$

*and*

$$-\frac{\partial h}{\partial t}(s,t;x,y) - \frac{y-x}{t-s}\frac{\partial h}{\partial y}(s,t;x,y) + \frac{1}{2}\frac{\partial^2 h}{\partial y^2}(s,t;x,y) = 0. \qquad (5.47)$$

*Proof.* This proof hinges upon the following lemma:

**Lemma 5.1.** *Let $M_{s,t}^f := f(s,t;B_s,B_t)$ $(s \leq t)$ for $f$ a regular function. Then, $(M_{s,t}^f, s \leq t)$ is a past-future martingale if and only if:*

*i) for fixed $t$,*

$$(M_{s,t}^f, s < t) \text{ is a } (\mathscr{F}_s^{(t)}, s < t) \text{ martingale, where: } \mathscr{F}_s^{(t)} = \mathscr{F}_s \vee \sigma(B_t) \quad (5.48)$$

*ii) for fixed $s$,*

$$(M_{s,t}^f, t > s) \text{ is a } \left({}^{(s)}\mathscr{F}_t^+, t > s\right) \text{ martingale, where:} \qquad (5.49)$$

$${}^{(s)}\mathscr{F}_t^+ = \sigma(B_s) \vee \mathscr{F}_t^+, \quad \text{and } \mathscr{F}_t^+ = \sigma(B_u, u \geq t)$$

*Proof.* Note that $(M_{s,t}^f, s \leq t)$ is a $(\mathscr{F}_{s,t}, s \leq t)$ martingale if and only if for every $\Phi_{s'} \in b(\mathscr{F}_{s'})$ (the space of bounded $\mathscr{F}_{s'}$-measurable r.v.'s), $\Psi_{t'} \in b(\mathscr{F}_{t'}^+)$, with $s' \leq s \leq t \leq t'$ one has:

$$\mathbb{E}\left[\Phi_{s'}M_{s,t}^f\Psi_{t'}\right] = \mathbb{E}\left[\Phi_{s'}M_{s',t'}^f\Psi_{t'}\right] \qquad (5.50)$$

We first prove that (5.50) implies conditions (*i*) and (*ii*) of Lemma 5.1

Indeed, taking $s' < s < t = t'$ and $\Psi_t = g(B_t)$ for a generic bounded function $g$, we have, from (5.50):

$$\mathbb{E}\left[\Phi_{s'}M_{s,t}^f g(B_t)\right] = \mathbb{E}\left[\Phi_{s'}M_{s',t}^f g(B_t)\right]$$

i.e. (5.48) and, by past-future symmetry, (5.50) also implies (5.49).

We now prove that conditions (*i*) and (*ii*) of Lemma 5.1 imply (5.50)

From the Markov property, the LHS of (5.50), say $L$, is equal to:

$$L = \mathbb{E}\left[\Phi_{s'}M_{s,t}^f \gamma(t,B_t)\right]$$

where

$$\gamma(t, B_t) = \mathbb{E}\left[\Psi_{t'} | \mathscr{F}_t\right] = \mathbb{E}\left[\Psi_{t'} | B_t\right].$$

From (5.48), we then get:

$$L = \mathbb{E}\left[\Phi_{s'} M_{s,t}^f \gamma(t, B_t)\right] = \mathbb{E}\left[\Phi_{s'} M_{s',t}^f \gamma(t, B_t)\right] = \mathbb{E}\left[\Phi_{s'} M_{s',t}^f \Psi_{t'}\right].$$

Again, from the Markov property, we get:

$$L = \mathbb{E}\left[\beta(s', B_{s'}) M_{s',t}^f \Psi_{t'}\right]$$

where $\beta(s', B_{s'}) = \mathbb{E}\left[\Phi_{s'} | B_{s'}\right]$. Now, using (5.49), we obtain:

$$L = \mathbb{E}\left[\beta(s', B_{s'}) M_{s',t}^f \Psi_{t'}\right] = \mathbb{E}\left[\beta(s', B_{s'}) M_{s',t'}^f \Psi_{t'}\right]$$

and Lemma 5.1 is proven.

$\square$

Let us remark that these arguments only make use of the Markov property, and not of the Brownian framework.

We now complete the proof of Theorem 5.2

For a fixed $t > 0$, $(B_s, s \leq t)$ is a $(\mathscr{F}_s^{(t)}, s \leq t)$ semi-martingale with:

$$B_s = \beta_s^{(t)} + \int_0^s \frac{B_t - B_u}{t - u} du \qquad (5.51)$$

where $(\beta_s^{(t)}, 0 \leq s \leq t)$ is a $(\mathscr{F}_s^{(t)}, 0 \leq s \leq t)$ Brownian motion (see [51]). We apply Itô's formula to $(M_{s,t}^f, s \leq t)$ in the filtration $(\mathscr{F}_s^{(t)}, s \leq t)$. We obtain, for $u < s < t$:

$$M_{s,t} = M_{u,t} + \int_u^s \frac{\partial f}{\partial s}(r, t; B_r, B_t) dr + \int_u^s \frac{\partial f}{\partial x}(r, t; B_r, B_t) \left(d\beta_r^{(t)} + \frac{B_t - B_r}{t - r} dr\right)$$
$$+ \frac{1}{2} \int_u^s \frac{\partial^2 f}{\partial x^2}(r, t; B_r, B_t) dr. \qquad (5.52)$$

Hence, the martingale property, from Point $(i)$ of Lemma 5.1, holds if and only if:

$$\frac{\partial f}{\partial s}(s, t; x, y) + \frac{y - x}{t - s} \frac{\partial f}{\partial x}(s, t; x, y) + \frac{1}{2} \frac{\partial^2 f}{\partial x^2}(s, t; x, y) = 0.$$

This is (5.46). We might obtain (5.47) using some similar arguments; however, a time inversion argument (Point $(ii)$ of Proposition 5.1) is much quicker and reduces the obtention of (5.47) to (5.46).

$\square$

### 5.2.3 Two Classes of PFH-Functions

We now introduce two classes of PFH-functions. Actually, we shall show that these two classes make it possible to describe all the PFH-functions.

**Theorem 5.3.** *Let $\nu$, $l$, $e$ and $d$ four reals. Then, the functions $f^{(l,\nu)}$ and $h^{(e,d)}$ defined by:*

$$f^{(l,\nu)}(s,t;x,y) = \exp\left\{-\frac{2}{t-s}(x+\nu s - l)(y+\nu t - l)\right\} \qquad (5.53)$$

*and*

$$h^{(e,d)}(s,t;x,y) = \exp\left\{\frac{x(e+dt)-y(e+ds)}{t-s} - \frac{e^2 + 2eds + d^2 st}{2(t-s)}\right\} \qquad (5.54)$$

$$= c(e,d)\exp\left\{\frac{2}{t-s}\left(\left(x-\frac{e+ds}{2}\right)\left(y+\frac{e+dt}{2}\right)-xy\right)\right\} \qquad (5.55)$$

*are PFH-functions.*

Note that, for $d = 0$:

$$h^{(e,0)}(s,t;x,y) = \exp\left(e\frac{x-y}{t-s} - \frac{e^2}{2(t-s)}\right), \qquad (5.56)$$

and for $e = 0$:

$$h^{(0,d)}(s,t;x,y) = \exp\left(d\frac{xt-ys}{t-s} - \frac{d^2 st}{2(t-s)}\right). \qquad (5.57)$$

We shall see in Subsection 5.2.5 how we came to think about these functions $f^{(l,\nu)}$ and $h^{(e,d)}$.

*Proof.* Elementary (but fastidious) computations show that $f^{(l,\nu)}$ and $h^{(e,d)}$ satisfy conditions (5.46) and (5.47), thus are PFH-functions.

□

### 5.2.4 Another Characterization of PFH-Functions

We shall say that a function $K : \mathbb{R}^+ \times \mathbb{R} \to \mathbb{R}^+$ is space-time harmonic for Brownian motion if $(K(t,B_t), t \geq 0)$ is a martingale (for the usual filtration $(\mathscr{F}_t, t \geq 0)$.) If $K$ is smooth, from Itô's formula, this condition is of course equivalent to:

$$\forall t \geq 0, \quad \mathbb{E}[|K(t,B_t)|] < +\infty \quad and \quad \frac{\partial K}{\partial t} + \frac{1}{2}\frac{\partial^2 K}{\partial x^2} = 0. \qquad (5.58)$$

It is known, from Widder's representation Theorem of positive space-time harmonic functions, that the condition (5.58) is also equivalent to the existence of a positive finite measure $\gamma$, carried by $\mathbb{R}$, such that:

$$K(t,x) = \int_{\mathbb{R}} \exp\left(\lambda x - \frac{\lambda^2 t}{2}\right) d\gamma(\lambda). \tag{5.59}$$

Our aim now is to obtain a representation formula like (5.59), but this time for PFH-functions. Of course, the functions that will play the role of $e_\lambda$ defined by $e_\lambda(t,x) := \exp\left(\lambda x - \frac{\lambda^2}{2} t\right)$ will be the functions $f^{(l,\nu)}$ and $h^{(e,d)}$ defined by (5.53) and (5.54).

### Theorem 5.4 (Another characterization of PFH-functions).
*A function h is a PFH-function if and only if:*

i) *For every $t > 0$ and $y \in \mathbb{R}$, there exists a space-time harmonic function $K^+_{(t,y)}$ for Brownian motion such that:*

$$h(s,t;x,y) = K^+_{(t,y)}\left(\frac{s}{t-s}, \frac{xt - ys}{(t-s)\sqrt{t}}\right), \tag{5.60}$$

*or equivalently, from (5.59):*
*For every $t > 0$ and $y \in \mathbb{R}$, there exists a positive finite measure $\gamma^+(t,y,d\eta)$ such that:*

$$h(s,t;x,y) = \int_{\mathbb{R}} \gamma^+(t,y,d\eta) \exp\left(\eta \frac{xt - ys}{(t-s)\sqrt{t}} - \frac{\eta^2}{2}\frac{s}{t-s}\right). \tag{5.61}$$

ii) *For every $s > 0$ and $x \in \mathbb{R}$, there exists a space-time harmonic function $K^-_{(\frac{1}{s},\frac{x}{s})}$ for Brownian motion such that:*

$$h(s,t;x,y) = K^-_{(\frac{1}{s},\frac{x}{s})}\left(\frac{s}{t-s}, \frac{(y-x)\sqrt{s}}{t-s}\right), \tag{5.62}$$

*or equivalently, from (5.59):*
*For every $s > 0$ and $x \in \mathbb{R}$, there exists a positive finite measure $\gamma^-(t,y,d\eta)$ such that:*

$$h(s,t;x,y) = \int_{\mathbb{R}} \gamma^-\left(\frac{1}{s}, \frac{x}{s}, d\eta\right) \exp\left(\eta \frac{(x-y)\sqrt{s}}{(t-s)} - \frac{\eta^2}{2}\frac{s}{t-s}\right). \tag{5.63}$$

*Proof.* We shall use Theorem 5.2, but, first of all, let us give another formulation of (5.46):

• For any given $t$ and $y$, the process $(h(s,t;b_s^{(y,t)},y), 0 \leq s < t)$ is a martingale, where $(b_s^{(y,t)}, 0 \leq s \leq t)$ is a Brownian bridge of length $t$ such that $b_t^{(y,t)} = y$ a.s. This results at once from Itô's formula, noting that $(b_s^{(y,t)}, 0 \leq s \leq t)$ is the solution of the SDE:

$$d\beta_s = \frac{y - \beta_s}{t - s} ds + dB_s. \tag{5.64}$$

Let us also remark also that, thanks to Proposition 5.1, an equivalent formulation of (5.47) is given by:

- Let $k$ defined by: $k(s,t;x,y) := h\left(\dfrac{1}{t},\dfrac{1}{s};\dfrac{y}{t},\dfrac{x}{s}\right)$. Then, $k$ satisfies (5.46).

Hence, we will have proven Theorem 5.4 if we can show the existence of a bijective correspondence $\Theta$ between:
- the set $\mathscr{H}_B$ of space-time harmonic functions for Brownian motion,
- the set $\mathscr{H}_{br,t}^{0\to y}$ of space-time harmonic functions for the Brownian bridge of length $t$ ending at $y$,
this correspondence being given by:

$$\Theta(K)(s,x) := h^{(t,y)}(s,x) := K\left(\frac{s}{t-s},\frac{xt-ys}{(t-s)\sqrt{t}}\right). \tag{5.65}$$

Let us show (5.65)
Let $K(u,z) : \mathbb{R}^+ \times \mathbb{R} \to \mathbb{R}^+$ and $h^{(t,y)}$ defined by:

$$h^{(t,y)}(s,x) := K\left(\frac{s}{t-s},\frac{xt-ys}{(t-s)\sqrt{t}}\right). \tag{5.66}$$

We also define: $L^{(t,y)}(\varphi) = \dfrac{\partial\varphi}{\partial s} + \dfrac{y-x}{t-s}\dfrac{\partial\varphi}{\partial x} + \dfrac{1}{2}\dfrac{\partial^2\varphi}{\partial x^2}$. Then, from (5.66), we deduce:

$$\frac{\partial h}{\partial s} = \frac{t}{(t-s)^2}\frac{\partial K}{\partial u} + \frac{(x-y)\sqrt{t}}{(t-s)^2}\frac{\partial K}{\partial z}$$

$$\frac{\partial h}{\partial x} = \frac{\sqrt{t}}{t-s}\frac{\partial K}{\partial z}, \qquad \frac{\partial^2 h}{\partial x^2} = \frac{t}{(t-s)^2}\frac{\partial^2 K}{\partial z^2}.$$

Hence:

$$\begin{aligned} L^{(t,y)}(h) &= \frac{t}{(t-s)^2}\frac{\partial K}{\partial u} + \frac{(x-y)\sqrt{t}}{(t-s)^2}\frac{\partial K}{\partial z} + \frac{y-x}{t-s}\frac{\sqrt{t}}{t-s}\frac{\partial K}{\partial z} + \frac{1}{2}\frac{t}{(t-s)^2}\frac{\partial^2 K}{\partial z^2} \\ &= \frac{t}{(t-s)^2}\left(\frac{\partial K}{\partial u} + \frac{1}{2}\frac{\partial^2 K}{\partial z^2}\right). \end{aligned} \tag{5.67}$$

It is therefore plain from (5.67) that $h$ is a space-time harmonic function for the Brownian bridge $b^{(y,t)}$ (i.e. $L^{(t,y)}(h) = 0$) if and only if $\dfrac{\partial K}{\partial u} + \dfrac{1}{2}\dfrac{\partial^2 K}{\partial z^2} = 0$, i.e. $K$ is space-time harmonic for Brownian motion.

$\square$

## 5.2.5 Description of Extremal PFH-Functions

From Theorem 5.4, to every PFH-function $h$, we can associate 2 families of positive finite measures $\gamma^+(t,y,d\eta)$ and $\gamma^-\left(\frac{1}{s},\frac{x}{s},d\eta\right)$. We now study the PFH-harmonic functions for which the measures $\gamma^+$ and $\gamma^-$ cannot be decomposed, i.e. such that the supports of $\gamma^+$ and $\gamma^-$ are each reduced to a point. More precisely:

**Definition 5.3.** A PFH-function $h$ is said to be extremal if the measures $\gamma^+(t,y,d\eta)$ and $\gamma^-\left(\frac{1}{s},\frac{x}{s},d\eta\right)$ have their supports each reduced to a point. In other terms, there exist two functions $\Psi, \Phi : \mathbb{R}^+ \times \mathbb{R} \to \mathbb{R}^+$ and two functions $\alpha, \beta : \mathbb{R}^+ \times \mathbb{R} \to \mathbb{R}$ such that:

$$\begin{cases} \gamma^+(t,y,d\eta) & = \Psi(t,y)\delta_{\alpha(t,y)}(d\eta), \\ \gamma^-\left(\frac{1}{s},\frac{x}{s},d\eta\right) & = \Phi(s,x)\delta_{\beta(s,x)}(d\eta). \end{cases} \tag{5.68}$$

**Theorem 5.5.** *Let $g$ an extremal PFH-function. Then, $g$ is one of the two functions $f^{(l,\nu)}$ or $h^{(e,d)}$ described by Theorem 5.3.*

*Proof.* i) Let $g$ an extremal PFH-function. First of all, with the same notations as in Theorem 5.4, there exist two families of measures $\gamma^+(t,y,d\eta)$ and $\gamma^-\left(\frac{1}{s},\frac{x}{s},d\eta\right)$ such that, for all $s \leq t$:

$$g(s,t;x,y) = \int_{\mathbb{R}} \gamma^+(t,y,d\eta)\exp\left(\eta\frac{xt-ys}{(t-s)\sqrt{t}} - \frac{\eta^2}{2}\frac{s}{t-s}\right) \tag{5.69}$$

$$= \int_{\mathbb{R}} \gamma^-\left(\frac{1}{s},\frac{x}{s},d\eta\right)\exp\left(\eta\frac{(x-y)\sqrt{s}}{(t-s)} - \frac{\eta^2}{2}\frac{s}{t-s}\right) \tag{5.70}$$

which leads, taking (5.68) into account, to the identity:

$$\Psi(t,y)\exp\left(\alpha\frac{xt-ys}{(t-s)\sqrt{t}} - \frac{\alpha^2}{2}\frac{s}{t-s}\right) = \Phi(s,x)\exp\left(\beta\frac{(x-y)\sqrt{s}}{(t-s)} - \frac{\beta^2}{2}\frac{s}{t-s}\right) \tag{5.71}$$

where, to simplify, we have written $\alpha$ for $\alpha(t,y)$ and $\beta$ for $\beta(s,x)$. Relation (5.71) being true for all $t \geq s$, we replace $t$ by $as$, with $a \geq 1$. After a few algebraic computations, (5.71) is seen to be equivalent to:

$$\Psi(as,y)\exp\left(-\frac{1}{2(a-1)}\left(\alpha^2 - 2\alpha x\sqrt{\frac{a}{s}} + \frac{2\alpha y}{\sqrt{as}}\right)\right)$$

$$= \Phi(s,x)\exp\left(-\frac{1}{2(a-1)}\left(\beta^2 - \frac{2\beta y}{\sqrt{s}} + \frac{2\beta x}{\sqrt{s}}\right)\right). \tag{5.72}$$

ii) We now study more carefully relation (5.72)
We take $x = 0$, and then $y = 0$ in (5.72). We obtain:

$$\Psi(as,y)\exp\left(-\frac{1}{2(a-1)}\left(\alpha^2 + \frac{2\alpha y}{\sqrt{as}}\right)\right)$$

$$= \Phi(s,0)\exp\left(-\frac{1}{2(a-1)}\left(\beta^2(s,0) - \frac{2\beta(s,0)y}{\sqrt{s}}\right)\right)$$

and

$$\Psi(as,0)\exp\left(-\frac{1}{2(a-1)}\left(\alpha^2(as,0)-2\alpha(as,0)x\sqrt{\frac{a}{s}}\right)\right)$$

$$= \Phi(s,x)\exp\left(-\frac{1}{2(a-1)}\left(\beta^2+\frac{2\beta x}{\sqrt{s}}\right)\right).$$

We plug back these two relations in (5.72):

$$\Phi(s,0)\exp\left(-\frac{1}{2(a-1)}\left(\beta^2(s,0)-\frac{2\beta(s,0)y}{\sqrt{s}}-2\alpha x\sqrt{\frac{a}{s}}\right)\right)$$

$$= \Psi(as,0)\exp\left(-\frac{1}{2(a-1)}\left(\alpha^2(as,0)-2\alpha(as,0)x\sqrt{\frac{a}{s}}-\frac{2\beta y}{\sqrt{s}}\right)\right). \quad (5.73)$$

Identifying the terms which depend on $x$ and $y$ in (5.73), we obtain:

$$\frac{\beta(s,0)y}{\sqrt{s}}+\alpha(as,y)x\sqrt{\frac{a}{s}}=\alpha(as,0)x\sqrt{\frac{a}{s}}+\frac{\beta(s,x)y}{\sqrt{s}} \quad (5.74)$$

and

$$\Phi(s,0)\exp\left(-\frac{1}{2(a-1)}\beta^2(s,0)\right)=\Psi(as,0)\exp\left(-\frac{1}{2(a-1)}\alpha^2(as,0)\right). \quad (5.75)$$

We then take the logarithm of (5.75), multiply by $a-1$ and make $a=1$. This yields:

$$\beta^2(s,0)=\alpha^2(s,0) \qquad \text{i.e.} \quad \beta(s,0)=\pm\alpha(s,0).$$

Furthermore, from (5.74), we deduce:

$$\beta(s,x)=\frac{\sqrt{s}}{y}\left(\beta(s,0)\frac{y}{\sqrt{s}}+\alpha(as,y)x\sqrt{\frac{a}{s}}-\alpha(as,0)x\sqrt{\frac{a}{s}}\right) \quad (5.76)$$

$$= \beta(s,0)+\frac{x}{y}\left(\alpha(as,y)-\alpha(as,0)\right)\sqrt{a}.$$

The LHS of this last relation does not depend on $y$, neither on $a$, so we get by making $y=a=1$:

$$\beta(s,x)=\beta(s,0)+x\left(\alpha(s,1)-\alpha(s,0)\right). \quad (5.77)$$

The same method yields, still from (5.74) but expressing this time $\alpha(as,y)$:

$$\alpha(as,y)=\frac{1}{x}\sqrt{\frac{s}{a}}\left(\alpha(as,0)x\sqrt{\frac{a}{s}}+\beta(s,y)\frac{y}{\sqrt{s}}-\beta(s,0)\frac{y}{\sqrt{s}}\right)$$

which reduces, after taking $x=1$ and $a=1$, to:

$$\alpha(s,y) = \alpha(s,0) + y(\beta(s,1) - \beta(s,0)). \tag{5.78}$$

Hence, from (5.77) and (5.78), with $x = y = 1$:

$$\alpha(s,1) - \alpha(s,0) = \beta(s,1) - \beta(s,0). \tag{5.79}$$

Plugging (5.77) and (5.78) back into (5.74), we obtain:

$$\frac{\beta(s,0)y}{\sqrt{s}} - \{\alpha(as,0) + y(\beta(as,1) - \beta(as,0))\} x\sqrt{\frac{a}{s}}$$

$$= \alpha(as,0)x\sqrt{\frac{a}{s}} + \frac{y}{\sqrt{s}}\{\beta(s,0) + x(\alpha(s,1) - \alpha(s,0))\}. \tag{5.80}$$

We identify in (5.80) the terms in $xy$, and then make $s = 1$:

$$\beta(a,1) - \beta(a,0) = \frac{\alpha(1,1) - \alpha(1,0)}{\sqrt{a}} =: \frac{c}{\sqrt{a}}. \tag{5.81}$$

Finally, gathering (5.76)–(5.81) yields:

$$\beta(s,x) = \beta(s,0) + \frac{cx}{\sqrt{s}}$$

$$\alpha(s,y) = \pm\beta(s,0) + \frac{cy}{\sqrt{s}},$$

or, denoting $l(s) = \beta(s,0)$ to simplify:

$$\beta(s,x) = l(s) + \frac{cx}{\sqrt{s}} \tag{5.82}$$

$$\alpha(s,y) = \pm l(s) + \frac{cy}{\sqrt{s}}. \tag{5.83}$$

*iii*) It remains to identify the possible values of $c$ and of the function $l$
To this end, we plug (5.82) and (5.83) back into (5.72):

$$\Psi(as,y)\exp\left\{-\frac{1}{2(a-1)}\left(\left(\pm l(as) + \frac{cy}{\sqrt{as}}\right)^2\right.\right.$$

$$\left.\left. -2x\sqrt{\frac{a}{s}}\left(\pm l(as) + \frac{cy}{\sqrt{as}}\right) + \frac{2y}{\sqrt{as}}\left(\pm l(as) + \frac{cy}{\sqrt{as}}\right)\right)\right\} \tag{5.84}$$

$$= \Phi(s,x)\exp\left\{-\frac{1}{2(a-1)}\left(\left(l(s) + \frac{cx}{\sqrt{s}}\right)^2\right.\right.$$

$$\left.\left. -\frac{2y}{\sqrt{s}}\left(l(s) + \frac{cx}{\sqrt{s}}\right) + \frac{2x}{\sqrt{s}}\left(l(s) + \frac{cx}{\sqrt{s}}\right)\right)\right\}.$$

We then take the logarithm of this last expression, multiply it by $a - 1$, and make $a = 1$:

$$\left(\pm l(s) + \frac{cy}{\sqrt{s}}\right)^2 - \frac{2x}{\sqrt{s}}\left(\pm l(s) + \frac{cy}{\sqrt{s}}\right) + \frac{2y}{\sqrt{s}}\left(\pm l(s) + \frac{cy}{\sqrt{s}}\right)$$

$$= \left(l(s) + \frac{cx}{\sqrt{s}}\right)^2 - \frac{2y}{\sqrt{s}}\left(l(s) + \frac{cx}{\sqrt{s}}\right) + \frac{2x}{\sqrt{s}}\left(l(s) + \frac{cx}{\sqrt{s}}\right). \quad (5.85)$$

But, in (5.85), the terms in $y^2$ and $x^2$ must be null, which leads to the equations:

$$\frac{c^2y^2}{s} + \frac{2cy^2}{s} = 0 \quad \text{i.e. } c = 0 \text{ or } c = -2,$$

$$\frac{c^2x^2}{s} + \frac{2cx^2}{s} = 0 \quad \text{i.e. } c = 0 \text{ or } c = -2. \quad (5.86)$$

The identification of the terms in $x$ and $y$ in (5.85) yields then:
- If $c = -2$,

$$\alpha(s,x) = l(s) - \frac{2x}{\sqrt{s}},$$

$$\beta(s,x) = l(s) - \frac{2x}{\sqrt{s}}.$$

- If $c = 0$,

$$\alpha(s,x) = -l(s),$$
$$\beta(s,x) = l(s).$$

iv) It remains to find $l$.
From (5.84), it is clear that the functions $\Psi$ and $\Phi$ are necessarily of the form:

$$\Psi(as,y) = \exp(i(as)y + j(as))$$
$$\Phi(s,x) = \exp(e(s)x + f(s)) \quad (5.87)$$

for four functions $i, j, e, f$. Then, according to the value of $c$ (and therefore to the sign of $\pm l(s)$), we can identify the coefficient in $x$ in (5.84) to get, for example when $c = -2$:

$$\frac{l(as)}{a-1}\sqrt{\frac{a}{s}} = -e(s) + \frac{l(s)}{a-1}\frac{1}{\sqrt{s}}.$$

Hence, for $s = 1$:

$$l(a) = -e(1)\frac{a-1}{\sqrt{a}} + \frac{l(1)}{\sqrt{a}} =: \frac{e}{\sqrt{a}} + d\sqrt{a} \quad (5.88)$$

with $d$ and $e$ two real constants.

*v*) To sum up, if *g* is an extremal PFH-function, we have necessarily:

- either $c = -2$ and:

$$\alpha(s,x) = \frac{e}{\sqrt{s}} + d\sqrt{s} - \frac{2x}{\sqrt{s}}$$

$$\beta(s,x) = \frac{e}{\sqrt{s}} + d\sqrt{s} - \frac{2x}{\sqrt{s}}$$

$$\Psi(s,y) = \exp\left(\frac{ey}{s} - \frac{e^2}{2s}\right) \tag{5.89}$$

$$\Phi(s,y) = \exp\left(-dx - \frac{d^2}{2}s\right),$$

- or $c = 0$ and:

$$\alpha(s,x) = \frac{e}{\sqrt{s}} + d\sqrt{s}$$

$$\beta(s,x) = -\frac{e}{\sqrt{s}} - d\sqrt{s}$$

$$\Psi(s,y) = \exp\left(-\frac{ey}{s} - \frac{e^2}{2s}\right) \tag{5.90}$$

$$\Phi(s,y) = \exp\left(dx - \frac{d^2}{2}s\right).$$

*vi*) Reverse study

So far, we have taken an extremal PFH-function *g*, and we have described the form it must necessarily take (these are relations (5.68), (5.69), (5.89) and (5.90)). But, do these relations actually give a PFH-function ? To check this, we shall adopt the reverse method. Starting from (5.89) and (5.90), we shall compute the corresponding functions *g* and verify that they are PFH-functions.

• The case $c = -2$

We have, computing *g* with the help of $\gamma^+$ as given by (5.68):

$$g(s,t;x,y) = \int_{\mathbb{R}} \exp\left(\eta \frac{xt - ys}{(t-s)\sqrt{t}} - \frac{\eta^2}{2}\frac{s}{t-s}\right) \Psi(t,y)\delta_{\alpha(t,y)}(d\eta)$$

$$= \exp\left(\frac{ey}{t} - \frac{e^2}{2t}\right) \exp\left(\left(\frac{e}{\sqrt{t}} + d\sqrt{t} - \frac{2y}{\sqrt{t}}\right) \frac{xt - ys}{(t-s)\sqrt{t}}\right.$$

$$\left. - \frac{s}{2(t-s)}\left(\frac{e}{\sqrt{t}} + d\sqrt{t} - \frac{2y}{\sqrt{t}}\right)^2\right)$$

$$= \exp\left(-\frac{de}{4}\right) \exp\left(-\frac{2}{t-s}\left(x - \frac{e+ds}{2}\right)\left(y - \frac{e+dt}{2}\right)\right).$$

The computation of $g$ performed this time with the help of $\gamma^-$ (as a verification) gives the same result. Hence, in the case $c = -2$, the function $g(s,t;x,y)$ we obtained is indeed a PFH-function: it is the function $f^{(l,\nu)}$ from Theorem 5.3 with $\nu = -\frac{d}{2}$ and $l = \frac{e}{2}$.

• The case $c = 0$

We have, still with $\gamma^+$:

$$g(s,t;x,y) = \exp\left(-\frac{ey}{t} - \frac{e^2}{2t}\right)$$

$$\cdot \exp\left(\left(\frac{e}{\sqrt{t}} + d\sqrt{t}\right)\frac{xt - ys}{(t-s)\sqrt{t}} - \frac{s}{2(t-s)}\left(\frac{e}{\sqrt{t}} + d\sqrt{t}\right)^2\right)$$

$$= \exp\left(\frac{x(e+dt) - y(e+ds)}{t-s} - \frac{e^2 + 2eds + d^2st}{2(t-s)}\right).$$

The computation of $g$ with the help of $\gamma^-$ gives also the same result. Hence, in the case $c = 0$, the function $g(s,t;x,y)$ we obtained is the function $h^{(e,d)}$ of Theorem 5.3. This ends the proof of Theorem 5.5.

□

**Corollary 5.1.** *A function $g$ is a positive PFH-function if and only if there exist two positive measures $\theta_1$ and $\theta_2$ carried by $\mathbb{R}^2$ such that:*

$$g(s,t;x,y) = \int_{\mathbb{R}^2} f^{(l,\nu)}(s,t;x,y)\theta_1(dl,d\nu) + \int_{\mathbb{R}^2} h^{(e,\delta)}(s,t;x,y)\theta_2(de,d\delta). \quad (5.91)$$

*Proof.* Let $\mathscr{C}_H$ denote the convex cone of positive PFH-functions, and denote by $X$ a base of $\mathscr{C}_H$. ($X$ is the intersection of $\mathscr{C}_H$ with a closed affine hyperplane, which does not contain 0, and intersects all the generators of the cone). An element $g$ of $X$ is said to be $X$-extremal if the relation $g = \frac{1}{2}(g_1 + g_2)$ with $g_1, g_2 \in X$ implies that $g_1 = g_2 = g$. It is clear that $X$-extremal elements are precisely the extremal PFH-functions in the sense of Definition 5.3. Indeed, if the measure $\gamma^+(t,y,d\eta)$ (resp. $\gamma^-(\frac{1}{s}, \frac{x}{s}, d\eta)$) is supported by more than one point (i.e. it is not a Dirac measure), we can always decompose it into the sum of two positive measures with disjoint supports. Then, relation (5.91) relies on an application of Choquet's representation Theorem, which expresses every element of $X$ as a barycenter of its extremal elements. We refer the interested reader to [16] for the details.

□

*Remark 5.4.* Looking carefully at Theorems 5.3 and 5.5, it seems natural to try to find all the PFH-functions having the form:

$$h(s,t;x,y) = \exp\left(a(s,t)xy + b(s,t)x + c(s,t)y + d(s,t)\right). \quad (5.92)$$

From Theorem 5.2, we must have:

$$\begin{cases} \dfrac{\partial a}{\partial s} = \dfrac{1}{t-s}a \\[2mm] a\left(\dfrac{1}{t-s} + \dfrac{1}{2}a\right) = 0 \\[2mm] \dfrac{\partial b}{\partial s} = \dfrac{1}{t-s}b \\[2mm] \dfrac{\partial c}{\partial s} + b\left(\dfrac{1}{t-s} + a\right) = 0 \\[2mm] \dfrac{\partial d}{\partial s} + \dfrac{1}{2}b^2 = 0. \end{cases} \qquad (5.93)$$

The second relation of (5.93) implies either $a = 0$ or $a = -\dfrac{2}{t-s}$. When $a = 0$, the solution of (5.93) is:

$$b(s,t) = \frac{e+dt}{t-s}, \quad c(s,t) = \frac{e+ds}{t-s}, \quad \text{and } d(s,t) = \frac{e^2 + 2eds + d^2st}{2(t-s)}.$$

The corresponding function $h$ is the function $h^{(e,d)}$ of Theorem 5.3. (This is the case $c = 0$, relation (5.90)). When $a = -\dfrac{2}{t-s}$, we find:

$$b(s,t) = \frac{l-\nu t}{t-s}, \quad c(s,t) = \frac{l-\nu s}{t-s}, \quad \text{and } d(s,t) = \frac{1}{2}\frac{(l-\nu s)(l-\nu t)}{t-s}.$$

The corresponding function $h$ is the function $f^{(l,\nu)}$ of Theorem 5.3. (This is the case $c = -2$, relation (5.89)).

**Exercise 5.2 (A few examples of past-future martingales).**
Let $(B_t, t \geq 0)$ be a standard Brownian motion.
1) Prove that $\left(\dfrac{B_t - B_s}{t-s}, s < t\right)$ is a past-future martingale.
2) Recall the initial enlargement formula (5.51):

$$B_s = \beta_s^{(t)} + \int_0^s \frac{B_t - B_u}{t-u}\,du \qquad (5.94)$$

where $(\beta_s^{(t)}, 0 \leq s \leq t)$ is a $(\mathscr{F}_s^{(t)} := \mathscr{F}_s \vee \sigma(B_t), 0 \leq s \leq t)$ Brownian motion (see [51]). Using (5.94), prove that, for $f \in L^2(\mathbb{R}^+)$:

$$\mathbb{E}\left[\int_0^\infty f(u)dB_u \middle| \mathscr{F}_{s,t}\right] = \int_0^s f(u)dB_u + \int_t^\infty f(u)dB_u + \frac{B_t - B_s}{t-s}\int_s^t f(u)du \qquad (5.95)$$

3) Deduce that:

$$\mathbb{E}\left[\exp\left(\int_0^\infty f(u)dB_u\right)|\mathscr{F}_{s,t}\right] = \exp\left(\int_0^s f(u)dB_u + \int_t^\infty f(u)dB_u\right.$$

$$\left. +\frac{B_t - B_s}{t-s}\int_s^t f(u)du + \frac{1}{2}\int_s^t f^2(u)du - \frac{1}{2(t-s)}\left(\int_s^t f(u)du\right)^2\right).$$

4) Let $a > 1/2$. Prove similarly that:

$$\mathbb{E}\left[\int_0^\infty e^{B_u - au}du|\mathscr{F}_{s,t}\right] = \int_0^s e^{B_u - au}du + \int_t^\infty e^{B_u - au}du$$

$$+ e^{B_s - as}\int_0^{t-s}\exp\left(\left(\frac{1}{2} + \frac{B_t - B_s}{t-s} - a\right)v - \frac{v^2}{2(t-s)}\right)dv.$$

<u>Note</u>: An integrable process $H$ such that $\left(\dfrac{H_t - H_s}{t-s}, s < t\right)$ is a past-future martingale is called a harness. It can be proven that a process $H$ is a harness if and only if for every $t > 0$, there exists a $(\mathscr{F}_s^{(t)})$-martingale $(M_s^{(t)}, s \geq 0)$ such that:

$$\forall s \leq t, \qquad H_s = M_s^{(t)} + \int_0^s \frac{H_t - H_u}{t-u}du,$$

see:

R. Mansuy and M. Yor [40, F]. Harnesses, Lévy bridges and Monsieur Jourdain. Stochastic Process. Appl., 115(2):329–338, 2005.

## 5.3 Notes and Comments

The assertion $(i)$ of Theorem 5.1 is due to A. Bentata and M. Yor ([6]). It has been developed in a course given by M. Yor in the Bachelier Séminaire in February 2008 at Institut H. Poincaré. "Yuri's formula" (Point $(iii)$ of Theorem 5.1) is due, in the particular case $\nu = -1/2$, to J. Akahori, Y. Imamura and Y. Yano ([1]). Section 5.1 is taken from D. Madan, B. Roynette and M. Yor ([48]). The notion of "past-future" martingales – Theorems 5.2 and 5.3 – is taken from the Bachelier Séminaire course at Institut H. Poincaré, already mentioned ([6]). The contents of Subsection 5.2.5 – which describes the space-time harmonic functions (see, e.g. Corollary 5.1) – are new.

# Chapter 6
# Put Option as Joint Distribution Function in Strike and Maturity

**Abstract** For a large class of $\mathbb{R}^+$-valued, continuous local martingales $(M_t, t \geq 0)$, with $M_0 = 1$ and $M_\infty = 0$, the put quantity: $\Pi_M(K,t) = \mathbb{E}\left[(K - M_t)^+\right]$ turns out to be the distribution function in both variables $K$ and $t$, for $K \leq 1$ and $t \geq 0$, of a probability $\gamma_M$ on $[0,1] \times [0,+\infty[$. We discuss in detail, in this Chapter, the case where $(M_t = \mathscr{E}_t := \exp(B_t - \frac{t}{2}), t \geq 0)$, for $(B_t, t \geq 0)$ a standard Brownian motion, and give an extension to the more general case of the semimartingale $\mathscr{E}_t^{\sigma,-\nu} := \exp(\sigma B_t - \nu t), (\sigma \neq 0, \nu > 0)$.

## 6.1 Put Option as a Joint Distribution Function and Existence of Pseudo-Inverses

### 6.1.1 Introduction

Throughout this Section 6.1, we consider a generic continuous local martingale $(M_t, t \geq 0)$ taking values in $\mathbb{R}^+$, and such that:

$$M_0 = 1, \quad \lim_{t \to \infty} M_t = 0. \tag{6.1}$$

To such a $(M_t, t \geq 0)$, we associate the function $\Pi_M : [0,1] \times \mathbb{R}^+ \to \mathbb{R}^+$ defined by:

$$\Pi_M(K,t) := \mathbb{E}\left[(K - M_t)^+\right], \quad (0 \leq K \leq 1, t \geq 0). \tag{6.2}$$

Note that this function is separately increasing in $K$ and $t$ (concerning the latter, since $(K-x)^+$ is convex, $((K - M_t)^+, t \geq 0)$ is a submartingale). Furthermore, we have:

i) $\Pi_M(K,0) = 0$ $(0 \leq K \leq 1)$ and, from the dominated convergence Theorem,

$$\Pi_M(K,+\infty) := \lim_{t \to \infty} \Pi_M(K,t) = K. \tag{6.3}$$

C. Profeta et al., *Option Prices as Probabilities*, Springer Finance,
DOI 10.1007/978-3-642-10395-7_6, © Springer-Verlag Berlin Heidelberg 2010

ii) From Theorem 2.1:

$$\Pi_M(K,t) = K\mathbb{P}(\mathscr{G}_K \le t) \tag{6.4}$$

with $\mathscr{G}_K := \sup\{t \ge 0; M_t = K\}$. In particular:

$$\Pi_M(1,t) = \mathbb{P}(\mathscr{G}_1 \le t). \tag{6.5}$$

Thus, $(\Pi_M(K,+\infty),\ K \le 1)$, (resp. $(\Pi_M(1,t),\ t \ge 0)$) is a distribution function on $[0,1]$, (resp. on $[0,+\infty[$) ; more precisely, these functions are, respectively, the distribution function of $U$, a standard uniform variable on $[0,1]$, and of $\mathscr{G}_1$.

To illustrate (6.5), let us recall that in the case $\left(M_t = \mathscr{E}_t := \exp\left(B_t - \frac{t}{2}\right),\ t \ge 0\right)$, with $(B_t, t \ge 0)$ a standard Brownian motion, it was shown in Chapter 1 that: $\mathscr{G}_1 \overset{(\text{law})}{=} 4B_1^2$; hence, (6.5) reads, in this case:

$$\Pi_{\mathscr{E}}(1,t) = \mathbb{P}(4B_1^2 \le t). \tag{6.6}$$

This formula may also be checked directly from the classical Black-Scholes formula.

## 6.1.2 Seeing $\Pi_M(K,t)$ as a Function of 2 Variables

It is thus a natural question to ask whether the function of $K$ and $t$: $(\Pi_M(K,t);\ K \le 1, t \ge 0)$ is the distribution function of a probability on $[0,1] \times [0,+\infty[$ which, assuming it exists, we shall denote by $\gamma\ (= \gamma_M)$. If so, we have:

$$\mathbb{E}\left[(K - M_t)^+\right] = \gamma\big([0,K] \times [0,t]\big), \quad (K \le 1,\ t \ge 0). \tag{6.7}$$

## 6.1.3 General Pattern of the Proof

Here is our strategy to attack this question. Note that by Fubini:

$$\mathbb{E}\left[(K - M_t)^+\right] = \int_0^K \mathbb{P}(M_t \le x)dx. \tag{6.8}$$

Assume that there exists, for every $x < 1$, a r.v. $Y_x \ge 0$ such that:

$$\mathbb{P}(M_t \le x) = \mathbb{P}(Y_x \le t) \quad (x < 1,\ t \ge 0). \tag{6.9}$$

**Definition 6.1.** We shall call this collection $(Y_x, x < 1)$ of r.v.'s (provided it exists) a decreasing pseudo-inverse of $(M_t,\ t \ge 0)$. (See Chapter 7, Definition 7.1 and 7.2 for more general definitions and some justifications of this terminology).

Let us go back to (6.8), and assume that $(M_t, t \geq 0)$ admits a decreasing pseudo-inverse $(Y_x, x < 1)$. Then, plugging (6.9) in (6.8), we find that $\gamma$ exists, and it is the probability:

$$\gamma(dx, dt) = dx \, \mathbb{P}(Y_x \in dt) \qquad \text{on } [0,1] \times [0, +\infty[.$$

Note that, a priori, we do not know the existence of $(Y_x, x < 1)$ as a process, that is a measurable function: $(x, \omega) \mapsto Y_x(\omega)$ ; if such a process exists, then $\gamma$ is the law of:

$$(U, Y_U) \tag{6.10}$$

where $U$ is uniform on $[0,1]$ and independent of $(Y_x, x < 1)$.

### 6.1.4 A Useful Criterion

In practice, most of the time, the function:

$$(K,t) \longmapsto \Pi_M(K,t)$$

is regular; if so, we find that $(M_t, t \geq 0)$ admits a decreasing pseudo-inverse if and only if:

$$\frac{\partial^2}{\partial K \partial t}\left(\Pi_M(K,t)\right) \geq 0 \tag{6.11}$$

and then:

$$\gamma(dK, dt) = dK \, \mathbb{P}(Y_K \in dt) = \left(\frac{\partial^2}{\partial K \partial t}\left(\Pi_M(K,t)\right)\right) dK \, dt. \tag{6.12}$$

### 6.1.5 Outline of the Following Sections

In Sections 6.2 and 6.3, we shall develop this program for

$$\left(M_t = \mathcal{E}_t := \exp\left(B_t - \frac{t}{2}\right), t \geq 0\right),$$

where $(B_t, t \geq 0)$ is a standard Brownian motion started at 0. In particular, we prove the existence of a decreasing pseudo-inverse for $(\mathcal{E}_t, t \geq 0)$. In Chapter 7, we shall study the existence of pseudo-inverses for Bessel and related processes, and more generally for linear diffusions in Chapter 8.

## 6.2 The Black-Scholes Paradigm

### 6.2.1 Statement of the Main Result

In this section, $(B_t,\ t \geq 0)$ denotes a standard Brownian motion started at 0 and $(\mathscr{E}_t,\ t \geq 0)$ is the exponential martingale defined by:

$$\mathscr{E}_t := \exp\left(B_t - \frac{t}{2}\right), \qquad (t \geq 0). \tag{6.13}$$

Note that $\mathscr{E}_0 = 1$ and $\mathscr{E}_t \underset{t \to \infty}{\longrightarrow} 0$ a.s. We define, for $0 \leq K \leq 1$ and $t \geq 0$:

$$\Pi_{\mathscr{E}}(K,t) := \mathbb{E}\left[(K - \mathscr{E}_t)^+\right]. \tag{6.14}$$

**Theorem 6.1.** *There exists a probability, which we shall denote by $\gamma$, on $[0,1] \times [0,+\infty[$ such that:*

$$\Pi_{\mathscr{E}}(K,t) = \gamma\big([0,K] \times [0,t]\big) \tag{6.15}$$

$\big($*see Point $(ii)$ of Proposition 6.3 for a description of the density of $\gamma$*$\big)$.

In order to prove Theorem 6.1 and to describe $\gamma$, we start with the following:

**Lemma 6.1.** *Denote by $\mathscr{N}$ the distribution function of the standard Gaussian variable:*

$$\mathscr{N}(x) := \frac{1}{\sqrt{2\pi}} \int_{-\infty}^{x} e^{-\frac{y^2}{2}} dy \qquad (x \in \mathbb{R}). \tag{6.16}$$

*Then:*

*i)  To any $a,b > 0$, one can associate a r.v. $Y_{a,b}$, taking values in $[0,+\infty[$, such that:*

$$\mathbb{P}(Y_{a,b} \leq t) = \mathscr{N}\left(a\sqrt{t} - \frac{b}{\sqrt{t}}\right), \qquad (t \geq 0). \tag{6.17}$$

*ii) The density $f_{Y_{a,b}}$ of $Y_{a,b}$ is given by:*

$$f_{Y_{a,b}}(t) = \frac{1}{\sqrt{2\pi}}\, e^{ab} \cdot \left(\frac{a}{2\sqrt{t}} + \frac{b}{2\sqrt{t^3}}\right) \exp\left(-\frac{1}{2}\left(a^2 t + \frac{b^2}{t}\right)\right). \tag{6.18}$$

*iii) Let us define:*

$$T_b^{(a)} := \inf\{t \geq 0\,;\, B_t + at = b\}, \tag{6.19}$$

$$G_b^{(a)} := \sup\{t \geq 0\,;\, B_t + at = b\}. \tag{6.20}$$

*Then:*

$$\mathbb{P}(Y_{a,b} \in dt) = \frac{1}{2}\left(\mathbb{P}(T_b^{(a)} \in dt) + \mathbb{P}(G_b^{(a)} \in dt)\right). \tag{6.21}$$

We note that this formula (6.21) allows to define the law of a process $(Y_{a,b}, b \geq 0)$ obtained as a fair coin toss of $(T_b^{(a)}, b \geq 0)$ on one hand and $(G_b^{(a)}, b \geq 0)$ on the other hand.

**Remark 6.1.**
**a)** It may be worth mentioning that Point $(i)$ of Lemma 6.1 admits a wide extension, since to any distribution function $F$ on $\mathbb{R}$, and any $a, b > 0$, we can associate a new distribution function $F_{a,b}$ on $\mathbb{R}^+$ via:

$$F_{a,b}(t) = F\left(a\sqrt{t} - \frac{b}{\sqrt{t}}\right) \quad (t \geq 0).$$

However, the particular case $F = \mathcal{N}$ fits extremely well with our discussion.

**b)** We note that $\dfrac{1}{Y_{a,b}} \overset{(law)}{=} Y_{b,a}$, which may be deduced from either (6.17), (6.18) or (6.21).

**Proof.**
**1)** Points $(i)$ and $(ii)$ of Lemma 6.1 are immediate since:

- $\displaystyle \lim_{t \downarrow 0} \mathcal{N}\left(a\sqrt{t} - \frac{b}{\sqrt{t}}\right) = 0$

- $\displaystyle \lim_{t \uparrow \infty} \mathcal{N}\left(a\sqrt{t} - \frac{b}{\sqrt{t}}\right) = 1$

- $\displaystyle \frac{\partial}{\partial t} \mathcal{N}\left(a\sqrt{t} - \frac{b}{\sqrt{t}}\right) = \frac{1}{\sqrt{2\pi}}\left(\frac{a}{2\sqrt{t}} + \frac{b}{2\sqrt{t^3}}\right) \exp\left(-\frac{1}{2}\left(a\sqrt{t} - \frac{b}{\sqrt{t}}\right)^2\right) \geq 0$

$$= f_{Y_{a,b}}(t).$$

Point $(iii)$ is then a direct consequence of (6.18), (1.11) and (1.12).

**2)** Another proof of Point $(iii)$ of Lemma 6.1:
Let us denote, for $\nu > 0$, $\left(\mathcal{E}_t^{(2\nu)} := \exp(2\nu B_t - 2\nu^2 t), t \geq 0\right)$. It is proven in Chapter 1, Theorem 1.3, that, for $A \geq 1$:

$$\mathbb{P}(T_{\log A}^{(\nu)} \leq t) = \mathbb{E}\left[\mathcal{E}_t^{(2\nu)} 1_{\{\mathcal{E}_t^{(2\nu)} > A^{2\nu}\}}\right] + A^{2\nu}\mathbb{P}(\mathcal{E}_t^{(2\nu)} > A^{2\nu})$$

and for $A \geq 0$ (Theorem 1.3, formula (1.54)):

$$\mathbb{P}(G_{\log A}^{(\nu)} \leq t) = \mathbb{E}\left[\mathcal{E}_t^{(2\nu)} 1_{\{\mathcal{E}_t^{(2\nu)} > A^{2\nu}\}}\right] - A^{2\nu}\mathbb{P}\left(\mathcal{E}_t^{(2\nu)} > A^{2\nu}\right).$$

Thus, by addition:

$$\frac{1}{2}\left(\mathbb{P}(T_{\log A}^{(\nu)} \leq t) + \mathbb{P}(G_{\log A}^{(\nu)} \leq t)\right)$$

$$= \mathbb{E}\left[\mathscr{E}_t^{(2\nu)} 1_{\{\mathscr{E}_t^{(2\nu)} > A^{2\nu}\}}\right]$$

$$= \mathbb{P}\left(e^{2\nu B_t + 2\nu^2 t} > A^{2\nu}\right) \qquad \text{(from Cameron-Martin formula (1.5))}$$

$$= \mathbb{P}(B_t > \log A - \nu t) = \mathbb{P}\left(B_1 > \frac{\log A - \nu t}{\sqrt{t}}\right)$$

$$= 1 - \mathscr{N}\left(\frac{\log A}{\sqrt{t}} - \nu\sqrt{t}\right) = \mathscr{N}\left(\nu\sqrt{t} - \frac{\log A}{\sqrt{t}}\right). \tag{6.22}$$

(6.18) is now an immediate consequence of (6.22), with $b = \log A$ and $\nu = a$, by derivation with respect to $t$.

$\square$

*Proof.* We have, for $K \leq 1$ and $t \geq 0$:

$$\Pi_{\mathscr{E}}(K,t) := \mathbb{E}\left[(K - \mathscr{E}_t)^+\right] = \int_0^K \mathbb{P}(\mathscr{E}_t \leq x)dx$$

$$= \int_0^K \mathscr{N}\left(\frac{\log x}{\sqrt{t}} + \frac{\sqrt{t}}{2}\right)dx \tag{6.23}$$

$$\left(\text{since } \mathbb{P}(\mathscr{E}_t \leq x) = \mathbb{P}(e^{B_t - \frac{t}{2}} \leq x) = \mathbb{P}\left(B_1 < \frac{\log x}{\sqrt{t}} + \frac{\sqrt{t}}{2}\right)\right)$$

$$= \int_0^K \mathbb{P}\left(Y_{\frac{1}{2}, \log \frac{1}{x}} \leq t\right)dx \qquad \text{(from Lemma 6.1).}$$

Hence:

$$\frac{\partial^2}{\partial K \partial t}\Pi_{\mathscr{E}}(K,t) = \frac{\partial}{\partial t}\mathbb{P}(Y_{\frac{1}{2}, \log \frac{1}{K}} \leq t) = f_{Y_{\frac{1}{2}, \log \frac{1}{K}}}(t) \geq 0.$$

This ends the proof of Theorem 6.1.

$\square$

*Remark 6.2.* More generally, if, for $x > 0$, $(M_t^{(x)}, t \geq 0)$ is a positive continuous local martingale such that $M_0^{(x)} = x$ a.s. and $\lim_{t \to \infty} M_t^{(x)} = 0$, then there exists a probability $\gamma_{M^{(x)}}$ on $[0,x] \times [0, +\infty[$ such that:

$$\frac{1}{x}\Pi_{M^{(x)}}(K,t) = \gamma_{M^{(x)}}\left([0,K] \times [0,t]\right), \qquad (K \leq x, t \geq 0)$$

if and only if $(M_t^{(x)}, t \geq 0)$ admits a decreasing pseudo-inverse $(Y_{x,y}, y < x)$.

## 6.2.2 Descriptions of the Probability γ

### 6.2.2.1 First Description of γ: Conditioning with Respect to $U$

**Proposition 6.1.** *The probability $\gamma$ on $[0, 1] \times [0, +\infty[$ is the law of the pair:*

$$(U, Y_{\frac{1}{2}, \log \frac{1}{U}}) \tag{6.24}$$

*where $U$ is uniform on $[0, 1]$ and independent of the process $(Y_{\frac{1}{2}, b}, b > 0)$.*

We now describe in words this probability viewed from (6.24): it is the law of a two components r.v.; the first component is the choice of a level out of the moneyness for a put, or the choice of a strike $K < 1$, uniformly on $[0, 1]$. Given this level, we construct the second variable on the outcome of a fair coin toss as either the first passage time of the stock price under the share measure to the level $\left(\frac{1}{K}\right)$, or the last passage time of the stock price under the share measure to level $\left(\frac{1}{K}\right)$.

*Proof.* We have, for $K \leq 1$ and $t \geq 0$:

$$\mathbb{P}(U \leq K, Y_{\frac{1}{2}, \log \frac{1}{U}} \leq t) = \int_0^K du\, \mathbb{P}(Y_{\frac{1}{2}, \log \frac{1}{u}} \leq t) \quad \text{(as explained just above)}$$

$$= \int_0^K \mathcal{N}\left(\frac{\sqrt{t}}{2} + \frac{\log u}{\sqrt{t}}\right) du$$

$$= \int_0^K \mathbb{P}(\mathcal{E}_t \leq u) du$$

$$\left(\text{since } \mathcal{E}_t \overset{\text{(law)}}{=} \exp\left(\sqrt{t}B_1 - \frac{t}{2}\right), \text{ for fixed } t\right)$$

$$= \mathbb{E}\left[(K - \mathcal{E}_t)^+\right] \quad \text{(from (6.8))}$$

$$= \gamma([0, K] \times [0, t]) \quad \text{(from (6.15))}.$$

The density of $\gamma$ with respect to the Lebesgue measure on $[0, 1] \times \mathbb{R}^+$ given by (6.33) (see Proposition 6.3 below) may also be obtained from the preceding relations. □

### 6.2.2.2 Second Description of γ: Conditioning with Respect to $\mathscr{G}_1$

**Proposition 6.2.** *The probability $\gamma$ on $[0, 1] \times [0, +\infty[$ is the law of the pair:*

$$\left(\exp(-2\mathfrak{e}) \vee \exp(-\sqrt{8\mathfrak{e}'B_1^2}), 4B_1^2\right) \tag{6.25}$$

*where $B_1, \mathfrak{e}, \mathfrak{e}'$ are independent, with $\mathfrak{e}$ and $\mathfrak{e}'$ two standard exponential variables.*

*Remark 6.3.* Upon comparing Propositions 6.1 and 6.2, it is quite natural to look for some understanding of the implied identities in laws between the first, resp. second, components of (6.25) and (6.24); precisely, we wish to check directly that:

$$\exp(-2\epsilon) \vee \exp(-\sqrt{8\epsilon'B_1^2}) \overset{(\text{law})}{=} U \tag{6.26}$$

and

$$Y_{\frac{1}{2},\log\frac{1}{U}} \overset{(\text{law})}{=} 4B_1^2 \quad \left( \overset{(\text{law})}{=} G_0^{\left(-\frac{1}{2}\right)} = \sup\left\{t \geq 0; B_t - \frac{t}{2} = 0\right\} \right). \tag{6.27}$$

**a)** We now prove (6.26)
• First, we have:

$$\sqrt{2\epsilon'B_1^2} \overset{(\text{law})}{=} \epsilon. \tag{6.28}$$

Indeed:

$$\mathbb{P}\left(\sqrt{2\epsilon'B_1^2} > x\right) = \mathbb{P}\left(\epsilon' > \frac{x^2}{2B_1^2}\right) = \mathbb{E}\left[\exp\left(-\frac{x^2}{2B_1^2}\right)\right]$$

$$\overset{(a)}{=} \mathbb{E}\left[\exp\left(-\frac{x^2}{2}T_1\right)\right] \overset{(b)}{=} \exp(-x)$$

since $T_1$, the first hitting time of 1 by $(B_t, t \geq 0)$ is distributed as $\dfrac{1}{B_1^2}$, hence $(a)$, and the Laplace transform of $T_1$ is well known to be given by $(b)$.
• Since $\exp(-2\epsilon) \overset{(\text{law})}{=} U^2$, we have, from (6.28):

$$\exp(-2\epsilon) \vee \exp(-\sqrt{8\epsilon'B_1^2}) \overset{(\text{law})}{=} U^2 \vee (U')^2$$

with $U$ and $U'$ uniform on $[0,1]$ and independent. But, for $y \in [0,1]$:

$$\mathbb{P}(U^2 \vee (U')^2 \leq y) = \left(\mathbb{P}(U^2 \leq y)\right)^2 = (\sqrt{y})^2 = y.$$

We have proven (6.26).
**b)** We now prove (6.27)
We have for every $t \geq 0$:

$$\mathbb{P}(Y_{\frac{1}{2},\log\frac{1}{U}} \leq t) = \int_0^1 \mathbb{P}(Y_{\frac{1}{2},\log\frac{1}{u}} \leq t)du \quad \text{(after conditioning by } U = u\text{)}$$

$$= \int_0^1 \mathcal{N}\left(\frac{\sqrt{t}}{2} + \frac{\log u}{\sqrt{t}}\right)du \quad \text{(from Lemma 6.1)}$$

$$= \mathbb{E}\left[(1 - \mathscr{E}_t)^+\right] \quad \text{(from (6.23))}$$

$$= \mathbb{P}(4B_1^2 \leq t) \quad \text{(from the Black-Scholes formula, see (1.19))}.$$

**We now prove Proposition 6.2**

Conditioning on $B_1^2$ and using the explicit formula for the density of $B_1^2$:

$$f_{B_1^2}(z) = \frac{1}{\sqrt{2\pi z}} e^{-\frac{z}{2}} 1_{\{z>0\}},$$

we have, for $K \leq 1$ and $t \geq 0$:

$$\mathbb{P}\left(\exp(-2\epsilon) \vee \left(\exp - \sqrt{8\epsilon' B_1^2}\right) \leq K, 4B_1^2 \leq t\right)$$

$$= \int_0^{\frac{t}{4}} \frac{dz}{\sqrt{2\pi z}} e^{-\frac{z}{2}} \mathbb{P}\left(\exp(-2\epsilon) \vee \left(\exp - \sqrt{8\epsilon' z}\right) \leq K\right)$$

$$= \int_0^{\frac{t}{4}} \frac{dz}{\sqrt{2\pi z}} e^{-\frac{z}{2}} \mathbb{P}\left(\exp(-2\epsilon) \leq K\right) \mathbb{P}\left(\epsilon' > \frac{(\log K)^2}{8z}\right)$$

$$= \sqrt{K} \int_0^{\frac{t}{4}} \frac{dz}{\sqrt{2\pi z}} e^{-\frac{z}{2}} \exp\left(-\frac{(\log K)^2}{8z}\right) \quad \left(\text{since } \exp(-\epsilon) \overset{\text{(law)}}{=} U\right)$$

$$= \sqrt{K} \mathbb{E}\left[1_{\{B_1^2 \leq \frac{t}{4}\}} \cdot \exp\left(-\frac{(\log K)^2}{8B_1^2}\right)\right]$$

$$= \mathbb{E}\left[(K - \mathcal{E}_t)^+\right] \tag{6.29}$$

where the last equality follows from Chapter 1, Theorem 1.4, which asserts that for $K \leq 1$:

$$\mathbb{E}\left[(K - \mathcal{E}_t)^+\right] = \sqrt{K} \mathbb{E}\left[1_{\{4B_1^2 \leq t\}} \exp\left(-\frac{(\log K)^2}{8B_1^2}\right)\right]. \tag{6.30}$$

$\square$

**Another proof of Proposition 6.2**

We have, from (1.14) with $\nu = -1/2$ and $K < 1$, since $\mathscr{G}_K = \sup\{t \geq 0; \mathscr{E}_t = K\} = G_{\log(K)}^{(-1/2)}$:

$$\mathbb{P}(\mathscr{G}_K \in ds) = \frac{1/2}{\sqrt{2\pi s}} \exp\left(-\frac{1}{2s}\left(\log K + \frac{s}{2}\right)^2\right) ds.$$

Hence:

$$K\mathbb{P}(\mathscr{G}_K \in ds) = \sqrt{K}\left(\exp\left(-\frac{(\log K)^2}{2s}\right)\right) \mathbb{P}(\mathscr{G}_1 \in ds)$$

$$= \exp\left(\frac{1}{2}\log K\right) \exp\left(-\frac{(\log K)^2}{2s}\right) \mathbb{P}(\mathscr{G}_1 \in ds)$$

$$= \mathbb{P}\left(\epsilon > -\frac{1}{2}\log K\right) \mathbb{P}\left(\epsilon' > \frac{(\log K)^2}{2s}\right) \mathbb{P}(\mathscr{G}_1 \in ds)$$

$$= \mathbb{P}\left(\exp(-2\epsilon) \vee \exp(-\sqrt{2s}\,\epsilon') < K\right) \mathbb{P}(\mathscr{G}_1 \in ds)$$

$$= \mathbb{P}\left(\exp(-2\epsilon) \vee \exp(-\sqrt{8B_1^2\epsilon'}) < K\right) \mathbb{P}(4B_1^2 \in ds)$$

since $\mathcal{G}_1 \overset{(\text{law})}{=} 4B_1^2$ (see Chapter 1, Theorem 1.2 and relation (1.19)). Hence, since from Theorem 2.1, $\mathbb{E}\left[(K - \mathcal{E}_s)^+\right] = K\mathbb{P}(\mathcal{G}_K \leq s)$, one has:

$$
\begin{aligned}
\gamma(dK, ds) &= \frac{\partial^2}{\partial K \partial s} \left(K\mathbb{P}(\mathcal{G}_K \leq s)\right) dK \, ds \\
&= \mathbb{P}\left(\exp(-2\mathfrak{e}) \vee \exp(-\sqrt{8B_1^2 \mathfrak{e}'}) \in dK\right) \mathbb{P}(4B_1^2 \in ds)
\end{aligned}
$$

which is Proposition 6.2.

$\square$

### 6.2.2.3 Third Description of $\gamma$: its Relation with Local Time-Space Calculus

Let us define $(L_s^K; K \geq 0, s \geq 0)$ the jointly continuous family of local times of the martingale $(\mathcal{E}_s, s \geq 0)$. This family is characterized by the occupation density formula:

$$
\int_0^t f(\mathcal{E}_s) d\langle \mathcal{E} \rangle_s = \int_0^\infty f(K) L_t^K \, dK
$$

for every Borel and positive function $f$. Here $(\langle \mathcal{E} \rangle_s, s \geq 0)$ denotes the bracket, i.e. the quadratic variation process of $(\mathcal{E}_s, s \geq 0)$ and we have:

$$
d\langle \mathcal{E} \rangle_s = \mathcal{E}_s^2 \, ds.
$$

The Itô-Tanaka formula yields, for $K \leq 1$:

$$
\mathbb{E}\left[(K - \mathcal{E}_t)^+\right] = \frac{1}{2}\mathbb{E}\left[L_t^K\right]. \tag{6.31}
$$

As a consequence of (6.31), we obtain Point ($i$) of the following:

**Proposition 6.3.**

i) *The probability $\gamma$ on $[0, 1] \times [0, +\infty[$ admits a density $f_\gamma$ and satisfies:*

$$
\gamma(dK, dt) = \frac{1}{2}\left(\frac{\partial^2}{\partial K \partial t}\mathbb{E}\left[L_t^K\right]\right) dK \, dt = f_\gamma(K, t) dK dt \qquad (0 \leq K \leq 1, \, t \geq 0). \tag{6.32}
$$

ii) *A closed form of $f_\gamma$ is:*

$$
f_\gamma(K, t) = \frac{1}{2\sqrt{2\pi Kt}}\left(\frac{1}{2} - \frac{\log K}{t}\right) \exp\left(-\frac{(\log K)^2}{2t} - \frac{t}{8}\right) 1_{[0,1]}(K) 1_{[0,+\infty[}(t). \tag{6.33}
$$

*Proof.* In Theorem 1.4, formula (1.64), we obtained the following explicit formula for $\mathbb{E}[L_t^K]$:

$$
\mathbb{E}\left[L_t^K\right] = \frac{\sqrt{K}}{\sqrt{2\pi}} \int_0^t \frac{ds}{\sqrt{s}} \exp\left(-\frac{(\log(K))^2}{2s} - \frac{s}{8}\right). \tag{6.34}
$$

Hence:

$$\frac{\partial^2}{\partial K \partial t} \mathbb{E}\left[L_t^K\right] = \frac{1}{\sqrt{2\pi t}} \frac{\partial}{\partial K} \left(\sqrt{K} \exp\left(-\frac{(\log K)^2}{2t} - \frac{t}{8}\right)\right) \tag{6.35}$$

and (6.33) is an easy consequence of (6.35) and (6.32).

□

### 6.2.2.4 Relation with a Result by N. Eisenbaum (see [22] and [23])

We now relate the above description of $\gamma$ as in Proposition 6.3 with the definition-formula established in [23]:

$$\int_{-\infty}^{\infty} \int_0^t f(K,s) d_{K,s}(L_s^K) = \int_0^t f(\mathscr{E}_s, s) d\mathscr{E}_s + \int_{1-t}^1 f(\mathscr{E}_{1-s}, 1-s) d_s \mathscr{E}_{1-s} \quad (t \leq 1). \tag{6.36}$$

This formula is the particular instance for $X_s = \mathscr{E}_s$ of the formula found in Theorem 2.2 of [23] for a general reversible semimartingale. Here, on the RHS of (6.36), the second stochastic integral is taken with respect to the natural filtration of $\widehat{\mathscr{E}}_s = \mathscr{E}_{1-s}$, which is, of course, that of $\widehat{B}_s = B_{1-s}$. We take $f$ bounded, Borel, with support in $[0,1]_K \times [0,1]_s$. In order to relate formula (6.36) with Proposition 6.3, we note that:

**a)** $\int_0^1 \int_0^1 f(K,s)\gamma(dK,ds) = \frac{1}{2} \mathbb{E}\left[\int_0^1 \int_0^1 f(K,s) d_{K,s}(L_s^K)\right]$ which follows from (6.31), and the monotone class Theorem.

**b)** From formula (6.36) and the fact that $(\mathscr{E}_t, t \geq 0)$ is a martingale, we deduce:

$$\mathbb{E}\left[\int_0^1 \int_0^1 f(K,s) d_{K,s}(L_s^K)\right] = \mathbb{E}\left[\int_0^1 f(\mathscr{E}_s, 1-s) d\mathscr{E}_s\right] \tag{6.37}$$

which we shall compute explicitly thanks to the semimartingale decomposition of $(\widehat{\mathscr{E}}_s, s \leq 1)$ in its own filtration. This is presented in the next:

**Proposition 6.4.**

i) *The canonical decomposition of $(B_t, t \leq 1)$ in the filtration* $\mathscr{B}_t^{(1)} := \sigma(B_s, s \leq t) \vee \sigma(B_1)$ *is:*

$$B_t = B_t^* + \int_0^t \frac{B_1 - B_s}{1-s} ds \tag{6.38}$$

*where $(B_t^*, t \leq 1)$ is a $(\mathscr{B}_t^{(1)}, t \leq 1)$ Brownian motion.*

ii) *The canonical decomposition of $\widehat{B}_t = B_{1-t}$ in its own filtration is:*

$$\widehat{B}_t = B_{1-t} = B_1 + \beta_t^* - \int_0^t \frac{B_{1-s}}{1-s} ds \tag{6.39}$$

*where $(\beta_t^*, t \leq 1)$ is a Brownian motion in $\{\widehat{\mathscr{B}}_t := \sigma(\widehat{B}_u, u \leq t), t \leq 1\}$.*

*iii) The canonical decomposition of $\widehat{\mathscr{E}}_t$ in $\widehat{\mathscr{B}}_t$ is:*

$$d\widehat{\mathscr{E}}_t = \widehat{\mathscr{E}}_t\left(d\beta_t^* + dt\left(1 - \frac{B_{1-t}}{1-t}\right)\right). \tag{6.40}$$

*Proof.* (*i*) is well-known, see, e.g., Jeulin-Yor [38] or Itô [34].
(*ii*) may be deduced from (*i*), when (*i*) is applied to $\beta_t = B_{1-t} - B_1$. Actually, formula (6.39) appears in [22], at the bottom of p. 308.
(*iii*) follows from (*ii*), thanks to Itô's formula.

$\qquad\qquad\qquad\qquad\qquad\qquad\qquad\qquad\qquad\qquad\qquad\qquad\qquad\qquad\qquad\qquad$ □

**Comments**
**a)** We are grateful to N. Eisenbaum (personal communication) for pointing out formula (6.39), which allowed to correct our original wrong derivation of the canonical decomposition of $\widehat{B}_t = B_{1-t}$:

$$\widehat{B}_t = B_{1-t} = B_1 + \widetilde{\beta}_t + \int_0^t \frac{du}{u}(\widehat{B}_u - B_1). \tag{6.41}$$

Indeed, by time-reversal from (6.38) in Proposition 6.4, there is the identity (6.41) where $\widetilde{\beta}_t = B_{1-t}^* - B_1^*$ is a Brownian motion, but (6.41) is <u>not</u> the canonical decomposition of $\widehat{B}_t$ in $\widehat{\mathscr{B}}_t$ (for a discussion of non-canonical decompositions of Brownian motion, see, e.g. Yor [95] and Hibino, Hitsuda, Muraoka [29]).

**b)** A slightly different derivation of (6.39) consists in remarking that $B_{1-t} = (1-t)B_1 + b(t)$ with $(b(t),\, 0 \le t \le 1)$ a standard Brownian bridge independent from $B_1$. From (*i*), this Brownian bridge admits the decomposition:

$$b(t) = \beta_t^* - \int_0^t \frac{b(s)}{1-s}\, ds.$$

Thus we obtain:

$$B_{1-t} = (1-t)B_1 + \beta_t^* - \int_0^t \frac{ds}{1-s}\left(b(s) + (1-s)B_1 - (1-s)B_1\right)$$

$$= (1-t)B_1 + \beta_t^* - \int_0^t \frac{B_{1-s}}{1-s}\, ds + tB_1$$

$$= B_1 + \beta_t^* - \int_0^t \frac{B_{1-s}}{1-s}\, ds$$

which is precisely (6.39). Now, plugging (6.40) in (6.37), and with the help of (*i*), we get:

$$\int_0^1 \int_0^1 f(K,s)\gamma(dK,ds) = \frac{1}{2}\,\mathbb{E}\left[\int_0^1 f(\mathscr{E}_s,s)\mathscr{E}_s\left(1 - \frac{B_s}{s}\right)ds\right]$$

$$= \frac{1}{2}\,\mathbb{E}\left[\int_0^1 \mathscr{E}_s f(\mathscr{E}_s,s)\left(\frac{1}{2} - \frac{\log\mathscr{E}_s}{s}\right)ds\right]$$

which matches perfectly with our previous formula (6.33). Thus, we have established a close link with local time-space calculus as developed in ([22]–[23]).

c) To summarize the argument in b): $(B_{1-t}, t \leq 1)$ is a Brownian bridge over the time interval $[0,1]$, starting with $B_1$ and ending at 0. Now, a Brownian bridge starting at $x$ and ending at $y$ solves:

$$X_t = x + \beta_t^* + \int_0^t \frac{y - X_s}{1-s} ds.$$

It remains to replace $x$ by $B_1$ and $y$ by 0 to recover (6.39).

We note that a similar remark appears on p. 563 of S. Tindel [85] in his stochastic calculus approach to spins systems.

### 6.2.3 An Extension of Theorem 6.1

In the next statement, we shall replace the standard Brownian martingale $(\mathscr{E}_t, t \geq 0)$ by the semimartingale $(\mathscr{E}_t^{\sigma,-\nu} := \exp(\sigma B_t - \nu t), t \geq 0)$ $(\sigma \neq 0, \nu > 0)$. Then we can show:

**Theorem 6.2.**

i) *There exists a probability on $[0,1] \times [0,+\infty[$, which we shall denote by $\gamma_{\sigma,\nu}$ such that:*

$$\Pi_{\sigma,\nu}(K,t) := \mathbb{E}\left[(K - \mathscr{E}_t^{\sigma,-\nu})^+\right] = \gamma_{\sigma,\nu}\left([0,K] \times [0,t]\right). \tag{6.42}$$

ii) *Moreover, $\gamma_{\sigma,\nu}$ is the law of:*

$$\left(U, Y_{\frac{\nu}{|\sigma|}, \frac{1}{|\sigma|} \log \frac{1}{U}}\right) \tag{6.43}$$

*where $U$ is uniform on $[0,1]$ and independent of $(Y_{a,b}, a,b > 0)$ as introduced in Lemma 6.1.*

*Proof.* We may choose $\sigma > 0$, since $\sigma B_t \overset{(law)}{=} -\sigma B_t$. Then, we write for $K \leq 1$, applying Fubini:

$$\mathbb{E}\left[(K - \mathscr{E}_t^{\sigma,-\nu})^+\right] = \int_0^K \mathbb{P}(\mathscr{E}_t^{\sigma,-\nu} \leq x) dx$$

$$= \int_0^K \mathbb{P}\left(B_1 < \frac{\nu\sqrt{t}}{\sigma} + \frac{\log x}{\sigma\sqrt{t}}\right) dx$$

$$= \int_0^K \mathscr{N}\left(\frac{\nu\sqrt{t}}{\sigma} + \frac{\log x}{\sigma\sqrt{t}}\right) dx$$

$$= \int_0^K \mathbb{P}\left(Y_{\frac{\nu}{\sigma}, \frac{1}{\sigma} \log \frac{1}{x}} \leq t\right) dx \qquad \text{(from Lemma 6.1)}$$

which implies points (*i*) and (*ii*) of Theorem 6.2.

$\square$

Note that (6.43) corresponds to the first description in Subsection 6.2.2 of the particular case $\sigma = 1$, $\nu = \dfrac{1}{2}$. We would like to see whether there is a second description of $\gamma_{\sigma,\nu}$ ; in particular, what is the law of $Y_{\frac{\nu}{\sigma},\frac{1}{\sigma}\log\frac{1}{U}}$ ? Let us denote by $f_{\sigma,\nu}$ the density of $Y_{\frac{\nu}{\sigma},\frac{1}{\sigma}\log\frac{1}{U}}$. Then, we have:

**Proposition 6.5.**

i) *The following formula holds:*

$$f_{\sigma,\nu}(t) = \frac{\sigma}{2\sqrt{2\pi t}} e^{-\frac{\nu^2 t}{2\sigma^2}} \left( 1 + \left( 2\frac{\nu}{\sigma^2} - 1 \right) \int_0^\infty e^{-\mu x - \frac{x^2}{2\sigma^2 t}} dx \right) \qquad (6.44)$$

*where $\mu = 1 - \dfrac{\nu}{\sigma^2}$,*

ii) *In particular, if $\dfrac{2\nu}{\sigma^2} = 1$, $\left(\text{this condition ensures that } (\mathscr{E}_t^{\sigma,-\nu}, t \geq 0) \text{ is a martingale!}\right)$, we have:*

$$f_{\sigma,\frac{\sigma^2}{2}}(t) = \frac{\sigma}{2\sqrt{2\pi t}} e^{-\frac{\sigma^2 t}{8}}. \qquad (6.45)$$

$\left( = f_{4B_1^2}(t) \text{ if } \sigma = 1 \text{ and } \nu = \dfrac{1}{2}, \text{ in agreement with Proposition 6.2, formula}\right.$
$(6.27)\Big)$.

*Proof.* From (6.21) and (6.43) we have:

$$f_{\sigma,\nu}(t)dt = \frac{1}{2}\int_0^1 du \left( \mathbb{P}\left( T_{\frac{1}{\sigma}\log\frac{1}{u}}^{(\frac{\nu}{\sigma})} \in dt \right) + \mathbb{P}\left( G_{\frac{1}{\sigma}\log\frac{1}{u}}^{(\frac{\nu}{\sigma})} \in dt \right) \right).$$

Making the change of variable $\log\left(\dfrac{1}{u}\right) = x$ and using (1.11) and (1.12), we obtain:

$$f_{\sigma,\nu}(t) = \frac{1}{2}\int_0^\infty e^{-x} \left( \left( \frac{x}{\sigma\sqrt{2\pi t^3}} + \frac{\nu}{\sigma\sqrt{2\pi t}} \right) e^{-\frac{(x-\nu t)^2}{2\sigma^2 t}} \right) dx$$

$$= \frac{\sigma}{2\sqrt{2\pi t}} \exp\left( -\frac{\nu^2 t}{2\sigma^2} \right) \int_0^\infty \left( \frac{x}{t\sigma^2} + \frac{\nu}{\sigma^2} \right) \exp\left( -x - \frac{x^2}{2\sigma^2 t} + \frac{x\nu}{\sigma^2} \right) dx.$$

We now introduce the parameter $\mu = 1 - \dfrac{\nu}{\sigma^2}$ and we compute the integral:

$$I := \int_0^\infty \left( \frac{x}{t\sigma^2} + \frac{\nu}{\sigma^2} \right) e^{-\mu x - \frac{x^2}{2\sigma^2 t}} dx$$

$$:= I_1 + \frac{\nu}{\sigma^2} I_2,$$

with:

$$I_1 = \int_0^\infty e^{-\mu x - \frac{x^2}{2\sigma^2 t}} \frac{x}{t\sigma^2} dx = 1 - \mu \int_0^\infty e^{-\mu x - \frac{x^2}{2\sigma^2 t}} dx$$

(after integrating by parts). Finally:

$$I = I_1 + \frac{\nu}{\sigma^2} I_2 = 1 - \mu I_2 + \frac{\nu}{\sigma^2} I_2$$
$$= 1 + \left(\frac{2\nu}{\sigma^2} - 1\right) \int_0^\infty e^{-\mu x - \frac{x^2}{2\sigma^2 t}} dx$$

and:

$$f_{\sigma,\nu}(t) = \frac{\sigma}{2\sqrt{2\pi t}} \left(1 + \left(\frac{2\nu}{\sigma^2} - 1\right) \int_0^\infty e^{-\mu x - \frac{x^2}{2\sigma^2 t}} dx\right) \exp\left(-\frac{\nu^2 t}{2\sigma^2}\right).$$

This proves (6.44).

□

## 6.2.4 $\gamma$ as a Signed Measure on $\mathbb{R}^+ \times \mathbb{R}^+$

In this paragraph, we extend the definition of $\gamma$ to $[0, +\infty[ \times [0, +\infty[$.

**Proposition 6.6.**

i) *There exists a signed measure $\gamma$ on $[0, +\infty[ \times [0, +\infty[$ such that:*

$$\Pi_{\mathscr{E}}(K, t) := \mathbb{E}\left[(K - \mathscr{E}_t)^+\right] = \gamma([0, K] \times [0, t]) \tag{6.46}$$

*holds for all values $K, t \geq 0$.*
ii) *$\gamma$ admits on $\mathbb{R}^+ \times \mathbb{R}^+$ the density $f_\gamma$ given by:*

$$f_\gamma(K, t) = \frac{1}{2\sqrt{2\pi K t}} \left(\frac{1}{2} - \frac{\log K}{t}\right) \exp\left(-\frac{(\log K)^2}{2t} - \frac{t}{8}\right) 1_{[0,+\infty[\times[0,+\infty[}(K, t) \tag{6.47}$$
$$= \frac{1}{2\sqrt{2\pi t}} \left(\frac{1}{2} - \frac{\log K}{t}\right) \exp\left(-\frac{1}{2t}\left(\log K + \frac{t}{2}\right)^2\right) 1_{[0,+\infty[\times[0,+\infty[}(K, t).$$

iii) *Consequently, if $\gamma = \gamma^+ - \gamma^-$ is the decomposition of $\gamma$ into its positive and negative parts, we have:*

$$\gamma^+(dK, dt) = 1_{\left\{K \leq e^{\frac{t}{2}}\right\}} \gamma(dK, dt). \tag{6.48}$$

*In particular:*

$$\gamma^+_{|(K,t\,;\,K \leq 1)} = \gamma_{|(K,t\,;\,K \leq 1)}$$

*is a probability.*
iv) *Formula (6.47) may be synthesized as follows:*
   • *if $K < 1$:*

$$\gamma(dK, dt) = \frac{dK}{2} \left\{\mathbb{P}\left(G^{(\frac{1}{2})}_{\log(\frac{1}{K})} \in dt\right) + \mathbb{P}\left(T^{(\frac{1}{2})}_{\log(\frac{1}{K})} \in dt\right)\right\}, \tag{6.49}$$

• *if K > 1:*

$$\gamma(dK,dt) = \frac{dK}{2} \left\{ \mathbb{P}(G^{(\frac{1}{2})}_{\log(\frac{1}{K})} \in dt) - \mathbb{P}(T^{(\frac{1}{2})}_{\log(\frac{1}{K})} \in dt) \right\}. \tag{6.50}$$

*Proof.* Points (*i*) and (*ii*) are easy to prove. Indeed, for every $K \geq 0$:

$$\mathbb{E}\left[(K - \mathscr{E}_t)^+\right] = \int_0^K \mathbb{P}(\mathscr{E}_t < x)dx = \int_0^K \mathscr{N}\left(\frac{\sqrt{t}}{2} + \frac{\log x}{\sqrt{t}}\right)dx.$$

This formula implies the existence of $\gamma$ and the density $f_\gamma$ of $\gamma$ is given by:

$$\begin{aligned}
f_\gamma(K,t) &= \frac{\partial^2}{\partial K \partial t} \int_0^K \mathscr{N}\left(\frac{\sqrt{t}}{2} + \frac{\log x}{\sqrt{t}}\right)dx \\
&= \frac{\partial}{\partial t} \mathscr{N}\left(\frac{\sqrt{t}}{2} + \frac{\log K}{\sqrt{t}}\right) \\
&= \frac{1}{2\sqrt{2\pi K t}}\left(\frac{1}{2} - \frac{\log K}{t}\right)\exp\left(-\frac{(\log K)^2}{2t} - \frac{t}{8}\right).
\end{aligned}$$

Point (*iii*) follows immediately. Formulae (6.49) and (6.50) are obtained as in (6.22). □

Finally, we note that an elementary change of variables allows to present $\gamma^+$ and $\gamma^-$ in a very simple manner.

**Proposition 6.7.** *For any $\varphi : \mathbb{R}^+ \times \mathbb{R}^+ \to \mathbb{R}^+$ Borel, the following 3 quantities are equal:*

• $$\int\int_{(K < e^{\frac{t}{2}})} \gamma^+(dK,dt)\, \varphi\left(\frac{t}{2} - \log K, t\right) \tag{6.51}$$

• $$\int\int_{(K > e^{\frac{t}{2}})} \gamma^-(dK,dt)\, \varphi\left(\log K - \frac{t}{2}, t\right) \tag{6.52}$$

• $$\frac{1}{2}\int_0^\infty dy\, \mathbb{E}\left[\varphi(y, T_y)\right] = \frac{1}{2}\int_0^\infty dy \int_0^\infty dt\, \frac{y}{\sqrt{2\pi t^3}}e^{-\frac{y^2}{2t}}\varphi(y,t) \tag{6.53}$$

*where $T_y = \inf\{t \geq 0; B_t = y\}$.*

*Proof.* This double identity follows easily, by obvious change of variables, from formula (6.47) for the density $f_\gamma$ of $\gamma$, and the facts that:

$$\gamma^+(dK,dt) = 1_{\left\{K \leq e^{\frac{t}{2}}\right\}}\gamma(dK,dt)\,;\; \gamma^-(dK,dt) = -1_{\left\{K \geq e^{\frac{t}{2}}\right\}}\gamma(dK,dt).$$

□

## 6.3 Notes and Comments

The idea to represent $\left(\mathbb{E}\left[(K-M_t)^+\right];0\leq K\leq 1, t\geq 0\right)$ as the cumulative distribution function of a probability $\gamma_M$ on $[0,1]\times[0,+\infty[$, i.e. $\mathbb{E}\left[(K-M_t)^+\right]=\gamma_M\left([0,K]\times[0,t]\right)$ $(0\leq K\leq 1, t\geq 0)$ appears, for $\left(M_t=\mathcal{E}_t:=\exp\left(B_t-\frac{t}{2}\right),t\geq 0\right)$ in the paper by D. Madan, B. Roynette and M. Yor ([49]). The construction of this probability $\gamma_M$ hinges upon the existence of a pseudo-inverse of the process $(\mathcal{E}_t, t\geq 0)$. This notion of a pseudo-inverse has also been introduced in the same paper ([49]), from which the descriptions of $\gamma_\mathcal{E}$ $(=\gamma)$ given by Propositions 6.1, 6.2 and 6.3 are taken. Links between Proposition 6.3 and the "local time-space calculus" (formula (6.37)) hinge upon some works by N. Eisenbaum (see [22] and [23]) whom we thank for telling us about an error in our first version of Proposition 6.4.

# Chapter 7
# Existence and Properties of Pseudo-Inverses for Bessel and Related Processes

**Abstract** In Chapter 6, we have shown the existence of a decreasing pseudo-inverse for the martingale $(M_t := \exp\left(B_t - \frac{t}{2}\right), t \geq 0)$. We shall now explore this notion in a more general framework, starting with the case of Bessel (and some related) processes. We show in particular that the tail probabilities of a Bessel process of index $\nu \geq 1/2$ increase with respect to time; in fact it is the distribution function of a random time which is related to first and last passage times of Bessel processes.

## 7.1 Introduction and Definition of a Pseudo-Inverse

### 7.1.1 Motivations

The aim of this Chapter 7 and of the following Chapter 8 is to give some mathematical meaning to the assertion:

"the stochastic process $(R_t, t \geq 0)$, with values in $\mathbb{R}^+$, has a tendency to increase"

$$(7.1)$$

and, more precisely, to "measure this tendency", by showing the existence of a "pseudo-inverse" of $R$ and by studying its properties. First, here are several possible interpretations of (7.1):

**a)** $(R_t, t \geq 0)$ is a.s. increasing; then, it admits an increasing inverse process.
**b)** $(R_t, t \geq 0)$ is a submartingale; under some adequate conditions, it admits a Doob-Meyer decomposition:

$$R_t = M_t + A_t \qquad (t \geq 0)$$

with $(A_t, t \geq 0)$ an increasing process and $(M_t, t \geq 0)$ a martingale. Since a martingale is a "well balanced" process, $(R_t, t \geq 0)$ has a tendency to increase, which is measured by $(A_t, t \geq 0)$.
**c)** $(\mathbb{E}[R_t], t \geq 0)$ is an increasing function.

C. Profeta et al., *Option Prices as Probabilities*, Springer Finance,
DOI 10.1007/978-3-642-10395-7_7, © Springer-Verlag Berlin Heidelberg 2010

**d)** $(R_t, t \geq 0)$ is "stochastically increasing", i.e.: for every $y \geq 0$, $\mathbb{P}(R_t \geq y)$ is an increasing function of $t$ (in the case $R_0 = x > 0$, we need to modify the previous assertion by taking $y \geq x$, to get a meaningful assertion).

We note that, trivially: (a) $\Longrightarrow$ (b) $\Longrightarrow$ (c) $\Longleftarrow$ (d) (when $x = 0$).

It is the assertion (d) which we have in mind throughout this study. It leads us to the definition of a pseudo-inverse of $(R_t, t \geq 0)$, which we now present.

## 7.1.2 Definitions and Examples

Let $\left(\Omega, (R_t, t \geq 0), \mathbb{P}_x, x \in \mathbb{R}^+\right)$ denote a process taking values on $\mathbb{R}^+$, which is a.s. continuous and such that $\mathbb{P}_x(R_0 = x) = 1$. In most of our applications, this process will be a diffusion, but to start with, we do not use the Markov property.

**Definition 7.1.** $(R_t, t \geq 0)$ admits an increasing pseudo-inverse, resp. a decreasing pseudo-inverse, if:

i)  for every $y > x$,

$$\lim_{t \to \infty} \mathbb{P}_x(R_t \geq y) = 1, \tag{7.2}$$

ii) for every $y > x$,

$$\text{the application from } \mathbb{R}^+ \text{ into } [0, 1] : t \to \mathbb{P}_x(R_t \geq y) \text{ is increasing.} \tag{7.3}$$

resp. if:

i)  for every $y < x$,

$$\lim_{t \to \infty} \mathbb{P}_x(R_t \leq y) = 1, \tag{7.2'}$$

ii) for every $y < x$,

$$\text{the application from } \mathbb{R}^+ \text{ into } [0, 1] : t \to \mathbb{P}_x(R_t \leq y) \text{ is increasing.} \tag{7.3'}$$

**Definition 7.2.** Assume that (7.2) and (7.3) $\left(\text{resp. (7.2') and (7.3')}\right)$ are satisfied. Then, there exists a family of positive r.v.'s $(Y_{x,y}, y > x)$ (resp. $Y_{x,y}, y < x$) such that:

$$\mathbb{P}_x(R_t \geq y) = \mathbb{P}(Y_{x,y} \leq t) \qquad (t \geq 0) \tag{7.4}$$

resp.

$$\mathbb{P}_x(R_t \leq y) = \mathbb{P}(Y_{x,y} \leq t) \qquad (t \geq 0). \tag{7.4'}$$

We call the family of positive r.v.'s $(Y_{x,y}, y > x)$ $\left(\text{resp. } (Y_{x,y}, y < x)\right)$ the increasing (resp. decreasing) pseudo-inverse of the process $(R_t, t \geq 0)$.

We note that, a priori, the family $(Y_{x,y}, y > x)$ $\left(\text{resp. } (Y_{x,y}, y < x)\right)$ is only a family of r.v.'s, and does not constitute a process. Note that formulae (7.4) and (7.4') may be read, either considering $x$ as fixed and $y$ varying or by considering the two parameters $x$ and $y$ as varying.

*Remark 7.1.*

**a)** If $(R_t, t \geq 0)$ admits an increasing (resp. decreasing) pseudo-inverse, then $R_t \xrightarrow[t \to \infty]{} +\infty$ in probability $\left(\text{resp. } R_t \xrightarrow[t \to \infty]{} 0 \text{ in probability}\right)$.

**b)** If $(R_t, t \geq 0)$ admits an increasing pseudo-inverse, then, for every $\alpha > 0$ (resp. for every $\alpha < 0$), the process $(R_t^\alpha, t \geq 0)$ admits an increasing (resp. decreasing) pseudo-inverse. If $(R_t, t \geq 0)$ admits a decreasing pseudo-inverse, then for every $\alpha > 0$ (resp. for every $\alpha < 0$), the process $(R_t^\alpha, t \geq 0)$ admits a decreasing (resp. increasing) pseudo-inverse.

**c)** Justifying the term "pseudo-inverse": Condition (7.3) indicates that the process $(R_t, t \geq 0)$ "has a tendency to increase". In case this process is indeed increasing, we introduce $(\tau_l, l \geq 0)$ its right-continuous inverse:

$$\tau_l := \inf\{t \geq 0; R_t > l\}.$$

Then, from (7.4), and for $y > x$;

$$\mathbb{P}_x(R_t \geq y) = \mathbb{P}_x(\tau_y \leq t) = \mathbb{P}(Y_{x,y} \leq t).$$

Thus $Y_{x,y} \overset{(\text{law})}{=} \tau_y$ under $\mathbb{P}_x$. Hence, in this case, we may choose for the family $(Y_{x,y}, y > x)$ the process $(\tau_y, y > x)$, and this justifies our terminology of pseudo-inverse.

### 7.1.3 Aim of this Chapter

In the set-up of the classical Black-Scholes formulae, i.e. when $M_t^{(x)} = \mathcal{E}_t^{(x)} :=$ $\exp\left(B_t - \dfrac{t}{2}\right)$, where under $\mathbb{P}_x, (B_t, t \geq 0)$ is a Brownian motion starting from $(\log x)$, we have proven in Chapter 6 the existence of a decreasing pseudo-inverse of $(M_t^{(x)}, t \geq 0)$ and thus established the existence of a probability $\gamma_{\mathcal{E}(x)}$ (see Theorem 6.1) characterized by:

$$\frac{1}{x}\mathbb{E}_x\left[(K - \mathcal{E}_t^{(x)})^+\right] = \gamma_{\mathcal{E}(x)}\left([0, K] \times [0, t]\right) \qquad (K \leq x, t \geq 0).$$

In Chapter 6, the study was done for $x = 1$, but reducing the study to the case $x = 1$ is easy. Note that we have also described this probability $\gamma_{\mathcal{E}(x)}$ in detail.

Now, let $(\Omega, (R_t, \mathcal{F}_t), t \geq 0; \mathbb{P}_x^{(\nu)}, x \geq 0)$ denote a Bessel process of index $\nu$ ($\nu \geq -1$). We shall prove in Section 7.2 (see Theorem 7.1) the following results:

- when $\nu \geq -\dfrac{1}{2}$, the existence of an increasing pseudo-inverse $(Y_{x,y}^{(\nu)}, y > x)$ for $(R_t, t \geq 0, \mathbb{P}_x^{(\nu)}, x \geq 0)$,

- when $\nu = -1$, the existence of a decreasing pseudo-inverse $(Y_{x,y}^{(-1)}, y < x)$ for $(R_t, t \geq 0, \mathbb{P}_x^{(-1)}, x \geq 0)$,
- when $\nu \in \left] -1, -\dfrac{1}{2} \right[ :$

  i) if $x = 0$, the existence of an increasing pseudo-inverse,

  ii) if $x > 0$, the non-existence of a pseudo-inverse.

In Section 7.3, we describe the laws of the r.v.'s $(Y_{x,y}^{(\nu)}, y > x)$. In particular, we obtain explicitly the Laplace transform of $(Y_{x,y}^{(\nu)}, y > x)$:

$$\mathbb{E}\left[ e^{-\lambda Y_{x,y}^{(\nu)}} \right] = \frac{I_\nu(x\sqrt{2\lambda})}{(x\sqrt{2\lambda})^\nu} (y\sqrt{2\lambda})^{\nu+1} K_{\nu+1}(y\sqrt{2\lambda}) \qquad \left( \nu \geq -\frac{1}{2} \right)$$

where $I_\nu$ and $K_\nu$ are the classical modified Bessel functions (see Appendix B.1). We also answer (partially) the question: are the r.v.'s $Y_{x,y}^{(\nu)}$ infinitely divisible ?

For $\nu > 0$, we apply (end of Section 7.2) the existence of a pseudo-inverse for the process $(R_t, t \geq 0)$ to the generalized Black-Scholes formula relative to the local martingale $(R_t^{-2\nu}, t \geq 0)$, when $(R_t, t \geq 0)$ denotes a $\delta = 2(\nu + 1)$ dimensional Bessel process.

In Section 7.4, we present two other families of Markovian positive submartingales which admit an increasing pseudo-inverse; they are:

i) the Bessel processes with drift, studied by S. Watanabe [89], with infinitesimal generator:

$$L^{(\nu,a)} = \frac{1}{2} \frac{d^2}{dx^2} + \left( \frac{2\nu+1}{2x} + a \frac{I_{\nu+1}}{I_\nu}(ax) \right) \frac{d}{dx}$$

for $\nu \geq -\dfrac{1}{2}$, and $a \geq 0$;

ii) the generalized squares of Ornstein-Uhlenbeck processes $\big($see, e.g., [66]$\big)$, also called Cox-Ingersoll-Ross (= CIR) processes in the Mathematical Finance literature, with infinitesimal generator:

$$\overline{L}^{(\nu,\beta)} = 2x \frac{d^2}{dx^2} + \left( 2\beta x + 2(\nu+1) \right) \frac{d}{dx}$$

again with $\nu \geq -\dfrac{1}{2}$, and $\beta \geq 0$.

Of course, letting $a \to 0$, (resp. $\beta \to 0$), in $(i)$, (resp. $(ii)$), we recover the result for the Bessel processes.

In Section 7.5, we exhibit a two parameter family $(Y_{x,y}^{(\nu,\alpha)}, \nu \geq 0, \alpha \in [0,1])$ of variables (indexed by $(x,y), x < y$) which extends the family $Y_{x,y}^{(\nu)}$, corresponding to $\alpha = 1$.

The following paragraph aims at relating the results of Section 7.3, which presents the properties of the r.v.'s $Y_{x,y}^{(\nu)}$, and those of Section 7.5, where the r.v.'s $Y_{x,y}^{(\nu,\alpha)}$ are defined and studied.

Consider $(R_t^{(\nu)}, t \geq 0)$ and $(R_t^{(\nu')}, t \geq 0)$ two independent Bessel processes starting from 0, and with respective indexes $\nu$ and $\nu'$, greater than $-1$, or dimensions: $d = 2(\nu + 1)$ and $d' = 2(\nu' + 1)$. From the additivity property of squares of Bessel processes, the process:

$$R_t^{(\nu+\nu'+1)} := \sqrt{(R_t^{(\nu)})^2 + (R_t^{(\nu')})^2} \qquad (t \geq 0),$$

is a Bessel process with index $\nu + \nu' + 1$ (or dimension: $d + d'$), starting from 0. Let, for $y > 0$:

$$G_y^{(\alpha)} := \sup\{t \geq 0; R_t^{(\alpha)} = y\}$$
$$T_y^{(\alpha)} := \inf\{t \geq 0; R_t^{(\alpha)} = y\} \qquad (\alpha = \nu, \nu + \nu' + 1).$$

It is then clear that:

$$T_y^{(\nu+\nu'+1)} \leq T_y^{(\nu)} \leq G_y^{(\nu)}, \qquad \text{and} \qquad G_y^{(\nu+\nu'+1)} \leq G_y^{(\nu)}.$$

These inequalities invite to look for some r.v.'s $Z_y^{(i)} (i = 1, 2, 3)$, such that:

$$G_y^{(\nu)} \overset{\text{(law)}}{=} G_y^{(\nu+\nu'+1)} + Z_y^{(1)} \tag{7.5}$$

$$G_y^{(\nu)} \overset{\text{(law)}}{=} T_y^{(\nu)} + Z_y^{(2)} \tag{7.6}$$

$$T_y^{(\nu)} \overset{\text{(law)}}{=} T_y^{(\nu+\nu'+1)} + Z_y^{(3)} \tag{7.7}$$

where the r.v.'s featured on the right-hand side of (7.5), (7.6) and (7.7) are independent.

The existence of these r.v.'s $Z_y^{(i)} (i = 1, 2, 3)$ is obtained in Sections 7.3 and 7.5 as a subproduct of the existence of pseudo-inverses for Bessel processes. Precisely, the identities (7.5), (7.6) and (7.7) are shown respectively as (7.117), in Proposition 7.4, resp. as (7.113), with $y = z$ and $z = 0$, resp. as (7.116) in Proposition 7.4. In Theorem 7.3 (Point $(iv)$ and $(v)$) and Theorem 7.7, other equalities in law of the type of (7.5), (7.6) and (7.7), are established, when the starting points of the Bessel processes differ from 0.

## 7.2 Existence of Pseudo-inverses for Bessel Processes

### 7.2.1 Statement of our Main Result

Let $\nu \geq -1$ and $(R_t, t \geq 0; \mathbb{P}_x^{(\nu)}, x \geq 0)$ denote a Bessel process with index $\nu$, i.e. with dimension $\delta = 2\nu + 2$, with $\delta \geq 0$.

**Theorem 7.1.**

i) *If* $\nu \geq -\dfrac{1}{2}, (R_t, t \geq 0)$ *admits an increasing pseudo-inverse; that is for every* $y > x \geq 0$, *there exists a positive r.v.* $Y_{x,y}^{(\nu)}$ *such that, for every* $t \geq 0$:

$$\mathbb{P}_x^{(\nu)}(R_t \geq y) = \mathbb{P}(Y_{x,y}^{(\nu)} \leq t). \tag{7.8}$$

ii) *Let* $\nu \in \left] -1, -\dfrac{1}{2} \right[$:

   – *if* $x > 0$, $(R_t, t \geq 0)$ *does not admit a pseudo-inverse,*

   – *if* $x = 0$, $(R_t, t \geq 0)$ *admits an increasing pseudo-inverse.*

iii) *If* $\nu = -1$, $(R_t, t \geq 0)$ *admits a decreasing pseudo-inverse, that is: there exists, for every* $x > y > 0$, *a positive r.v.* $Y_{x,y}^{(-1)}$ *such that, for every* $t \geq 0$:

$$\mathbb{P}_x^{(-1)}(R_t \leq y) = \mathbb{P}(Y_{x,y}^{(-1)} \leq t). \tag{7.9}$$

Thus, in Theorem 7.1, the value $\nu = -\frac{1}{2}$ appears as a critical value. This is, in fact, rather natural. Indeed, for $\nu \geq -\frac{1}{2}$, $(R_t, t \geq 0)$ is a submartingale - therefore it is, in a way, increasing in $t$ - and it may be written as:

$$R_t = x + B_t + \frac{2\nu + 1}{2} \int_0^t \frac{ds}{R_s} \qquad \text{if } \nu > -\frac{1}{2}$$

where $(B_t, t \geq 0)$ denotes a Brownian motion starting from 0, and for $\nu = -\frac{1}{2}$, $R_t = |x + B_t| = |x| + \beta_t + L_t^{-x}$, where $(L_t^{-x}, t \geq 0)$ denotes the local time of $B$ at level $-x$ and $(\beta_t, t \geq 0)$ is a Brownian motion.

For $\nu \in \left] -1, -\dfrac{1}{2} \right[$, such a representation is no longer true. In fact:

$$R_t = x + \beta_t + \frac{2\nu + 1}{2} k_t$$

with

$$k_t := \text{p.v.} \int_0^t \frac{ds}{R_s} := \int_0^\infty a^{2\nu}(L_t^a - L_t^0)da$$

where $(L_t^a, t \geq 0, a \geq 0)$ denotes the jointly continuous family of diffusion local times associated with $(R_t, t \geq 0)$ $\left(\text{see [70], Exercise 1.26, Chap.XI, p.451}\right)$; in this case, $(R_t, t \geq 0)$ is no longer a semimartingale, but it is a Dirichlet process. $\left(\text{See}\right.$ Bertoin [7] for a deep study of excursion theory for the process $\left.\left((R_t, k_t), t \geq 0\right)\right)$. In order to prove Theorem 7.1, we gather a few results about Bessel processes, see also Appendices B.2 and B.3.

## 7.2.2 A Summary of some Results About Bessel Processes

### 7.2.2.1 The Density of the Bessel Semi-Group $\left(\text{see [70], Chap. XI}\right)$

We denote by $p^{(\nu)}(t,x,y)$ the density of $R_t$ under $\mathbb{P}_x^{(\nu)}$ $(\nu > -1)$; one has:

$$p^{(\nu)}(t,x,y) = \frac{y}{t}\left(\frac{y}{x}\right)^{\nu}\exp\left(-\frac{x^2+y^2}{2t}\right)I_{\nu}\left(\frac{xy}{t}\right) \qquad (\nu > -1) \qquad (7.10)$$

if $x > 0$, whereas for $x = 0$,

$$p^{(\nu)}(t,0,y) = 2^{-\nu}t^{-(\nu+1)}\Gamma(\nu+1)y^{2\nu+1}\exp\left(-\frac{y^2}{2t}\right) \qquad (\nu > -1) \qquad (7.11)$$

where $I_{\nu}$ denotes the modified Bessel function with index $\nu$ (see Appendix B.1).

### 7.2.2.2 Density of Last Passage Times $\left(\text{See [25] and [65]}\right)$

Let $y \geq 0$ and denote by $G_y$ the last passage time of $R$ at level $y$:

$$G_y := \sup\{t \geq 0;\ R_t = y\}, \qquad (7.12)$$
$$(= 0 \text{ if this set } \{t \geq 0;\ R_t = y\} \text{ is empty}).$$

Then, for $x \leq y$, applying Theorem 2.4:

$$\mathbb{P}_x^{(\nu)}(G_y \in dt) = \frac{\nu}{y}p^{(\nu)}(t,x,y)dt \qquad (\nu > 0). \qquad (7.13)$$

It follows easily from (7.13) and (7.11) that, under $\mathbb{P}_0^{(\nu)}$, the law of $G_y$ is that of $\dfrac{y^2}{2\gamma_{\nu}}$, where $\gamma_{\nu}$ is a gamma r.v. with parameter $\nu$ $(\nu > 0)$:

$$G_y \overset{\text{(law)}}{=} \frac{y^2}{2\gamma_{\nu}} \qquad (G_y \text{ being considered under } \mathbb{P}_0^{(\nu)}) \qquad (7.14)$$

and:

$$\mathbb{E}_x^{(\nu)}\left[e^{-\lambda G_y}\right] = 2\nu\left(\frac{y}{x}\right)^{\nu} I_{\nu}(x\sqrt{2\lambda})K_{\nu}(y\sqrt{2\lambda}) \qquad (0 < x < y,\ \nu > 0),$$

$$\mathbb{E}_0^{(\nu)}\left[e^{-\lambda G_y}\right] = \frac{1}{2^{\nu-1}\Gamma(\nu)}(y\sqrt{2\lambda})^{\nu} K_{\nu}(y\sqrt{2\lambda}) \qquad (y > 0,\ \nu > 0).$$

### 7.2.2.3 Laplace Transform of First Hitting Times (See [41] and [65])

Let $y \geq 0$ and denote by $T_y$ the first hitting time of $R$ at level $y$:

$$T_y := \inf\{t \geq 0; R_t = y\} \tag{7.15}$$
$$(= +\infty \text{ if this set } \{t \geq 0; R_t = y\} \text{ is empty}).$$

Then:

$$\mathbb{E}_x^{(\nu)}\left[e^{-\lambda T_y}\right] = \left(\frac{y}{x}\right)^{\nu} \frac{I_{\nu}(x\sqrt{2\lambda})}{I_{\nu}(y\sqrt{2\lambda})} \qquad (x \leq y,\ \lambda \geq 0). \tag{7.16}$$

In particular, for $x = 0$

$$\mathbb{E}_0^{(\nu)}\left[e^{-\lambda T_y}\right] = \frac{1}{2^{\nu}\Gamma(\nu+1)} \frac{(y\sqrt{2\lambda})^{\nu}}{I_{\nu}(y\sqrt{2\lambda})}. \tag{7.17}$$

### 7.2.2.4 Resolvent Kernel (see [12])

Let $u_{\lambda}^{(\nu)}$ denote the density of the potential kernel of $(R_t, t \geq 0)$ under $\mathbb{P}^{(\nu)}$:

$$u_{\lambda}^{(\nu)}(x,y) = \int_0^{\infty} e^{-\lambda t} p^{(\nu)}(t,x,y)dt \qquad (\lambda \geq 0).$$

Then, for every positive Borel function $f$:

$$\int_x^{\infty} u_{\lambda}^{(\nu)}(x,y)f(y)dy = \frac{2}{x^{\nu}}I_{\nu}(x\sqrt{2\lambda}) \int_x^{\infty} y^{\nu+1} K_{\nu}(y\sqrt{2\lambda})f(y)dy \tag{7.18}$$

and

$$u_{\lambda}^{(\nu)}(x,y) = 2y\left(\frac{y}{x}\right)^{\nu} I_{\nu}(x\sqrt{2\lambda})K_{\nu}(y\sqrt{2\lambda}) \qquad (x < y)$$

where $K_{\nu}$ denotes the Bessel-McDonald function with index $\nu$ (see Appendix B.1).

### 7.2.2.5 Bessel Realizations of the Hartman-Watson Laws (See [94] and [65])

It was established by Hartman-Watson [28] that, for any $\nu \geq 0$ and $r > 0$, the ratio

$$\lambda \longrightarrow \frac{I_{\sqrt{\lambda+\nu^2}}}{I_{\nu}}(r)$$

is the Laplace transform of a probability, say $\theta_r^{(\nu)}$ on $\mathbb{R}^+$, which may then be called Hartman-Watson distribution. In [94] and [65] these laws were shown to be those of the Bessel clocks $\int_0^t \dfrac{ds}{R_s^2}$ for Bessel bridges $(R_s, s \le t)$ conditioned at both ends. Precisely, for $\nu > -1$ and $\mu \ne 0$, one has:

$$\mathbb{E}_x^{(\nu)}\left[\exp\left(-\frac{\mu^2}{2}\int_0^t \frac{ds}{R_s^2}\right)\bigg| R_t = y\right] = \frac{I_{\sqrt{\mu^2+\nu^2}}}{I_\nu}\left(\frac{xy}{t}\right) \qquad (x, y \ge 0). \qquad (7.19)$$

This formula may be obtained from the particular case $\nu = 0$:

$$\mathbb{E}_x^{(0)}\left[\exp\left(-\frac{\mu^2}{2}\int_0^t \frac{ds}{R_s^2}\right)\bigg| R_t = y\right] = \frac{I_{|\mu|}}{I_0}\left(\frac{xy}{t}\right)$$

together with the absolute continuity relationship:

$$\mathbb{P}_{x|\mathscr{R}_t \cap \{t < T_0\}}^{(\nu)} = \left(\frac{R_t}{x}\right)^\nu \exp\left(-\frac{\nu^2}{2}\int_0^t \frac{ds}{R_s^2}\right) \cdot \mathbb{P}_{x|\mathscr{R}_t}^{(0)} \qquad (7.20)$$

where $(\mathscr{R}_t, t \ge 0)$ denotes the natural filtration of $(R_t, t \ge 0)$. If $\nu > 0$, $\mathscr{R}_t \cap \{t < T_0\}$ may be replaced by $\mathscr{R}_t$ in formula (7.20) since then $T_0 = \infty$ a.s., whereas if $\nu \in \, ]-1, 0[$, then, for $\mu^2 > 0$:

$$\mathbb{E}_x^{(\nu)}\left[\exp\left(-\frac{\mu^2}{2}\int_0^t \frac{ds}{R_s^2}\right)\bigg| R_t = y\right] = \mathbb{E}_x^{(\nu)}\left[\exp\left(-\frac{\mu^2}{2}\int_0^t \frac{ds}{R_s^2}\right)1_{\{T_0 > t\}}\bigg| R_t = y\right]$$

since, if $T_0 < t$, $\int_0^t \dfrac{ds}{R_s^2} = +\infty$.

We make several remarks:

**a)** For $\nu < 0$, letting $\mu \to 0$ in (7.19), we get, using (7.20):

$$\mathbb{P}_x^{(\nu)}(T_0 > t | R_t = y) = \frac{I_{-\nu}}{I_\nu}\left(\frac{xy}{t}\right). \qquad (7.21)$$

**b)** For $\nu > -\dfrac{1}{2}$, formula (7.19) becomes, taking $\mu^2 = 2\nu + 1$:

$$\mathbb{E}_x^{(\nu)}\left[\exp\left\{-\left(\nu + \frac{1}{2}\right)\int_0^t \frac{ds}{R_s^2}\right\}\bigg| R_t = y\right] = \frac{I_{\nu+1}}{I_\nu}\left(\frac{xy}{t}\right) \qquad \left(\nu > -\frac{1}{2}\right). \quad (7.22)$$

From (7.22), we deduce that, for $\nu \ge -\dfrac{1}{2}$ and $z \ge 0$:

$$I_{\nu+1}(z) \le I_\nu(z). \qquad (7.23)$$

$$\left( (7.23) \text{ is an immediate consequence of (7.22) for } \nu > -\frac{1}{2} \text{ and for } \nu = -\frac{1}{2}: \right.$$

$$\left. I_{\frac{1}{2}}(z) = \sqrt{\frac{2}{\pi z}} \sinh(z) \le I_{-\frac{1}{2}}(z) = \sqrt{\frac{2}{\pi z}} \cosh(z); \text{ see (B.6)} \right).$$

### 7.2.2.6 The Hirsch-Song Formula $\big($see [30]$\big)$

**Proposition 7.1.** *Let* $\nu > -\dfrac{1}{2}$.

i) *The following Hirsch-Song formula between transition densities holds:*

$$\frac{\partial}{\partial x} p^{(\nu)}(t,x,y) = -\frac{\partial}{\partial y} \left( \frac{x}{y} p^{(\nu+1)}(t,x,y) \right). \tag{7.24}$$

ii) *Let H denote the operator defined on the space of* $\mathscr{C}^1$ *functions, with derivative equal to 0 at 0, and which are bounded, as well as their derivative:*

$$Hf(x) = \frac{1}{x} f'(x). \tag{7.25}$$

*Let* $(P_t^{(\nu)}, t \ge 0)$ *denote the Bessel semi-group with index* $\nu$. *Then:*

$$HP_t^{(\nu)} = P_t^{(\nu+1)} H. \tag{7.26}$$

iii) *Let* $(Q_t^{(\nu)}, t \ge 0)$ *denote the semi-group of the squared Bessel process, with index* $\nu$. *Then:*

$$DQ_t^{(\nu)} = Q_t^{(\nu+1)} D \tag{7.27}$$

*where D denotes the differentiation operator:* $Df(x) = f'(x)$, *with domain the space of* $\mathscr{C}^1$ *functions, bounded as well as their derivative.*

It is easily verified that (7.26) and (7.27) are equivalent $\big($since $Q_t^{(\nu)} f(z) = P_t^{(\nu)}(\tilde{f})(\sqrt{z})$, with $\tilde{f}(z) = f(z^2)\big)$. On the other hand, denoting by $\overline{L}^{(\nu)}$ the infinitesimal generator of the semi-group $Q_t^{(\nu)}$:

$$\overline{L}^{(\nu)} f(x) = 2x f''(x) + 2(\nu+1) f'(x), \tag{7.28}$$

an easy computation allows to obtain $D\overline{L}^{(\nu)} = \overline{L}^{(\nu+1)} D$, hence (7.27), (see Subsubsection 7.2.2.7 below about the Fokker-Planck formula).

The intertwining relation (7.27) shall be generalized in Chapter 8, Remark 8.9. (See also Hirsch-Yor [31].)

We show (7.24), following [30]
For $f$ regular with compact support, one has:

$$P_t^{(\nu)} f(x) = \int_0^\infty p^{(\nu)}(t,x,y) f(y) dy,$$

hence:

$$\frac{\partial}{\partial x}P_t^{(\nu)}f(x) = \int_0^\infty \frac{\partial}{\partial x}p^{(\nu)}(t,x,y)f(y)dy. \tag{7.29}$$

On the other hand, with obvious notations:

$$\frac{\partial}{\partial x}P_t^{(\nu)}f(x) = \frac{\partial}{\partial x}\mathbb{E}^{(\nu)}\left[f(R_t^x)\right] = \mathbb{E}^{(\nu)}\left[f'(R_t^x)\frac{\partial R_t^x}{\partial x}\right].$$

However, since:

$$R_t^x = x + B_t + \frac{2\nu+1}{2}\int_0^t \frac{ds}{R_s^x} \qquad \left(\nu > -\frac{1}{2}\right),$$

we obtain, by differentiation with respect to $x$:

$$\frac{\partial R_t^x}{\partial x} = 1 - \frac{2\nu+1}{2}\int_0^t \frac{ds}{(R_s^x)^2}\frac{\partial R_s^x}{\partial x} \tag{7.30}$$

(see also Vostrikova [88]).

The linear equation (7.30) may then be integrated, and we obtain:

$$\frac{\partial R_t^x}{\partial x} = \exp\left(-\frac{2\nu+1}{2}\int_0^t \frac{ds}{(R_s^x)^2}\right). \tag{7.31}$$

Thus:

$$\int_0^\infty \frac{\partial}{\partial x}p^{(\nu)}(t,x,y)f(y)dy = \mathbb{E}_x^{(\nu)}\left[f'(R_t)\exp\left(-\frac{2\nu+1}{2}\int_0^t \frac{ds}{R_s^2}\right)\right]$$

$$= \int_0^\infty \mathbb{E}_x^{(\nu)}\left[\exp\left(-\frac{2\nu+1}{2}\int_0^t \frac{ds}{R_s^2}\right)\Big| R_t = y\right]f'(y)p^{(\nu)}(t,x,y)dy$$

$$= \int_0^\infty p^{(\nu)}(t,x,y)f'(y)\frac{I_{\nu+1}}{I_\nu}\left(\frac{xy}{t}\right)dy \qquad \text{(from (7.22))}$$

$$= -\int_0^\infty f(y)\frac{\partial}{\partial y}\left(\frac{I_{\nu+1}}{I_\nu}\left(\frac{xy}{t}\right)p^{(\nu)}(t,x,y)\right)dy \qquad \text{(after an integration by parts)}$$

$$= -\int_0^\infty f(y)\frac{\partial}{\partial y}\left(\frac{x}{y}p^{(\nu+1)}(t,x,y)\right)dy \qquad \text{(from (7.10))}. \tag{7.32}$$

The comparison of (7.32) and (7.29) then implies Point $(i)$ of Proposition 7.1.

### 7.2.2.7 The Fokker-Planck Formula

The infinitesimal generator $L^{(\nu)}$ of the Bessel semi-group $(P_t^{(\nu)}, t \geq 0)$ with index $\nu$ is given by:

$$L^{(\nu)}f(x) = \frac{1}{2}f''(x) + \frac{2\nu+1}{2x}f'(x). \tag{7.33}$$

Its domain is the space of functions $f$ such that $L^{(\nu)}f$ are bounded continuous and satisfy: $\lim_{x \to 0} x^{2\nu+1}f'(x) = 0$. One has:

$$\frac{\partial}{\partial t} p^{(\nu)}(t,x,y) = L^{(\nu)}\big(p^{(\nu)}(t,\cdot,y)\big)(x)$$
$$= L^{(\nu)*}\big(p^{(\nu)}(t,x,\cdot)\big)(y) \qquad \text{(Fokker-Planck)} \qquad (7.34)$$

where the operator $L^{(\nu)*}$, the adjoint of $L^{(\nu)}$, is defined by:

$$L^{(\nu)*} f(x) = \frac{1}{2} f''(x) - \frac{\partial}{\partial x}\left(\frac{2\nu+1}{2x} f(x)\right). \qquad (7.35)$$

The infinitesimal generator $\overline{L}^{(\nu)}$ of the semi-group $(Q_t^{(\nu)}, t \geq 0)$ of the squared Bessel process with index $\nu$, is given by:

$$\overline{L}^{(\nu)} f(x) = 2x f''(x) + 2(\nu+1) f'(x) \qquad (7.36)$$

and its adjoint $\overline{L}^{(\nu)*}$ is defined by:

$$\overline{L}^{(\nu)*} f(x) = \frac{\partial^2}{\partial x^2}\big(2x f(x)\big) - 2(\nu+1) f'(x). \qquad (7.37)$$

### 7.2.2.8 Sojourn Time Below Level $y$ $\big($see [65], also [52], Th. 11.6, p.180$\big)$

Let:

$$A_y^- = \int_0^\infty 1_{\{R_t \leq y\}} dt \qquad \left(= \int_0^{G_y} 1_{\{R_t \leq y\}} dt\right).$$

Then:

$$\mathbb{E}_y^{(\nu)}\left[e^{-\lambda A_y^-}\right] = \frac{2\nu}{y\sqrt{2\lambda}} \frac{I_\nu(y\sqrt{2\lambda})}{I_{\nu-1}(y\sqrt{2\lambda})} \qquad (\nu > 0, \, \lambda > 0). \qquad (7.38)$$

## 7.2.3 Proof of Theorem 7.1

### 7.2.3.1 A Useful Lemma

**Lemma 7.1.** *Let $I_\nu$ denote the modified Bessel function with index $\nu$.*

*i) If $\nu \geq -\dfrac{1}{2}$, then, for every $z \geq 0$,*

$$I_\nu(z) \geq I_{\nu+1}(z). \qquad (7.39)$$

*ii) If $\nu = -1$, then, for every $z \geq 0$,*

$$I_1(z) = I_{-1}(z) \leq I_0(z). \qquad (7.40)$$

*iii) If $\nu \in\, ]-1, -\frac{1}{2}[$,*

   a) *for z small enough, $I_\nu(z) > I_{\nu+1}(z)$,*
   b) *whereas for z large enough, $I_\nu(z) < I_{\nu+1}(z)$.*

*Proof.* Point $(i)$ has already been proven: it is relation (7.23).
Then, Point $(ii)$ follows from the Point $(i)$ since $I_{-1} = I_1$.

We prove Point $(iii)$
In the neighborhood of 0, one has:

$$\frac{I_\nu(z)}{I_{\nu+1}(z)} \underset{z\to 0}{\sim} \frac{(z/2)^\nu}{\Gamma(\nu+1)} \frac{\Gamma(\nu+2)}{(z/2)^{\nu+1}} = \frac{2}{z}(\nu+1) \underset{z\to 0}{\longrightarrow} +\infty \qquad (\nu > -1) \qquad (7.41)$$

In the neighborhood of $+\infty$, one has $\big($see [46], p. 122 and 123$\big)$:

$$I_\mu(z) \underset{z\to\infty}{=} \frac{e^z}{\sqrt{2\pi z}}\left(1 - \frac{4\mu^2-1}{4}\frac{1}{2z} + o\left(\frac{1}{z^2}\right)\right).$$

Thus, with $\mu = \nu$ and $\nu \in\, ]-1, -\frac{1}{2}[$:

$$I_\nu(z) \underset{z\to\infty}{=} \frac{e^z}{\sqrt{2\pi z}}\left(1 - \frac{4\nu^2-1}{4}\frac{1}{2z} + o\left(\frac{1}{z^2}\right)\right),$$

$$I_{\nu+1}(z) \underset{z\to\infty}{=} \frac{e^z}{\sqrt{2\pi z}}\left(1 - \frac{4(\nu+1)^2-1}{4}\frac{1}{2z} + o\left(\frac{1}{z^2}\right)\right).$$

Now, for $\nu \in\, ]-1, -\frac{1}{2}[$, one has $4\nu^2 - 1 \geq 4(\nu+1)^2 - 1$; indeed, this is equivalent to $\nu^2 \geq 1 + \nu^2 + 2\nu$, i.e. $\nu \leq -\dfrac{1}{2}$.

   Hence, the ratio $\dfrac{I_\nu(z)}{I_{\nu+1}(z)} \underset{z\to\infty}{\longrightarrow} 1^-$ as $z \to +\infty$. As a conclusion, from this point and (7.41), for $\nu \in\, ]-1, -\frac{1}{2}[$:

- for z large enough, $\dfrac{I_\nu(z)}{I_{\nu+1}(z)} < 1$,

- for z small enough, $\dfrac{I_\nu(z)}{I_{\nu+1}(z)} > 1$.

$\square$

### 7.2.3.2 The Case $x = 0$

When $x = 0$ and $\nu > -1$, we shall show the existence of an increasing process $(J_t, t \geq 0)$ such that $J_t \underset{t\to\infty}{\longrightarrow} +\infty$ a.s. and

$$\mathbb{P}_0^{(\nu)}(R_t \geq y) = \mathbb{P}(J_t \geq y). \qquad (7.42)$$

This formula obviously implies the existence of a process which is a pseudo-inverse of $(R_t, t \geq 0)$. Let us prove (7.42), which hinges in fact simply on the scaling property of $(R_t, t \geq 0)$. Indeed, one has:

$$\mathbb{P}_0^{(\nu)}(R_t \geq y) = \mathbb{P}_0^{(\nu)}\left(\sqrt{t} \geq \frac{y}{R_1}\right) \qquad \text{(by scaling)}$$

$$= \mathbb{P}_0^{(\nu)}\left(t \geq \frac{y^2}{R_1^2}\right)$$

$$= \mathbb{P}\left(t \geq \frac{y^2}{2\gamma_{\nu+1}}\right)$$

$\left(\text{since } R_1^2 \overset{\text{(law)}}{=} 2\gamma_{\nu+1} \text{ under } \mathbb{P}_0^{(\nu)}, \text{ where } \gamma_{\nu+1} \text{ is a gamma r.v. with parameter } \nu+1\right)$

$$= \mathbb{P}_0^{(\nu+1)}(G_y \leq t) \qquad \text{(from (7.14))}$$

$$= \mathbb{P}_0^{(\nu+1)}\left(\inf_{u \geq t} R_u \geq y\right) = \mathbb{P}_0^{(\nu+1)}(J_t \geq y),$$

with $J_t := \inf_{u \geq t} R_u$. Clearly $(J_t, t \geq 0)$ is an increasing process and $J_t \xrightarrow[t \to \infty]{} +\infty \ \mathbb{P}_0^{(\nu+1)}$ a.s. since $\nu + 1 > 0$.

*Remark 7.2.* In a general set-up, it was proven by Kamae-Krengel [39], that to a real-valued process $(X_t, t \geq 0)$ which admits an increasing pseudo-inverse, one can always associate an increasing process $(A_t, t \geq 0)$ such that, for every $t \geq 0$, $X_t \overset{\text{(law)}}{=} A_t$.

### 7.2.3.3 We now Prove Point $(i)$ of Theorem 7.1

From the comparison theorem, (applied to Bessel processes) for $\nu \geq -\frac{1}{2}$, one has:

$$\mathbb{P}_x^{(\nu)}(R_t \geq y) \geq \mathbb{P}_0^{(-\frac{1}{2})}(R_t \geq y)$$

$$= \mathbb{P}_0(|B_t| \geq y) = \mathbb{P}_0\left(|B_1| > \frac{y}{\sqrt{t}}\right) \xrightarrow[t \to \infty]{} 1$$

where $(B_t, t \geq 0)$ is a Brownian motion started from 0.

It suffices then, in order to prove Point $(i)$ of Theorem 7.1 to show that: $t \longmapsto \mathbb{P}_x^{(\nu)}(R_t \geq y)$ is an increasing function of $t$. Now, one has:

$$\frac{\partial}{\partial t}\mathbb{P}_x^{(\nu)}(R_t \geq y) = \int_y^\infty \frac{\partial}{\partial t}p^{(\nu)}(t,x,z)dz$$

$$= \int_y^\infty \left[\frac{1}{2}\frac{\partial^2}{\partial z^2}p^{(\nu)}(t,x,z) - \frac{\partial}{\partial z}\left(\frac{2\nu+1}{2z}p^{(\nu)}(t,x,z)\right)\right]dz$$

$$\text{(from Fokker-Planck, (7.34))}$$

$$= -\frac{1}{2}\frac{\partial}{\partial y}p^{(\nu)}(t,x,y) + \frac{2\nu+1}{2y}p^{(\nu)}(t,x,y).$$

Now, from (7.10):

$$\frac{\partial}{\partial y} p^{(\nu)}(t,x,y) = p^{(\nu)}(t,x,y) \left[ \frac{\nu+1}{y} - \frac{y}{t} + \frac{x}{t} \frac{I'_\nu(\frac{xy}{t})}{I_\nu(\frac{xy}{t})} \right].$$

Hence:

$$\frac{\partial}{\partial t} \mathbb{P}_x^{(\nu)}(R_t \geq y) = p^{(\nu)}(t,x,y) \left[ \frac{\nu}{2y} + \frac{y}{2t} - \frac{x}{2t} \frac{I'_\nu(\frac{xy}{t})}{I_\nu(\frac{xy}{t})} \right]$$

$$= p^{(\nu)}(t,x,y) \left[ \frac{\nu}{2y} + \frac{y}{2t} - \frac{x}{2t} \left( \frac{I_{\nu+1}}{I_\nu} \left( \frac{xy}{t} \right) + \frac{\nu t}{xy} \right) \right]$$

$$\left( \text{since } \frac{I'_\nu(z)}{I_\nu(z)} = \frac{I_{\nu+1}}{I_\nu}(z) + \frac{\nu}{z} \quad \text{(see (B.5))} \right)$$

$$= \frac{p^{(\nu)}(t,x,y)}{2t \, I_\nu(\frac{xy}{t})} \left[ y I_\nu \left( \frac{xy}{t} \right) - x I_{\nu+1} \left( \frac{xy}{t} \right) \right]. \tag{7.43}$$

It now follows from Lemma 7.1 that $\frac{\partial}{\partial t} \mathbb{P}_x^{(\nu)}(R_t \geq y)$ is positive, since $y > x$ and $I_\nu(z) \geq I_{\nu+1}(z)$.

### 7.2.3.4  We now Prove Point $(ii)$ of Theorem 7.1

This follows immediately from (7.43) and from point $(iii)$ of Lemma 7.1: for $x \neq 0$, $\frac{\partial}{\partial t} \mathbb{P}_x^{(\nu)}(R_t \geq y)$ does not have a fixed sign, hence $t \longmapsto \mathbb{P}_x^{(\nu)}(R_t \geq y)$ cannot be a monotone function.

### 7.2.3.5  We now Prove Point $(iii)$ of Theorem 7.1

We need to show that, for every $x > 0$ and $0 < y < x$, $\mathbb{P}_x^{(-1)}(R_t \leq y)$ is an increasing function of $t$. We may of course replace $R_t$ by $R_t^2$. Now, the law of $R_t^2$, when $R_0^2 = x$, is given by:

$$\mathbb{P}(R_t^2 \in dy) = \exp\left( -\frac{x}{2t} \right) \delta_0(dy) + q(t,x,y)dy$$

with

$$q(t,x,y) := \frac{1}{2t} \sqrt{\frac{x}{y}} \exp\left( -\frac{x+y}{2t} \right) I_1 \left( \frac{xy}{t} \right). \tag{7.44}$$

Then, we need to show that, for every $y < x$, one has:

$$\frac{\partial}{\partial t} \left( \exp -\frac{x}{2t} \right) + \frac{\partial}{\partial t} \int_0^y q(t,x,z)dz \geq 0.$$

But, since $\dfrac{\partial}{\partial t}\left(\exp-\dfrac{x}{2t}\right) \geq 0$, it suffices to show that:

$$\frac{\partial}{\partial t}\int_0^y q(t,x,z)dz \geq 0.$$

Now, one has:

$$\frac{\partial}{\partial t}\int_0^y q(t,x,z)dz = \int_0^y \frac{\partial}{\partial t}q(t,x,z)dz$$

$$= \int_0^y \frac{\partial^2}{\partial z^2}\left(2zq(t,x,z)\right)dz \quad \text{(from Fokker-Planck, with } \nu = -1)$$

$$= \frac{\partial}{\partial y}\left(2yq(t,x,y)\right)$$

$$= 2q(t,x,y) + 2yq(t,x,y)\left[-\frac{1}{2y} - \frac{1}{2t} + \frac{1}{2t}\sqrt{\frac{x}{y}}\frac{I_1'}{I_1}\left(\frac{\sqrt{xy}}{t}\right)\right]$$

$$= q(t,x,y)\left[1 - \frac{y}{t} + \frac{\sqrt{xy}}{t}\frac{I_0}{I_1}\left(\frac{\sqrt{xy}}{t}\right) - \frac{t}{\sqrt{xy}}\right]$$

$$\left(\text{since } \frac{I_1'}{I_1}(z) = \frac{I_0}{I_1}(z) - \frac{1}{z}\right)$$

$$= \frac{\sqrt{y}}{t}q(t,x,y)\left[\sqrt{x}\frac{I_0}{I_1}\left(\frac{\sqrt{xy}}{t}\right) - \sqrt{y}\right] \geq 0$$

from Point $(ii)$ of Lemma 7.1 and since $y < x$. This ends the proof of Theorem 7.1.

$\square$

*Remark 7.3.*
**a)** The property: $t \to \mathbb{P}_x^{(\nu)}(R_t \geq y)$ is an increasing function of $t$ (for $y > x$) is equivalent to: for every Borel increasing function $\varphi$ with support in $]x, +\infty[$, $t \to \mathbb{E}_x^{(\nu)}\left[\varphi(R_t)\right]$ is an increasing function of $t$. Indeed, we have, for such a $\varphi$:

$$\mathbb{E}_x^{(\nu)}\left[\varphi(R_t)\right] = \mathbb{E}_x^{(\nu)}\left[\int_x^{R_t}d\varphi(y)\right] = \mathbb{E}_x^{(\nu)}\left[\int_x^\infty d\varphi(y)1_{\{R_t \geq y\}}\right]$$

$$= \int_x^\infty d\varphi(y)\mathbb{P}_x^{(\nu)}(R_t \geq y).$$

**b)** Formula (7.43)

$$\frac{\partial}{\partial t}\mathbb{P}_x^{(\nu)}(R_t \geq y) = \frac{p^{(\nu)}(t,x,y)}{2t I_\nu(\frac{xy}{t})}\left[yI_\nu\left(\frac{xy}{t}\right) - xI_{\nu+1}\left(\frac{xy}{t}\right)\right]$$

may still be written, with the help of (7.10):

$$\frac{\partial}{\partial t}\mathbb{P}(Y_{x,y}^{(\nu)} \leq t) = \frac{\partial}{\partial t}\mathbb{P}_x^{(\nu)}(R_t \geq y) = \frac{y}{2t}p^{(\nu)}(t,x,y) - \frac{x^2}{2ty}p^{(\nu+1)}(t,x,y). \quad (7.45)$$

We denote by $\theta^{(\nu)}(t,x,y)$ the RHS of (7.45). We now check directly that this positive function of $t$ integrates to 1 i.e., it is a density of probability:

$$\int_0^\infty \theta^{(\nu)}(t,x,y)dt = 1. \tag{7.46}$$

For this purpose, we use the classical Lipschitz-Hankel formula $\big($see, e.g. [90], p. 384$\big)$, for $\nu > 0$:

$$\nu \int_0^\infty \frac{du}{u} \exp(-au)I_\nu(u) = \frac{1}{(a+\sqrt{a^2-1})^\nu} \qquad (a > 1). \tag{7.47}$$

Differentiating both sides with respect to $a$, we obtain:

$$\int_0^\infty \exp(-au)I_\nu(u)du = \frac{1}{(a+\sqrt{a^2-1})^\nu}\frac{1}{\sqrt{a^2-1}} \qquad (a > 1,\ \nu > 0). \tag{7.48}$$

In order to check (7.46), we first prove that:

$$\Lambda_{x,y}^{(\nu)} := \int_0^\infty \frac{dt}{t}p^{(\nu)}(t,x,y) = \frac{2y}{y^2-x^2} \qquad (y > x). \tag{7.49}$$

Note that (7.49) (which we assume for a moment) shows that $\Lambda_{x,y}^{(\nu)}$ does not depend on $\nu$ ! We deduce from (7.49) that:

$$\int_0^\infty \left[\frac{y}{2t}p^{(\nu)}(t,x,y) - \frac{x^2}{2ty}p^{(\nu+1)}(t,x,y)\right]dt = \frac{y}{2}\Lambda_{x,y}^{(\nu)} - \frac{x^2}{2y}\Lambda_{x,y}^{(\nu+1)}$$
$$= \left(\frac{y}{2} - \frac{x^2}{2y}\right)\frac{2y}{y^2-x^2} = 1$$

which is the desired result (7.46).

**c)** We now prove (7.49). From (7.10), we get:

$$\Lambda_{x,y}^{(\nu)} = \left(\frac{y}{x}\right)^\nu y \int_0^\infty \frac{dt}{t^2} \exp\left(-\frac{x^2+y^2}{2t}\right)I_\nu\left(\frac{xy}{t}\right)$$
$$= \left(\frac{y}{x}\right)^\nu y \int_0^\infty du \exp\left(-\frac{x^2+y^2}{2}u\right)I_\nu(xyu)$$

(after the change of variable $\frac{1}{t} = u$)

$$= \left(\frac{y}{x}\right)^\nu \frac{y}{xy} \int_0^\infty dz\exp\left(-\frac{x^2+y^2}{2xy}z\right)I_\nu(z)$$

(after the change of variable $xyu = z$)

$$= \left(\frac{y}{x}\right)^\nu \frac{1}{x}\frac{1}{(a+\sqrt{a^2-1})^\nu}\frac{1}{\sqrt{a^2-1}}$$

with the help of (7.48), where $a := \dfrac{x^2 + y^2}{2xy} > 1$ since $(x < y)$. On the other hand, there are the elementary identities:

$$\sqrt{a^2 - 1} = \frac{y^2 - x^2}{2xy} \quad \text{and} \quad a + \sqrt{a^2 - 1} = \frac{y}{x},$$

$$\text{which imply: } \Lambda_{x,y}^{(\nu)} = \frac{1}{x} \frac{1}{\frac{y^2 - x^2}{2xy}} = \frac{2y}{y^2 - x^2}$$

which is the desired result (7.49). We shall come back to this computation, in Section 7.3, in order to establish certain properties of the r.v.'s $Y_{x,y}^{(\nu)}$.

### 7.2.4 Interpretation in Terms of the Local Martingales $(R_t^{-2\nu}, t \geq 0)$

We assume here $\nu > 0$. We know that $(R_t, t \geq 0)$ solves the S.D.E.:

$$R_t = r + B_t + \frac{2\nu + 1}{2} \int_0^t \frac{ds}{R_s} \tag{7.50}$$

where $(B_t, t \geq 0)$ is a Brownian motion starting from 0. Let $M_t^{(\nu)} := R_t^{-2\nu}$ and denote $M_0^{(\nu)} = x \, (= r^{-2\nu})$. Itô's formula yields:

$$M_t^{(\nu)} = x - 2\nu \int_0^t (M_s^{(\nu)})^{\frac{2\nu + 1}{2\nu}} dB_s. \tag{7.51}$$

We denote by $\mathbb{Q}_x^{(\nu)}$ the law of the solution of (7.51). Thus $(M_t^{(\nu)}, t \geq 0; \mathbb{Q}_x^{(\nu)}, x \geq 0)$ is the family of laws of a Markov process and $(M_t^{(\nu)}, t \geq 0)$ is a positive local martingale, hence, a supermartingale, which converges a.s. to 0 as $t \to \infty$. Gathering the results of Theorem 7.1, Point $(b)$ of Remark 7.1 and Remark 6.2, we have obtained:

**Theorem 7.2.**

i)   *For every $\nu > 0$ and $x > 0$, there exists a probability measure $\gamma_{M^{(\nu)}}$ on $[0, x] \times [0, +\infty[$ such that:*

$$\frac{1}{x} \mathbb{Q}_x^{(\nu)} \left[ (K - M_t^{(\nu)})^+ \right] = \gamma_{M^{(\nu)}} \left( [0, K] \times [0, t] \right) \qquad (K \leq x, \, t \geq 0). \tag{7.52}$$

ii)  *There exists a family of positive r.v.'s $(\overline{Y}_{x,y}^{(\nu)}, y < x)$ such that:*

$$\mathbb{Q}_x^{(\nu)}(M_t^{(\nu)} \leq y) = \mathbb{P}(\overline{Y}_{x,y}^{(\nu)} \leq t) \qquad (y < x, \, t \geq 0). \tag{7.53}$$

*Remark 7.4.*
**a)** Since:

$$Q_x^{(\nu)}(M_t^{(\nu)} \leq y) = \mathbb{P}_{x^{-\frac{1}{2\nu}}}^{(\nu)}\left(\frac{1}{R_t^{2\nu}} \leq y\right) = \mathbb{P}_{x^{-\frac{1}{2\nu}}}^{(\nu)}(R_t \geq y^{-\frac{1}{2\nu}}) \tag{7.54}$$

we have:

$$\overline{Y}_{x,y}^{(\nu)} \overset{\text{(law)}}{=} Y_{x^{-\frac{1}{2\nu}}, y^{-\frac{1}{2\nu}}}^{(\nu)} \qquad (x > y). \tag{7.55}$$

In Section 7.3, we shall study in details the laws of the r.v.'s $(Y_{x,y}^{(\nu)}, y > x)$ which, thanks to (7.55), allows to obtain easily the corresponding properties of the r.v.'s $(\overline{Y}_{x,y}^{(\nu)}, y < x)$.

**b)** When $\nu \in ]-1, 0[, (R_t^{-2\nu}, t \geq 0)$ is no longer a local martingale, but is a sub-martingale $\big($see [19]$\big)$ such that:

$$R_t^{-2\nu} = N_t^{(\nu)} + L_t^{(\nu)} \tag{7.56}$$

where $(N_t^{(\nu)}, t \geq 0)$ is a martingale and $(L_t^{(\nu)}, t \geq 0)$ an increasing process such that $dL_t^{(\nu)} = 1_{\{R_t = 0\}} dL_t^{(\nu)}$.

## 7.3  Some Properties of the r.v.'s $(Y_{x,y}^{(\nu)}, y > x)\left(\nu \geq -\frac{1}{2}\right)$

We recall that these r.v.'s $Y_{x,y}^{(\nu)}$ (or, more exactly, their laws) are defined (see Theorem 7.1) via:

$$\mathbb{P}_x^{(\nu)}(R_t \geq y) = \mathbb{P}(Y_{x,y}^{(\nu)} \leq t). \tag{7.57}$$

### 7.3.1  The Main Theorem

**Theorem 7.3.** *Here are a few properties (of the laws) of* $Y_{x,y}^{(\nu)}\left(\nu \geq -\frac{1}{2}, y > x\right)$:

i) *Laplace transform of* $Y_{x,y}^{(\nu)}$

$$\mathbb{E}\left[e^{-\lambda Y_{x,y}^{(\nu)}}\right] = \frac{I_\nu(x\sqrt{2\lambda})}{(x\sqrt{2\lambda})^\nu}(y\sqrt{2\lambda})^{\nu+1} K_{\nu+1}(y\sqrt{2\lambda}) \quad (x > 0, \ \lambda \geq 0)$$

$$= (y\sqrt{2\lambda})\left(\frac{y}{x}\right)^\nu I_\nu(x\sqrt{2\lambda}) K_{\nu+1}(y\sqrt{2\lambda}). \tag{7.58}$$

*For x = 0, we get:*

$$\mathbb{E}\left[e^{-\lambda Y_{0,y}^{(\nu)}}\right] = \frac{1}{2^{\nu}\Gamma(\nu+1)}(y\sqrt{2\lambda})^{\nu+1}K_{\nu+1}(y\sqrt{2\lambda}).$$ (7.59)

*ii) Scaling property*

The law of $\dfrac{Y_{x,y}^{(\nu)}}{xy}$ depends only on the ratio $\dfrac{y}{x}(>1)$. In particular:

$$Y_{x,y}^{(\nu)} = x^2 Y_{1,\frac{y}{x}}^{(\nu)} \qquad (y > x)$$ (7.60)

*and*

$$\mathbb{E}\left[e^{-\frac{\lambda Y_{x,y}^{(\nu)}}{xy}}\right] = \sqrt{2\lambda}\,b^{\nu+\frac{1}{2}}I_{\nu}\left(\sqrt{\frac{2\lambda}{b}}\right)K_{\nu+1}(\sqrt{2\lambda b})$$ (7.61)

*where* $b = \dfrac{y}{x} > 1$.

*iii) Further results about the laws of $Y_{x,y}^{(\nu)}$*

- $Y_{0,y}^{(\nu)} \overset{\text{(law)}}{=} \dfrac{y^2}{2\gamma_{\nu+1}} \overset{\text{(law)}}{=} G_y^{(\nu+1)}$ (7.62)

*where* $G_y^{(\nu+1)}$ *denotes the r.v.* $G_y$ *under* $\mathbb{P}_0^{(\nu+1)}$.
　*The following relations hold:*

- $\mathbb{P}(Y_{x,y}^{(\nu)} \leq t) = \mathbb{P}\left(\gamma_{\nu+1} \geq \dfrac{y^2}{2t}\right) + \displaystyle\int_0^x \dfrac{z}{y}p^{(\nu+1)}(t,z,y)dz$ (7.63)

- $\dfrac{\partial}{\partial x}\mathbb{P}(Y_{x,y}^{(\nu)} \leq t) = \dfrac{x}{y}p^{(\nu+1)}(t,x,y) = \dfrac{x}{\nu+1}f_{G_{x,y}^{(\nu+1)}}(t)$ (7.64)

*where* $f_{G_{x,y}^{(\nu+1)}}$ *denotes the density of the r.v.* $G_y$ *under* $\mathbb{P}_x^{(\nu+1)}$, $(x < y)$.
- The density $f_{Y_{x,y}^{(\nu)}}$ of $Y_{x,y}^{(\nu)}$ is given by:

$$f_{Y_{x,y}^{(\nu)}}(t) = \dfrac{1}{2t}\left[yp^{(\nu)}(t,x,y) - \dfrac{x^2}{y}p^{(\nu+1)}(t,x,y)\right].$$ (7.65)

- $\mathbb{E}\left[Y_{x,y}^{(\nu)}\right] = \dfrac{y^2}{2\nu} - \dfrac{x^2}{2(\nu+1)} \qquad (\nu > 0).$ (7.66)

*iv) An equation satisfied by $Y_{x,y}^{(\nu)}$*

$$Y_{x,z}^{(\nu)} \overset{\text{(law)}}{=} T_{x,y}^{(\nu)} + Y_{y,z}^{(\nu)} \qquad (x < y < z).$$ (7.67)

*In particular:*

$$Y_{0,z}^{(\nu)} \stackrel{\text{(law)}}{=} T_{0,y}^{(\nu)} + Y_{y,z}^{(\nu)} \qquad (0 < y < z). \tag{7.68}$$

*The r.v.'s which occur on the RHS of (7.67) and (7.68) are independent and $T_{x,y}^{(\nu)}$ is the first hitting time of level $y$ by the process $(R_t, t \geq 0)$ starting from $x$.*

v) $$T_{0,x}^{(\nu)} + Y_{x,y}^{(\nu)} \stackrel{\text{(law)}}{=} G_y^{(\nu+1)} \qquad (0 < x < y). \tag{7.69}$$

*The r.v.'s which occur on the LHS of (7.69) are independent $\big($see (7.12), (7.13) and (7.14) for the definition of $G_y^{(\nu+1)}\big)$.*

*Proof.*

*i)* We compute the Laplace transform of $Y_{x,y}^{(\nu)}$

From (7.57), we deduce:

$$\int_0^\infty e^{-\lambda t} \mathbb{P}_x^{(\nu)}(R_t \geq y)dt = \int_0^\infty dt \int_y^\infty e^{-\lambda t} p^{(\nu)}(t,x,z)dz$$

$$= \int_y^\infty dz \int_0^\infty e^{-\lambda t} p^{(\nu)}(t,x,z)dt$$

$$= \int_y^\infty u_\lambda^{(\nu)}(x,z)dz \qquad \text{(see Subsubsection 7.2.2.4)}$$

$$= \mathbb{E}\left[\int_0^\infty e^{-\lambda t} 1_{\{t \geq Y_{x,y}^{(\nu)}\}} dt\right] = \frac{1}{\lambda} \mathbb{E}\left[e^{-\lambda Y_{x,y}^{(\nu)}}\right].$$

Hence:

$$\mathbb{E}\left[e^{-\lambda Y_{x,y}^{(\nu)}}\right] = \lambda \int_y^\infty u_\lambda^{(\nu)}(x,z)dz \tag{7.70}$$

$$= \frac{2\lambda}{x^\nu} I_\nu(x\sqrt{2\lambda})(2\lambda)^{-\frac{\nu}{2}-1} \int_{y\sqrt{2\lambda}}^\infty K_\nu(h)h^{\nu+1}dh \qquad \big(\text{from (7.18)}\big).$$

Now, since:

$$-z^{\nu+1} K_\nu(z) = \frac{\partial}{\partial z}\left(z^{\nu+1} K_{\nu+1}(z)\right) \qquad \text{(see (B.4))}$$

we obtain:

$$\mathbb{E}\left[e^{-\lambda Y_{x,y}^{(\nu)}}\right] = (2\lambda)^{-\frac{1}{2}} \frac{I_\nu(x\sqrt{2\lambda})}{x^\nu}(y\sqrt{2\lambda})^{\nu+1} K_{\nu+1}(y\sqrt{2\lambda})$$

$$= \frac{I_\nu(x\sqrt{2\lambda})}{(x\sqrt{2\lambda})^\nu}(y\sqrt{2\lambda})^{\nu+1} K_{\nu+1}(y\sqrt{2\lambda}).$$

Formula (7.59) may be obtained by letting $x$ tend to 0 in (7.58) and (7.66) follows from (7.58) by differentiation.

*ii*) Proof of the scaling property

• It is an immediate consequence of (7.58). The fact that the law of $\dfrac{Y_{x,y}^{(\nu)}}{xy}$ depends only on the ratio $\left(\dfrac{y}{x}\right)\left(\dfrac{y}{x} > 1, x > 0\right)$ may also be obtained with the help of (7.45). Indeed, we deduce from (7.45) and (7.10) that:

$$\mathbb{P}\left(\frac{Y_{x,y}^{(\nu)}}{xy} \leq t\right) = \mathbb{P}\left(\frac{xy}{Y_{x,y}^{(\nu)}} \geq \frac{1}{t}\right) = \int_0^{txy} \frac{du}{2u}\left[y\,p^{(\nu)}(u,x,y) - \frac{x^2}{y}\,p^{(\nu+1)}(u,x,y)\right]$$

or, equivalently:

$$\mathbb{P}\left(\frac{xy}{Y_{x,y}^{(\nu)}} \geq t\right) = \int_0^{\frac{xy}{t}} \frac{du}{2u}\left[y\,p^{(\nu)}(u,x,y) - \frac{x^2}{y}\,p^{(\nu+1)}(u,x,y)\right].$$

Given (7.10), this formula yields the density $f_{\frac{xy}{Y_{x,y}^{(\nu)}}}$ of the r.v. $\dfrac{xy}{Y_{x,y}^{(\nu)}}$:

$$f_{\frac{xy}{Y_{x,y}^{(\nu)}}}(u) = \frac{1}{2}\left(\frac{y}{x}\right)^{\nu+1} e^{-a(x,y)u}\left(I_\nu(u) - \frac{x}{y}I_{\nu+1}(u)\right)$$

where $a(x,y) := \dfrac{x^2 + y^2}{2xy} = \dfrac{1}{2}\left(\dfrac{x}{y} + \dfrac{y}{x}\right)$.

Thus, the law of $\dfrac{xy}{Y_{x,y}^{(\nu)}}$, hence that of $\dfrac{Y_{x,y}^{(\nu)}}{xy}$ only depends on the ratio $\left(\dfrac{y}{x}\right)$.

• We may also prove this scaling property as a direct consequence of the scaling property of the Bessel process. Indeed, we have:

$$\mathbb{P}(Y_{x,y}^{(\nu)} \leq txy) = \mathbb{P}_x^{(\nu)}(R_{txy} \geq y)$$

$$= \mathbb{P}_x^{(\nu)}\left(\frac{1}{\sqrt{xy}}R_{txy} \geq \frac{y}{\sqrt{xy}}\right)$$

$$= \mathbb{P}_{\frac{x}{\sqrt{xy}}}^{(\nu)}\left(R_t \geq \sqrt{\frac{y}{x}}\right) \qquad \text{(by scaling of the Bessel process)}$$

$$= \mathbb{P}_{\sqrt{\frac{x}{y}}}^{(\nu)}\left(R_t \geq \sqrt{\frac{y}{x}}\right).$$

*iii*) Proof of (7.62)
This formula (7.62) has been obtained during the proof of Theorem 7.1 (Subsection 7.2.3.2, in the case $x = 0$). We note that one can deduce (7.59) from (7.62). Indeed, from (7.62), since $\mathbb{P}(Y_{0,y}^{(\nu)} \leq t) = \mathbb{P}\left(\dfrac{y^2}{2\gamma_{\nu+1}} \leq t\right)$, one has:

$$\int_0^\infty e^{-\lambda t} \mathbb{P}(Y_{0,y}^{(\nu)} \leq t) dt = \int_0^\infty e^{-\lambda t} \mathbb{P}\left(\frac{y^2}{2\gamma_{\nu+1}} \leq t\right) dt, \quad \text{i.e.:}$$

$$\mathbb{E}\left[e^{-\lambda Y_{0,y}^{(\nu)}}\right] = \mathbb{E}\left[e^{-\frac{\lambda}{2}\frac{y^2}{\gamma_{\nu+1}}}\right]$$

$$= \frac{1}{\Gamma(\nu+1)} \int_0^\infty z^\nu \exp\left\{-\frac{1}{2}\left(2z + \frac{\lambda y^2}{z}\right)\right\} dz$$

$$= \frac{2^{\frac{1-\nu}{2}}}{\Gamma(\nu+1)} \lambda^{\frac{\nu+1}{2}} y^{\nu+1} K_{\nu+1}(y\sqrt{2\lambda}) \quad \text{(see (B.9))}$$

$$= \frac{1}{2^\nu \Gamma(\nu+1)} (y\sqrt{2\lambda})^{\nu+1} K_{\nu+1}(y\sqrt{2\lambda}).$$

*iv)* We now prove (7.63) and (7.64)
From the Hirsch-Song formula (7.24), one has:

$$\frac{\partial}{\partial x} p^{(\nu)}(t,x,y) = -\frac{\partial}{\partial y}\left(\frac{x}{y} p^{(\nu+1)}(t,x,y)\right).$$

Then:

$$\frac{\partial}{\partial x} \mathbb{P}(Y_{x,y}^{(\nu)} \leq t) = \frac{\partial}{\partial x} \mathbb{P}_x^{(\nu)}(R_t \geq y)$$

$$= \int_y^\infty \frac{\partial}{\partial x} p^{(\nu)}(t,x,z) dz$$

$$= -\int_y^\infty \frac{\partial}{\partial z}\left(\frac{x}{z} p^{(\nu+1)}(t,x,z)\right) dz$$

$$= \frac{x}{y} p^{(\nu+1)}(t,x,y)$$

and the relation:

$$\frac{x}{y} p^{(\nu+1)}(t,x,y) = \frac{x}{\nu+1} f_{G_{x,y}^{(\nu+1)}}(t)$$

follows from (7.13). Formula (7.63) is obtained by integration of (7.64) (with respect to $x$):

$$\mathbb{P}(Y_{x,y}^{(\nu)} \leq t) = \mathbb{P}(Y_{0,y}^{(\nu)} \leq t) + \int_0^x \frac{\partial}{\partial z} \mathbb{P}(Y_{z,y}^{(\nu)} \leq t) dz$$

$$= \mathbb{P}\left(\gamma_{\nu+1} \geq \frac{y^2}{2t}\right) + \int_0^x \frac{z}{y} p^{(\nu+1)}(t,z,y) dz$$

from (7.62) and (7.64). We note that (7.63) may also be written, from (7.64):

$$\mathbb{P}(Y_{x,y}^{(\nu)} \leq t) = \mathbb{P}\left(\gamma_{\nu+1} \geq \frac{y^2}{2t}\right) + \int_0^x \frac{z}{\nu+1} f_{G_{z,y}^{(\nu+1)}}(t) dz. \tag{7.71}$$

We also remark that the computation of $\mathbb{E}\left[Y_{x,y}^{(\nu)}\right]$, given by (7.66) when $\nu > 0$, may be obtained from (7.71). Indeed we deduce from (7.71) that:

$$\mathbb{P}(Y_{x,y}^{(\nu)} \ge t) = \mathbb{P}\left(\gamma_{\nu+1} \le \frac{y^2}{2t}\right) - \int_0^x \frac{z}{\nu+1} f_{G_{z,y}^{(\nu+1)}}(t)dz.$$

Thus, integrating this relation in $t$ from 0 to $+\infty$, for $\nu > 0$, we obtain:

$$\int_0^\infty \mathbb{P}(Y_{x,y}^{(\nu)} \ge t)dt = \int_0^\infty dt \int_0^{\frac{y^2}{2t}} \frac{1}{\Gamma(\nu+1)} e^{-z} z^\nu dz - \frac{x^2}{2(\nu+1)}$$

$$= \frac{1}{\Gamma(\nu+1)} \int_0^\infty e^{-z} z^\nu \frac{y^2}{2z} dz - \frac{x^2}{2(\nu+1)} = \frac{y^2}{2\nu} - \frac{x^2}{2(\nu+1)}.$$

$v$) Formula (7.65) is an immediate consequence of (7.45).

$vi$) We now prove that $Y_{x,y}^{(\nu)}$ satisfies equation (7.67)
Indeed, this follows from a simple application of the Markov property. Let $x < y < z$. Since the process $(R_t, t \ge 0)$ starting from $x$ needs to pass through $y$ to reach $z$, we obtain:

$$\mathbb{P}_x^{(\nu)}(R_t \ge z) = \mathbb{P}_x^{(\nu)}(T_y < \infty; R_t \ge z)$$

$$= \mathbb{P}_x^{(\nu)}\left(1_{\{T_y \le t\}} \widehat{\mathbb{P}}_y^{(\nu)}(\widehat{R}_{t-T_y} \ge z)\right)$$

where in the expression $\widehat{\mathbb{P}}_y^{(\nu)}(\widehat{R}_{t-T_y} \ge z)$ the term $T_y$ is frozen. Hence, conditioning with respect to $T_y = u$, we obtain:

$$\mathbb{P}_x^{(\nu)}(R_t \ge z) = \int_0^t \mathbb{P}_x^{(\nu)}(T_y \in du)\mathbb{P}_y^{(\nu)}(R_{t-u} \ge z), \quad \text{i.e.}:$$

$$\mathbb{P}(Y_{x,z}^{(\nu)} \le t) = \int_0^t \mathbb{P}_x^{(\nu)}(T_y \in du)\mathbb{P}(Y_{y,z}^{(\nu)} \le t - u)$$

hence:

$$Y_{x,z}^{(\nu)} \overset{\text{(law)}}{=} T_{x,y}^{(\nu)} + Y_{y,z}^{(\nu)}.$$

It is also possible to obtain (7.67) by using the Laplace transforms of $Y_{x,y}^{(\nu)}$ and $T_{x,y}^{(\nu)}$. Indeed, from (7.58):

$$\mathbb{E}\left[e^{-\lambda Y_{x,z}^{(\nu)}}\right] = \frac{I_\nu(x\sqrt{2\lambda})}{(x\sqrt{2\lambda})^\nu}(z\sqrt{2\lambda})^{\nu+1} K_{\nu+1}(z\sqrt{2\lambda})$$

whereas, from (7.58) and (7.16):

$$\mathbb{E}\left[e^{-\lambda(T_{x,y}^{(\nu)}+Y_{y,z}^{(\nu)})}\right] = \left(\frac{y}{x}\right)^\nu \frac{I_\nu(x\sqrt{2\lambda})}{I_\nu(y\sqrt{2\lambda})} \cdot \frac{I_\nu(y\sqrt{2\lambda})}{(y\sqrt{2\lambda})^\nu}(z\sqrt{2\lambda})^{\nu+1} K_{\nu+1}(z\sqrt{2\lambda})$$

$$= \frac{I_\nu(x\sqrt{2\lambda})}{(x\sqrt{2\lambda})^\nu}(z\sqrt{2\lambda})^{\nu+1} K_{\nu+1}(z\sqrt{2\lambda}) = \mathbb{E}\left[e^{-\lambda Y_{x,z}^{(\nu)}}\right].$$

*vii*) The proof of (7.69) hinges on the same arguments as previously and, from (7.14), on:

$$\mathbb{E}\left[e^{-\lambda G_y^{(\nu+1)}}\right] = \frac{1}{2^\nu \Gamma(\nu+1)}(y\sqrt{2\lambda})^{\nu+1}K_{\nu+1}(y\sqrt{2\lambda}).$$

This ends the proof of Theorem 7.3.

        □

## 7.3.2  Some Further Relations

**1)** We note that the r.v.'s $(Y_{x,y}^{(\nu)}, y > x)$ are not the only ones which satisfy equation (7.67). Indeed, let, for $x < y$:

$$A_{x,y}^{(\nu),-} := \int_0^\infty 1_{\{R_x^{(\nu)}(s) \leq y\}}\,ds \quad \left(= \int_0^{G_y} 1_{\{R_x^{(\nu)}(s) \leq y\}}\,ds\right)$$

where here $\left(R_x^{(\nu)}(s), s \geq 0\right)$ denotes the Bessel process with index $\nu$ starting from $x$. Then, an application of the Markov property yields, for $x < y < z$:

$$A_{x,z}^{(\nu),-} \overset{(\text{law})}{=} T_{x,y}^{(\nu)} + A_{y,z}^{(\nu),-}. \tag{7.72}$$

We note that, although the r.v.'s $(A_{x,y}^{(\nu),-}, y > x)$ satisfy the same equation as the r.v.'s $(Y_{x,y}^{(\nu)}, y > x)$, they do not have the same law. Indeed, for $x = 0$, from Ciesielski-Taylor $([17])$

$$A_{0,y}^{(\nu),-} \overset{(\text{law})}{=} T_{0,y}^{(\nu-1)} \qquad (\nu > 0) \tag{7.73}$$

whereas $Y_{0,y}^{(\nu)} \overset{(\text{law})}{=} G_y^{(\nu+1)}$ (from (7.62)) and the laws of $T_{0,y}^{(\nu-1)}$ and $G_y^{(\nu+1)}$ differ; indeed $\mathbb{E}\left[e^{-\lambda T_{0,y}^{(\nu-1)}}\right] = \frac{1}{2^{\nu-1}\Gamma(\nu)}\frac{(y\sqrt{2\lambda})^{\nu-1}}{I_{\nu-1}(y\sqrt{2\lambda})}$, from (7.17), whereas, from (7.59) and (7.62),

$$\mathbb{E}\left[e^{-\lambda Y_{0,y}^{(\nu)}}\right] = \mathbb{E}_0^{(\nu+1)}\left[e^{-\lambda G_y}\right] = \frac{1}{2^\nu \Gamma(\nu+1)}(y\sqrt{2\lambda})^{\nu+1}K_{\nu+1}(y\sqrt{2\lambda}).$$

**2)** For $\nu > 0$, the equality:

$$\mathbb{E}\left[Y_{x,y}^{(\nu)}\right] = \frac{y^2}{2\nu} - \frac{x^2}{2(\nu+1)}$$

may be obtained in a different manner than the previously developed ones, by making use this time of the r.v.'s $A_{x,y}^{(\nu),-}$. Indeed, for $x < y$:

$$\mathbb{E}\left[Y_{x,y}^{(\nu)}\right] = \int_0^\infty \mathbb{P}(Y_{x,y}^{(\nu)} \geq t)dt = \mathbb{E}_x^{(\nu)}\left[\int_0^\infty 1_{\{R_t \leq y\}}dt\right] = \mathbb{E}\left[A_{x,y}^{(\nu),-}\right]. \qquad (7.74)$$

Thus, we need to compute $\mathbb{E}\left[A_{x,y}^{(\nu),-}\right]$. Itô's formula, for $\nu > 0$, implies, from (7.50):

$$(R_t \wedge y)^2 = x^2 + 2\int_0^t R_s 1_{\{R_s < y\}}\left(dB_s + \frac{2\nu+1}{2R_s}ds\right) - yL_t^y + \int_0^t 1_{\{R_s < y\}}ds \qquad (7.75)$$

where $(L_t^y, t \geq 0)$ denotes the local time of $(R_t, t \geq 0)$ at level $y$. Hence, taking expectation in (7.75) and letting $t$ tend to $+\infty$, we obtain:

$$y^2 = x^2 + (2\nu + 2)\mathbb{E}\left[A_{x,y}^{(\nu),-}\right] - y\mathbb{E}_x^{(\nu)}[L_\infty^y]. \qquad (7.76)$$

However:

$$\begin{aligned}
\mathbb{E}_x^{(\nu)}[L_\infty^y] = \mathbb{E}_y^{(\nu)}[L_\infty^y] &= \int_0^\infty p^{(\nu)}(t,y,y)dt \\
&= \int_0^\infty \frac{y}{t}\exp\left(-\frac{y^2}{t}\right)I_\nu\left(\frac{y^2}{t}\right)dt \qquad \text{(from (7.10))} \\
&= y\int_0^\infty I_\nu(z)e^{-z}\frac{dz}{z} \qquad \left(\text{after the change of variable } \frac{y^2}{t} = z\right) \\
&= \frac{y}{\nu} \qquad\qquad\qquad\qquad\qquad\qquad\qquad\qquad\qquad (7.77)
\end{aligned}$$

from the Lipschitz-Hankel formula (see (7.47)). Finally, from (7.77), (7.76) and (7.74):

$$\mathbb{E}\left[Y_{x,y}^\nu\right] = \mathbb{E}\left[A_{x,y}^{(\nu),-}\right] = \frac{1}{2\nu+2}\left(y^2\left(1 + \frac{1}{\nu}\right) - x^2\right) = \frac{y^2}{2\nu} - \frac{x^2}{2(\nu+1)}.$$

**3)** It may be of interest to express the law of the r.v. $Y_{x,y}^{(\nu)}$ in terms of the only process $(R_t, t \geq 0; \mathbb{P}_x^{(\nu)})$. Here is our result.

**Proposition 7.2.** *Let $\nu > 0$ and $h$ be a generic positive Borel function. Then:*

$$\mathbb{E}\left[h(Y_{x,y}^{(\nu)})\right] = \frac{y^2}{2\nu}\mathbb{E}_x^{(\nu)}\left[\frac{h(G_y)}{G_y}\left(1 - \frac{x}{y}\exp\left\{-\left(\nu + \frac{1}{2}\right)\int_0^{G_y}\frac{ds}{R_s^2}\right\}\right)\right] \qquad (7.78)$$

$$\begin{aligned}
&= \frac{y(y-x)}{2\nu}\mathbb{E}_x^{(\nu)}\left[\frac{h(G_y)}{G_y}\right] \\
&\quad + \frac{xy}{2\nu}\mathbb{E}_x^{(\nu)}\left[\frac{h(G_y)}{G_y}\left(1 - \exp\left\{-\left(\nu + \frac{1}{2}\right)\int_0^{G_y}\frac{ds}{R_s^2}\right\}\right)\right]. \qquad (7.79)
\end{aligned}$$

*Proof.* The density of the r.v. $Y_{x,y}^{(\nu)}$, $f_{Y_{x,y}^{(\nu)}}$, equals, from (7.43) and (7.22):

$$f_{Y_{x,y}^{(\nu)}}(t) = \frac{p^{(\nu)}(t,x,y)}{2t}\left(y - x\frac{I_{\nu+1}}{I_\nu}\left(\frac{xy}{t}\right)\right) \tag{7.80}$$

$$= \frac{y\,p^{(\nu)}(t,x,y)}{2t}\left(1 - \frac{x}{y}\mathbb{E}_x^{(\nu)}\left[\exp\left\{-\left(\nu+\frac{1}{2}\right)\int_0^t \frac{ds}{R_s^2}\right\}\middle| R_t = y\right]\right).$$

On the other hand, from (7.13), the density $f_{G_{x,y}^{(\nu)}}$ of $G_y$ under $\mathbb{P}_x^{(\nu)}$ equals:

$$f_{G_{x,y}^{(\nu)}}(t) = \frac{\nu}{y}p^{(\nu)}(t,x,y). \tag{7.81}$$

Thus, since for every positive and predictable process $H$, one has ([24]):

$$\mathbb{E}_x^{(\nu)}\left[H_{G_y}\right] = \int_0^\infty \mathbb{P}_x^{(\nu)}(G_y \in dt)\mathbb{E}_x^{(\nu)}\left[H_t|R_t = y\right], \tag{7.82}$$

we derive from (7.80) that:

$$\mathbb{E}\left[h(Y_{x,y}^{(\nu)})\right]$$

$$= \int_0^\infty \frac{h(t)}{2t}p^{(\nu)}(t,x,y)y\left(1 - \frac{x}{y}\mathbb{E}_x^{(\nu)}\left[\exp\left\{-\left(\nu+\frac{1}{2}\right)\int_0^t \frac{ds}{R_s^2}\right\}\middle| R_t = y\right]\right)dt$$

$$= \frac{y^2}{2\nu}\int_0^\infty \frac{h(t)}{t}\mathbb{P}_x^{(\nu)}(G_y \in dt)\left(1 - \frac{x}{y}\mathbb{E}_x^{(\nu)}\left[\exp\left\{-\left(\nu+\frac{1}{2}\right)\int_0^t \frac{ds}{R_s^2}\right\}\middle| R_t = y\right]\right)$$

$$= \frac{y^2}{2\nu}\mathbb{E}_x^{(\nu)}\left[\frac{h(G_y)}{G_y}\left(1 - \frac{x}{y}\exp\left\{-\left(\nu+\frac{1}{2}\right)\int_0^{G_y}\frac{ds}{R_s^2}\right\}\right)\right]$$

from (7.81). This is formula (7.78), which, after some mild rearrangement, yields formula (7.79).

$\square$

We note that both terms in (7.79):

$$\frac{y(y-x)}{2\nu}\mathbb{E}_x^{(\nu)}\left[\frac{h(G_y)}{G_y}\right] \quad \text{and} \quad \frac{xy}{2\nu}\mathbb{E}_x^{(\nu)}\left[\frac{h(G_y)}{G_y}\left(1 - \exp\left\{-\left(\nu+\frac{1}{2}\right)\int_0^{G_y}\frac{ds}{R_s^2}\right\}\right)\right]$$

are positive measures (viewed via the integration of $h$). The first one has total mass:

$$\frac{y(y-x)}{2\nu}\mathbb{E}_x^{(\nu)}\left[\frac{1}{G_y}\right] = \frac{y(y-x)}{2\nu}\int_0^\infty \frac{1}{ty}\nu p^{(\nu)}(t,x,y)dt \quad \text{(from (7.81))}$$

$$= \frac{y(y-x)}{2\nu}\frac{\nu}{y}\frac{2y}{y^2-x^2} = \frac{y}{y+x} \quad \text{(from (7.49))}$$

whereas the second has total mass $\dfrac{x}{y+x}$. In particular, for $x = 0$, we recover:

$$\mathbb{E}\left[h(Y_{0,y}^{(\nu)})\right] = \frac{y^2}{2\nu}\mathbb{E}_0^{(\nu)}\left[\frac{h(G_y)}{G_y}\right]$$

and the second term in (7.79) vanishes.

**4)** The r.v.'s $\overline{Y}_{x,y}^{(-1)}(x > y)$ In the same way that Point $(i)$ of Theorem 7.1 allows to define the r.v.'s $\left(Y_{x,y}^{(\nu)}, y > x; \nu \geq -\frac{1}{2}\right)$, Point $(iii)$ of this theorem allows to define the positive r.v.'s $(\overline{Y}_{x,y}^{(-1)}, x > y)$ characterized by:

$$\mathbb{Q}_x^{(-1)}(M_t \leq y) = \mathbb{P}(\overline{Y}_{x,y}^{(-1)} \leq t)$$

where $(M_t, t \geq 0)$ under $\mathbb{Q}_x^{(-1)}$ is a Bessel square process with index $-1$ started from $x$. With arguments similar to those used to prove Theorem 7.3, we obtain:

$$\mathbb{E}\left[e^{-\lambda \overline{Y}_{x,y}^{(-1)}}\right] = K_1(\sqrt{2\lambda x})\left[\sqrt{2\lambda x} + \frac{y}{\sqrt{x}}\sqrt{2\lambda}I_2(\sqrt{2\lambda y})\right]. \qquad (7.83)$$

**5)** Use of additivity property of squares of Bessel processes

We have shown, in Theorem 7.1, that for $\nu \geq -\frac{1}{2}$ and for $y > 1: t \to \mathbb{P}_1^{(\nu)}(R_t \geq y)$ is an increasing function of $t$. It is the distribution function of the r.v. $Y_{1,y}^{(\nu)}$ (the general case, with $x < y$, may be deduced from this one, by scaling). On the other hand, it is well known $\left(\text{see, e.g., [70], Chap. XI}\right)$ that $\mathbb{Q}_x^{(d)} * \mathbb{Q}_{x'}^{(d')} = \mathbb{Q}_{x+x'}^{(d+d')}$, where $\mathbb{Q}_x^{(d)}$ (resp. $\mathbb{Q}_{x'}^{(d')}$) denotes the law of a squared Bessel process, with dimension $d = 2\nu + 2$, starting from $x$ (resp. with dimension $d' = 2\nu' + 2$ starting from $x'$). Hence:

$$\mathbb{P}_1^{(\nu)}(R_t \geq \sqrt{y}) = \mathbb{P}_1^{(\nu)}(R_t^2 \geq y) = \mathbb{P}\left(R_0^2(t) + X(t) \geq y\right)$$

where $(R_0^2(t), t \geq 0)$ is a squared Bessel process with dimension $d = 2\nu + 2$ starting from 0 and $(X_t, t \geq 0)$ is a squared Bessel process with dimension 0 started from 1. Then:

$$\mathbb{P}_1^{(\nu)}(R_t \geq \sqrt{y}) = \mathbb{P}\left(\gamma_{\frac{d}{2}} \geq \frac{1}{2t}(y - X_t)^+\right) \qquad (7.84)$$

since $R_0^2(t) \overset{(\text{law})}{=} t R_0^2(1)$ by scaling and $R_0^2(1) \overset{(\text{law})}{=} 2\gamma_{\frac{d}{2}}$. Consequently:

$$\mathbb{P}_1^{(\nu)}(R_t \geq \sqrt{y}) = \frac{1}{\Gamma(\frac{d}{2})}\int_0^\infty \mathbb{P}\left(\frac{1}{2t}(y - X_t)^+ \leq z\right) z^{\frac{d}{2}-1}e^{-z}dz. \qquad (7.85)$$

Therefore, from (7.84), Theorem 7.1 hinges only upon the properties of the process $(X_t, t \geq 0)$, which is a squared Bessel process of dimension 0 started from 1. For $d = 2$ (i.e. $\nu = 0$), (7.85) becomes:

$$\mathbb{P}_1^{(0)}(R_t \geq \sqrt{y}) = \mathbb{E}\left[\exp\left\{-\frac{1}{2t}(y - X_t)^+\right\}\right] \qquad (7.86)$$

since $\gamma_{\frac{d}{2}} = \gamma_1$ is then a standard exponential variable. We note that, Theorem 7.1, applied for $\nu = 0 \left(> -\tfrac{1}{2}!\right)$ implies that the RHS of (7.86) is a distribution function (i.e.: that of the r.v. $Y_{1,\sqrt{y}}^{(0)}$).

For $d = 1 \left(\text{i.e. } \nu = -\tfrac{1}{2}\right)$, (7.85) becomes:

$$\mathbb{P}_1^{(-\frac{1}{2})}(R_t \geq \sqrt{y}) = \sqrt{\frac{2}{\pi}} \mathbb{E}\left[\int_{\sqrt{\frac{(y-X_t)^+}{t}}}^{\infty} e^{-\frac{u^2}{2}} du\right] \tag{7.87}$$

which is also a distribution function with respect to $t$, that of the r.v. $Y_{1,\sqrt{y}}^{(-\frac{1}{2})}$. This remark invites to ask the following (open) questions:

- Which are the probabilities $\pi$ on $\mathbb{R}^+$ such that: $\mathbb{E}\left[\pi\left(\frac{1}{2t}(y - X_t)^+\right)\right]$ is a distribution function in $t$, with $\overline{\pi}(x) = \pi(]x, +\infty[)$ ? Theorem 7.1 implies that it is the case when $\pi$ is a mixture of gamma laws, with parameter $\frac{d}{2} \geq \frac{1}{2}$ but that it is not true if $\frac{d}{2} < \frac{1}{2}$.

- More generally, which are the properties of the process $\left(Z_t = \frac{1}{2t}(y - X_t)^+, t \geq 0\right)$, with $y > 1$, which may explain the above increase property ?

**6) Around infinite divisibility properties for $Y_{x,y}^{(\nu)}$**

**Proposition 7.3.**

i) *Let $x > 0$ and $V_x$ a positive, $\tfrac{1}{2}$ stable r.v. such that $\mathbb{E}\left[e^{-\lambda V_x}\right] = \exp(-x\sqrt{2\lambda})(\lambda \geq 0)$. If $\nu > \frac{1}{2}$, then $V_x + Y_{x,y}^{(\nu)}$ (with $V_x$ and $Y_{x,y}^{(\nu)}$ independent) is infinitely divisible.*

ii) *If $\nu > \frac{1}{2}$ and $\frac{y}{x}$ is large enough, then $Y_{x,y}^{(\nu)}$ is infinitely divisible.*

*Proof.*
i) We prove (i)
By scaling we may suppose $x = 1$ and $y > 1$. From (7.58) we have:

$$\mathbb{E}\left[e^{-\lambda Y_{1,y}^{(\nu)}}\right] = y^{\nu+1}\sqrt{2\lambda}I_\nu(\sqrt{2\lambda})K_{\nu+1}(y\sqrt{2\lambda}) \tag{7.88}$$

$$:= \exp\left(-h(\lambda)\right).$$

Hence:

$$h'(\lambda) = -\frac{1}{2\lambda} - \frac{1}{\sqrt{2\lambda}}\frac{I_\nu'(\sqrt{2\lambda})}{I_\nu(\sqrt{2\lambda})} - \frac{y}{\sqrt{2\lambda}}\frac{K_{\nu+1}'(y\sqrt{2\lambda})}{K_{\nu+1}(y\sqrt{2\lambda})}$$

$$= -\frac{1}{\sqrt{2\lambda}}\frac{I_{\nu+1}}{I_\nu}(\sqrt{2\lambda}) + \frac{y}{\sqrt{2\lambda}}\frac{K_\nu}{K_{\nu+1}}(y\sqrt{2\lambda}) \tag{7.89}$$

(see Appendix B.1). On the other hand $\big($see Ismail [32]$\big)$:

$$\frac{1}{\sqrt{2\lambda}}\frac{K_\nu(\sqrt{2\lambda})}{K_{\nu+1}(\sqrt{2\lambda})} = \frac{4}{\pi^2}\int_0^\infty \frac{1}{2\lambda+t^2}\frac{dt}{t\big(J_{\nu+1}^2(t)+Y_{\nu+1}^2(t)\big)} \tag{7.90}$$

and

$$\frac{I_{\nu+1}(\sqrt{2\lambda})}{I_\nu(\sqrt{2\lambda})} = \sqrt{2\lambda}\sum_{n=1}^\infty \frac{2}{2\lambda+j_{\nu,n}^2} \tag{7.91}$$

where $J_{\nu+1},Y_{\nu+1}$ are the Bessel functions with index $\nu$ (see [46], p.98) and $(j_{\nu,n},n\geq 1)$ is the increasing sequence of the positive zeroes of $J_\nu$. Hence:

$$\frac{1}{2}h'(\lambda) = \frac{1}{\pi}\int_0^\infty \frac{y^2}{2\lambda y^2+t^2}\frac{2}{\pi t}\frac{dt}{J_{\nu+1}^2(t)+Y_{\nu+1}^2(t)} - \sum_{n\geq 1}\frac{1}{2\lambda+j_{\nu,n}^2}. \tag{7.92}$$

Let now $Z_y := V_1 + Y_{1,y}^{(\nu)}(y>1)$. We have, from (7.92):

$$\mathbb{E}\left[e^{-\lambda Z_y}\right] = \exp\big(-h(\lambda)-\sqrt{2\lambda}\big) := \exp\big(-g(\lambda)\big)$$

with:

$$\frac{1}{2}g'(\lambda) = \frac{1}{2}h'(\lambda) + \frac{1}{2\sqrt{2\lambda}} \tag{7.93}$$

$$= \left[\frac{1}{\pi}\int_0^\infty \frac{y^2}{2\lambda y^2+t^2}\frac{2}{\pi t}\frac{dt}{J_{\nu+1}^2(t)+Y_{\nu+1}^2(t)}\right]$$

$$+ \left[\frac{1}{\pi}\int_0^\infty \frac{dt}{2\lambda+t^2} - \sum_{n\geq 1}^\infty \frac{1}{2\lambda+j_{\nu,n}^2}\right] \tag{7.94}$$

$$:=(1)+(2) \qquad \left(\text{we used } \frac{1}{\sqrt{2\lambda}} = \frac{2}{\pi}\int_0^\infty \frac{dt}{2\lambda+t^2}\right).$$

One needs, to prove $(i)$, to show that $g'$ is completely monotone. For (1), it is clear since it is the Stieltjes transform of a positive measure. For (2), one has:

$$\frac{1}{\pi}\int_0^\infty \frac{dt}{2\lambda+t^2} - \sum_{n\geq 1}^\infty \frac{1}{2\lambda+j_{\nu,n}^2} = \frac{1}{\pi}\left(\sum_{n=1}^\infty \int_{j_{\nu,n-1}}^{j_{\nu,n}}\frac{dt}{2\lambda+t^2} - \pi\sum_{n=1}^\infty \frac{1}{2\lambda+j_{\nu,n}^2}\right)$$

$$= \frac{1}{\pi}\sum_{n=1}^\infty \left\{\int_{j_{\nu,n-1}}^{j_{\nu,n}}\left(\frac{1}{2\lambda+t^2}-\frac{1}{2\lambda+j_{\nu,n}^2}\right)dt + (j_{\nu,n}-j_{\nu,n-1}-\pi)\frac{1}{2\lambda+j_{\nu,n}^2}\right\}$$

$$= \frac{1}{\pi}\sum_{n=1}^\infty \left\{\int_{j_{\nu,n-1}}^{j_{\nu,n}}\frac{j_{\nu,n}^2-t^2}{(2\lambda+t^2)(2\lambda+j_{\nu,n}^2)}dt + (j_{\nu,n}-j_{\nu,n-1}-\pi)\frac{1}{2\lambda+j_{\nu,n}^2}\right\}$$

which is clearly completely monotone since $\left(\text{see [32], p. 357}\right)$, $j_{\nu,n} - j_{\nu,n-1} > \pi$ as $\nu > \dfrac{1}{2}$.

*ii*) We now prove (*ii*)

One needs to show that $h'(\lambda)$, as given by (7.92) is the Laplace transform, for $y$ large enough, of a positive measure. Now, one has:

$$\frac{1}{2}h'(\lambda) = \frac{1}{\pi}\int_0^\infty \frac{1}{2\lambda + s^2}\frac{2}{\pi s}\frac{ds}{J_{\nu+1}^2(ys) + Y_{\nu+1}^2(ys)} - \sum_{n\geq 1}\frac{1}{2\lambda + j_{\nu,n}^2} \tag{7.95}$$

$$= \frac{1}{\pi}\sum_{n=1}^\infty \int_{j_{\nu,n-1}}^{j_{\nu,n}} \left(\frac{1}{2\lambda + s^2} - \frac{1}{2\lambda + j_{\nu,n}^2}\right)\frac{2}{\pi s}\frac{ds}{J_{\nu+1}^2(ys) + Y_{\nu+1}^2(ys)}$$

$$+ \frac{1}{\pi}\sum_{n=1}^\infty \left(\int_{j_{\nu,n-1}}^{j_{\nu,n}} \frac{2}{\pi s}\frac{ds}{J_{\nu+1}^2(ys) + Y_{\nu+1}^2(ys)} - \pi\right)\frac{1}{2\lambda + j_{\nu,n}^2}. \tag{7.96}$$

But, from Watson $([90], \text{p. 449})$:

$$J_{\nu+1}^2(z) + Y_{\nu+1}^2(z) \underset{z\to\infty}{\sim} \frac{2}{\pi z}.$$

Hence:

$$\int_{j_{\nu,n-1}}^{j_{\nu,n}} \frac{2}{\pi s}\frac{ds}{J_{\nu+1}^2(ys) + Y_{\nu+1}^2(ys)} \underset{y\to+\infty}{\sim} \int_{j_{\nu,n-1}}^{j_{\nu,n}} \frac{2}{\pi s}\frac{\pi sy}{2}ds$$

$$\underset{y\to+\infty}{\sim} y(j_{\nu,n} - j_{\nu,n-1}) \qquad \text{uniformly in } n.$$

Thus, for $y$ large enough, and since $\nu > \dfrac{1}{2}$ implies $j_{\nu,n} - j_{\nu,n-1} > \pi$, one has:

$$\frac{1}{2}h'(\lambda) = \frac{1}{\pi}\sum_{n=1}^\infty \int_{j_{\nu,n-1}}^{j_{\nu,n}} \frac{j_{\nu,n}^2 - s^2}{(2\lambda + s^2)(2\lambda + j_{\nu,n}^2)}\frac{2}{\pi s}\frac{ds}{J_{\nu+1}^2(ys) + Y_{\nu+1}^2(ys)}$$

$$+ \sum_{n=1}^\infty \alpha_n(y)\frac{1}{2\lambda + j_{\nu,n}^2}$$

where the $\alpha_n(y)$ are all positive. It is then clear that $h'(\lambda)$ is completely monotone.

$\square$

**7) On the negative moments of $Y_{x,y}^{(\nu)}$ $(x < y)$**

For every $m > 0$, we have:

$$\mathbb{E}\left[\frac{1}{(Y_{x,y}^{(\nu)})^m}\right] < \infty. \tag{7.97}$$

Indeed, we have, for $m > 0$:

$$\mathbb{E}\left[\frac{1}{(Y_{x,y}^{(\nu)})^m}\right] = \frac{1}{\Gamma(m)}\int_0^\infty \mathbb{E}\left[e^{-\lambda Y_{x,y}^{(\nu)}}\right]\lambda^{m-1}d\lambda$$

and, from (7.58), $\mathbb{E}\left[e^{-\lambda Y_{x,y}^{(\nu)}}\right] \underset{\lambda\to+\infty}{\sim} \frac{1}{2}\left(\frac{y}{x}\right)^\nu \frac{1}{\sqrt{xy}}e^{-(y-x)\sqrt{2\lambda}}$   (see Appendix B.1).

## 7.4 Two Extensions of Bessel Processes with Increasing Pseudo-Inverses

### 7.4.1 Bessel Processes with Index $\nu \geq -\frac{1}{2}$ and Drift $a > 0$

S. Watanabe $\big($see, in particular [89], p.117 and 118, see also [65], especially Sections 7 and 8$\big)$ introduced the Bessel processes $\big((R_t, t \geq 0), \mathbb{P}_x^{(\nu,a)}, x \geq 0\big)$ with index $\nu$ and drift $a$, whose extended infinitesimal generator is given by:

$$\frac{1}{2}\frac{d^2}{dx^2} + \left(\frac{2\nu+1}{2x} + a\frac{I_{\nu+1}}{I_\nu}(ax)\right)\frac{d}{dx}. \tag{7.98}$$

(This is the expression (7.8) in [89], where we replaced $\alpha$ by $(2\nu+1)$ and $\sqrt{2c}$ by $a$). We recall that, for integer dimensions $d = 2(\nu+1)$, these processes may be obtained by taking $|\overrightarrow{B}_t + \overrightarrow{m}\cdot t|$, where $(\overrightarrow{B}_t)$ is a $d$-dimensional Brownian motion, starting from 0, and $a = |\overrightarrow{m}|$, for some $\overrightarrow{m} \in \mathbb{R}^d$.

Here is a first generalization of our Theorem 7.1 (recovered by letting $a \to 0$).

**Theorem 7.4.** For $\nu \geq -\frac{1}{2}$, and $a > 0$, the process $\big((R_t, t \geq 0), (\mathbb{P}_x^{(\nu,a)}, x \geq 0)\big)$ admits an increasing pseudo-inverse.

**Sketch of the proof of Theorem 7.4** (It is very similar to that of Theorem 7.1).

We need to show that, for $y > x$, $\frac{\partial}{\partial t}\mathbb{P}_x^{(\nu,a)}(R_t \geq y) \geq 0$. From Fokker-Planck formula (7.34), we get:

$$\frac{\partial}{\partial t}\mathbb{P}_x^{(\nu,a)}(R_t \geq y) = -\frac{1}{2}\frac{\partial}{\partial y}p(t,x,y) + \left(\frac{2\nu+1}{2y} + a\frac{I_{\nu+1}(ay)}{I_\nu(ay)}\right)p(t,x,y) \tag{7.99}$$

where $p(t,x,y)$, the density with respect to the Lebesgue measure of $R_t$ under $\mathbb{P}_x^{(\nu,a)}$, equals, from $\big([89], \text{p.117-118}\big)$:

$$p(t,x,y) = y\frac{I_\nu(ay)}{I_\nu(ax)}\frac{e^{-\frac{a^2 t}{2}}}{t}\exp\left(-\left(\frac{x^2+y^2}{2t}\right)\right)I_\nu\left(\frac{xy}{t}\right). \tag{7.100}$$

Hence:

$$\frac{\partial}{\partial y} p(t,x,y) = p(t,x,y) \left[ \frac{1}{y} + a \frac{I'_\nu(ay)}{I_\nu(ay)} - \frac{y}{t} + \frac{x}{t} \frac{I'_\nu\left(\frac{xy}{t}\right)}{I_\nu\left(\frac{xy}{t}\right)} \right].$$

Then, we deduce from:

$$\frac{I'_\nu(z)}{I_\nu(z)} = \frac{I_{\nu+1}(z)}{I_\nu(z)} + \frac{\nu}{z},$$

and from (7.99) that:

$$\frac{\frac{\partial}{\partial t} \mathbb{P}_x^{(\nu,a)}(R_t \ge y)}{p(t,x,y)} = -\frac{1}{2y} - \frac{a}{2} \left( \frac{I_{\nu+1}(ay)}{I_\nu(ay)} + \frac{\nu}{ay} \right) + \frac{y}{2t} - \frac{x}{2t} \left( \frac{I_{\nu+1}\left(\frac{xy}{t}\right)}{I_\nu\left(\frac{xy}{t}\right)} + \frac{\nu t}{xy} \right)$$

$$+ \frac{2\nu+1}{2y} + a \frac{I_{\nu+1}(ay)}{I_\nu(ay)}$$

$$= \frac{a}{2} \frac{I_{\nu+1}(ay)}{I_\nu(ay)} + \frac{1}{2t} \left[ y - x \frac{I_{\nu+1}\left(\frac{xy}{t}\right)}{I_\nu\left(\frac{xy}{t}\right)} \right] \ge 0 \qquad (7.101)$$

since $y > x$ and $I_{\nu+1}(z) \le I_\nu(z)$, from Lemma 7.1.

$\square$

## 7.4.2 Squares of Generalized Ornstein-Uhlenbeck Processes, also Called CIR Processes in Mathematical Finance

Let $\left( (R_t, t \ge 0), (\mathbb{Q}_x^{(\nu,\beta)}, x \ge 0) \right)$ denote the square of a generalized Ornstein-Uhlenbeck process, with infinitesimal generator:

$$2x \frac{d^2}{dx^2} + (2\beta x + 2(\nu+1)) \frac{d}{dx}. \qquad (7.102)$$

For $d = 2(\nu+1)$ an integer, this process may be constructed as the square of the Euclidean norm of the $d$-dimensional Ornstein-Uhlenbeck, with parameter $\beta > 0$, that is the solution of:

$$X_t = x_0 + B_t + \beta \int_0^t X_s ds$$

where $(B_t, t \ge 0)$ denotes a $d$-dimensional Brownian motion. See, e.g., Pitman-Yor [66] for results about this family of diffusions.

**Theorem 7.5.** *For $\nu \ge -\frac{1}{2}$, and $\beta \ge 0$, the process $\left( (R_t, t \ge 0); (\mathbb{Q}_x^{(\nu,\beta)}, x \ge 0) \right)$ admits an increasing pseudo-inverse.*

**Sketch of the proof of Theorem 7.5.** We need to show that $\frac{\partial}{\partial t} \mathbb{Q}_x^{(\nu,\beta)}(R_t \ge y) \ge 0$ for every $y > x$. From the Fokker-Planck formula (7.34), we get:

$$\frac{\partial}{\partial t} \mathbb{Q}_x^{(\nu,\beta)}(R_t \ge y) = -2y \frac{\partial}{\partial y} p(t,x,y) + (2\beta y + 2\nu) p(t,x,y) \qquad (7.103)$$

where $p(t,x,y)$, the density with respect to the Lebesgue measure of $R_t$ under $\mathbb{Q}_x^{(\nu,\beta)}$, is given by (see [66]):

$$p(t,x,y) = \frac{\beta}{2\sinh(\beta t)} \left(\frac{y}{x}\right)^{\frac{\nu}{2}} \exp\left\{-\beta\left((1-\nu)t + \frac{xe^{\beta t} + ye^{-\beta t}}{2\sinh(\beta t)}\right)\right\} I_\nu\left(\frac{\beta\sqrt{xy}}{\sinh(\beta t)}\right).$$

(7.104)

Hence:

$$\frac{\frac{\partial}{\partial y}p(t,x,y)}{p(t,x,y)} = \frac{\nu}{2y} - \beta\frac{e^{-\beta t}}{2\sinh(\beta t)} + \frac{1}{2}\sqrt{\frac{x}{y}}\frac{\beta}{\sinh(\beta t)}\frac{I_\nu'}{I_\nu}\left(\frac{\beta\sqrt{xy}}{\sinh(\beta t)}\right).$$

(7.105)

Thus, from (7.105) and (7.103):

$$\frac{\frac{\partial}{\partial t}\mathbb{Q}_x^{(\nu,\beta)}(R_t \geq y)}{p(t,x,y)} = 2\beta y + \nu + \beta y\frac{e^{-\beta t}}{\sinh(\beta t)} - \frac{\beta}{\sinh(\beta t)}\sqrt{xy}\frac{I_\nu'}{I_\nu}\left(\frac{\beta\sqrt{xy}}{\sinh(\beta t)}\right)$$

and, using again the relation:

$$\frac{I_\nu'}{I_\nu}(z) = \frac{I_{\nu+1}}{I_\nu}(z) + \frac{\nu}{z},$$

we have:

$$\Delta_t := \frac{\frac{\partial}{\partial t}\mathbb{Q}_x^{(\nu,\beta)}(R_t \geq y)}{p(t,x,y)}$$

$$= 2\beta y + \nu + \beta y\frac{e^{-\beta t}}{\sinh(\beta t)} - \frac{\beta}{\sinh(\beta t)}\sqrt{xy}\left(\frac{I_{\nu+1}}{I_\nu}\left(\frac{\beta\sqrt{xy}}{\sinh(\beta t)}\right) + \frac{\nu\sinh(\beta t)}{\beta\sqrt{xy}}\right)$$

$$= 2\beta y + \frac{\beta y e^{-\beta t}}{\sinh(\beta t)} - \frac{\beta}{\sinh(\beta t)}\sqrt{xy}\frac{I_{\nu+1}}{I_\nu}\left(\frac{\beta\sqrt{xy}}{\sinh(\beta t)}\right).$$

Denoting $z := \frac{\beta\sqrt{xy}}{\sinh(\beta t)}$, and using $y > x \Longrightarrow z \leq \frac{\beta y}{\sinh(\beta t)}$, we have:

$$\Delta_t \geq 2z\sinh(\beta t) + z\sqrt{\frac{y}{x}}e^{-\beta t} - z\frac{I_{\nu+1}}{I_\nu}(z) \quad \text{(since } y \geq x\text{)}$$

$$\geq z\left(e^{\beta t} - e^{-\beta t} + e^{-\beta t} - \frac{I_{\nu+1}}{I_\nu}(z)\right)$$

$$= z\left(e^{\beta t} - \frac{I_{\nu+1}}{I_\nu}(z)\right) \geq 0$$

as $\beta \geq 0$ implies $e^{\beta t} \geq 1 \geq \frac{I_{\nu+1}}{I_\nu}(z)$, from Lemma 7.1.

□

*Remark 7.5.*

**a)** When $d$ is an integer, the previous computation may be obtained in a simpler manner, using the fact that $R_t$ is the square of the norm of a $d$-dimensional Gaussian variable with mean $x_0 \exp(\beta t)$, and (common) variance $\dfrac{e^{2\beta t} - 1}{2\beta}$.

**b)** We also deduce, from the comparison theorem for SDE's, that, for $\nu' \geq \nu$ and $\beta' \geq \beta$:

$$Q_x^{(\nu',\beta')}(R_t \geq y) \geq Q_x^{(\nu,\beta)}(R_t \geq y) \qquad (y \geq x)$$

hence, with obvious notation, for $x \leq y$, and $t \geq 0$:

$$\mathbb{P}(Y_{x,y}^{(\nu',\beta')} \leq t) \geq \mathbb{P}(Y_{x,y}^{(\nu,\beta)} \leq t)$$

which states that the r.v.'s. $Y_{x,y}^{(\nu,\beta)}$ are stochastically decreasing in the parameters $\nu$ and $\beta$. These variables $Y_{x,y}^{(\nu,\beta)}$ are different from those discussed in Section 7.5.

### 7.4.3 A Third Example

Let us consider the process:

$$X_t^{(\nu)} = \int_0^t ds \exp 2\left((B_t - B_s) + \nu(t - s)\right), \qquad (t \geq 0)$$

Here, $(B_t, t \geq 0)$ denotes a 1-dimensional Brownian motion, starting from 0. The process $(X_t^{(\nu)}, t \geq 0)$ is easily shown to be Markov, since, thanks to Itô's formula, we obtain:

$$dX_t^{(\nu)} = \left(2(\nu+1)X_t^{(\nu)} + 1\right)dt + 2X_t^{(\nu)}dB_t.$$

**Theorem 7.6.** *For $\nu + 1 \geq 0$, the submartingale $(X_t^{(\nu)}, t \geq 0)$ admits an increasing pseudo-inverse.*

*Proof.* Using time reversal (from $t$), we get, for any fixed $t \geq 0$:

$$X_t^{(\nu)} \overset{\text{(law)}}{=} A_t^{(\nu)} := \int_0^t du \exp\left(2(B_u + \nu u)\right).$$

Now, the process $(A_t^{(\nu)}, t \geq 0)$ is increasing, and, if we denote by $(\tau_y^{(\nu)}, y \geq 0)$ its inverse, we obtain:

$$\mathbb{P}(X_t^{(\nu)} \geq y) = \mathbb{P}(A_t^{(\nu)} \geq y) = \mathbb{P}(\tau_y^{(\nu)} \leq t) \qquad (y \geq 0).$$

We note that, here, we may define the pseudo-inverse process $(Y_y^{(\nu)}, y > 0)$ of the process $(X_t^{(\nu)}, t \geq 0)$ as a "time" process, precisely the process $(\tau_y^{(\nu)}, y \geq 0)$.

$\square$

## 7.5 The More General Family $(Y_{x,y}^{(\nu,\alpha)}; x < y, \nu \geq 0, \alpha \in [0,1])$

In Section 7.2, we introduced the r.v.'s $\left(Y_{x,y}^{(\nu)}, x < y; \nu \geq -\dfrac{1}{2}\right)$ and studied their properties in Section 7.3. We shall now introduce further positive r.v.'s $\left(Y_{x,y}^{(\nu,\alpha)}; x < y, \nu \geq 0, \alpha \in [0,1]\right)$ which extend the family $(Y_{x,y}^{(\nu)})$, corresponding to $\alpha = 1$, and we shall describe some of their properties. Let us insist again that these variables have nothing to do with those introduced in Remark 7.5 following Theorem 7.5.

### 7.5.1 Some Useful Formulae

For ease of the reader, we recall some notation and formulae which we have already used and which shall be useful to us below.

$T_{x,y}^{(\nu)}$: a r.v. whose law, under $\mathbb{P}_x^{(\nu)}$, is that of $\inf\{t \geq 0; R_t = y\}$,

$T_y^{(\nu)} = T_{0,y}^{(\nu)}$,

$G_{x,y}^{(\nu)}$: a r.v. whose law, under $\mathbb{P}_x^{(\nu)}$, is that of $\sup\{t \geq 0; R_t = y\}$,

$G_y^{(\nu)} = G_{0,y}^{(\nu)}$.

Then:

$$\mathbb{E}\left[e^{-\lambda T_{x,y}^{(\nu)}}\right] = \left(\frac{y}{x}\right)^{\nu} \frac{I_{\nu}(x\sqrt{2\lambda})}{I_{\nu}(y\sqrt{2\lambda})} \qquad (x < y) \tag{7.106}$$

$$\mathbb{E}\left[e^{-\lambda T_y^{(\nu)}}\right] = \frac{1}{2^{\nu}\Gamma(\nu+1)} \frac{(y\sqrt{2\lambda})^{\nu}}{I_{\nu}(y\sqrt{2\lambda})} \qquad (y > 0) \tag{7.107}$$

$$\mathbb{E}\left[e^{-\lambda G_{x,y}^{(\nu)}}\right] = 2\nu \left(\frac{y}{x}\right)^{\nu} I_{\nu}(x\sqrt{2\lambda})K_{\nu}(y\sqrt{2\lambda}) \qquad (x < y) \tag{7.108}$$

$$\mathbb{E}\left[e^{-\lambda G_y^{(\nu)}}\right] = \frac{1}{2^{\nu-1}\Gamma(\nu)}(y\sqrt{2\lambda})^{\nu}K_{\nu}(y\sqrt{2\lambda}) \qquad (0 < y, \nu > 0) \tag{7.109}$$

$$\mathbb{E}\left[e^{-\lambda Y_{x,y}^{(\nu)}}\right] = \frac{I_{\nu}(x\sqrt{2\lambda})}{(x\sqrt{2\lambda})^{\nu}}(y\sqrt{2\lambda})^{\nu+1}K_{\nu+1}(y\sqrt{2\lambda}). \tag{7.110}$$

From these formulae, we deduced, in Theorem 7.3:

$$T_x^{(\nu)} + Y_{x,y}^{(\nu)} \overset{(\text{law})}{=} G_y^{(\nu+1)} \tag{7.111}$$

$$Y_{x,z}^{(\nu)} \overset{(\text{law})}{=} T_{x,y}^{(\nu)} + Y_{y,z}^{(\nu)} \qquad (x < y < z). \tag{7.112}$$

On the other hand, it is obvious that:

$$T_{x,y}^{(\nu)} + G_{y,z}^{(\nu)} \overset{(\text{law})}{=} G_{x,z}^{(\nu)} \qquad (x < y < z). \tag{7.113}$$

## 7.5.2 Definition of $(G_y^{(\nu+\theta,\nu)}, y > 0, \nu, \theta \ge 0)$ and $(T_y^{(\nu+\theta,\nu)}, y > 0, \nu, \theta \ge 0)$

From Ismail-Kelker [33], or Pitman-Yor $\big([65], \text{p.336, formula } (9.b.1)\big)$, there exists, for every $y > 0, \nu, \theta \ge 0$, a positive r.v. $G_y^{(\nu+\theta,\nu)}$ such that:

$$\mathbb{E}\left[e^{-\lambda G_y^{(\nu+\theta,\nu)}}\right] = \frac{\Gamma(\nu+\theta)}{\Gamma(\nu)} \frac{2^\theta}{(y\sqrt{2\lambda})^\theta} \frac{K_\nu(y\sqrt{2\lambda})}{K_{\nu+\theta}(y\sqrt{2\lambda})}. \tag{7.114}$$

Likewise from [33], or [65], p.336, formula (9.a.1), there exists, for every $y > 0$, $\nu, \theta \ge 0$, a positive r.v. $T_y^{(\nu+\theta,\nu)}$ such that:

$$\mathbb{E}\left[e^{-\lambda T_y^{(\nu+\theta,\nu)}}\right] = \frac{\Gamma(\nu+\theta+1)}{\Gamma(\nu+1)} \frac{2^\theta}{(y\sqrt{2\lambda})^\theta} \frac{I_{\nu+\theta}(y\sqrt{2\lambda})}{I_\nu(y\sqrt{2\lambda})}. \tag{7.115}$$

From the relations (7.107) and (7.115) on one hand, and relations (7.109) and (7.114) on the other hand, we immediately deduce the following Proposition which completes the results of Pitman-Yor [65].

**Proposition 7.4.**

i)  *For every $y > 0$, $\nu \ge -\dfrac{1}{2}$ and $\theta \ge 0$:*

$$T_y^{(\nu+\theta)} + T_y^{(\nu+\theta,\nu)} \stackrel{(\text{law})}{=} T_y^{(\nu)}. \tag{7.116}$$

ii)  *For every $y > 0$, $\nu > 0$ and $\theta \ge 0$:*

$$G_y^{(\nu+\theta)} + G_y^{(\nu+\theta,\nu)} \stackrel{(\text{law})}{=} G_y^{(\nu)}. \tag{7.117}$$

Of course, it is desirable to give a "more probabilistic" proof of (7.116) and (7.117). Here is such a proof for relation (7.116).

**Another proof of (7.116)**

To simplify, we take $y = 1$. Let $(R_t^{(\nu)}, t \ge 0)$ and $(R_t^{(\theta-1)}, t \ge 0)$ two Bessel processes starting from 0, independent, with respective indexes $\nu$ and $\theta - 1$ (i.e. with dimension resp. $2\nu + 2$ and $2\theta$). Let $\left(R_t^{(\nu+\theta)} := \sqrt{(R_t^{(\nu)})^2 + (R_t^{(\theta-1)})^2}, t \ge 0\right)$. $(R_t^{(\nu+\theta)}, t \ge 0)$ is a Bessel process with index $\nu + \theta$, i.e. with dimension $2\nu + 2\theta + 2$, started at 0. Let $T_1^{(\nu+\theta)} := \inf\{t \ge 0; R_t^{(\nu+\theta)} = 1\}$. It is clear that $T_1^{(\nu+\theta)} \le T_1^{(\nu)}$ (with $T_1^{(\nu)} := \inf\{t \ge 0; R_t^{(\nu)} = 1\}$) and that $R_{T_1^{(\nu+\theta)}}^{(\nu)} \le 1$. Thus:

$$T_1^{(\nu)} = T_1^{(\nu+\theta)} + \inf\left\{u \geq 0; R^{(\nu)}_{T_1^{(\nu+\theta)}+u} = 1\right\}. \tag{7.118}$$

On the other hand, it follows from the intertwining properties of the Bessel semi-groups $\big($see [14] or [18]$\big)$ that:

$$\mathbb{E}_0^{(\nu+\theta)}\left[f(R_t^{(\nu)})\,\big|\,\mathscr{R}_t^{(\nu+\theta)}\right] = \mathbb{E}\left[f\left(r\sqrt{\beta_{\nu+1,\theta}}\right)\right] \tag{7.119}$$

where on the right-hand side of (7.119) $r = R_t^{(\nu+\theta)}$ and $\beta_{\nu+1,\theta}$ is a beta variable with parameters $\nu + 1$ and $\theta$. We then deduce from (7.119) (which is valid for the $(\mathscr{R}_t^{(\nu+\theta)})$ stopping time $T = T_1^{(\nu+\theta)}$) that:

$$R^{(\nu)}_{T_1^{(\nu+\theta)}} \overset{(\text{law})}{=} \sqrt{\beta_{\nu+1,\theta}} \tag{7.120}$$

and that $T_1^{(\nu+\theta)}$ and $(R^{(\nu)}_{T_1^{(\nu+\theta)}+u}, u \geq 0)$ are independent. It then follows from (7.118) that:

$$\mathbb{E}\left[e^{-\lambda T_1^{(\nu)}}\right] = \mathbb{E}\left[e^{-\lambda T_1^{(\nu+\theta)}}\right]\int_0^1 \frac{x^\nu(1-x)^{\theta-1}}{B(\nu+1,\theta)}\mathbb{E}\left[e^{-\lambda T^{(\nu)}_{\sqrt{x},1}}\right]dx. \tag{7.121}$$

Taking (7.106) into account, we shall have proven (7.116) once we establish:

$$\int_0^1 \frac{x^\nu(1-x)^{\theta-1}}{B(\nu+1,\theta)}\left(\frac{1}{\sqrt{x}}\right)^\nu \frac{I_\nu(\sqrt{2\lambda x})}{I_\nu(\sqrt{2\lambda})}dx = \frac{2^\theta\Gamma(\nu+\theta+1)}{\Gamma(\nu+1)}\frac{1}{(\sqrt{2\lambda})^\theta}\frac{I_{\nu+\theta}(\sqrt{2\lambda})}{I_\nu(\sqrt{2\lambda})}. \tag{7.122}$$

Now (7.122) is easily established, by using the series expansion:

$$I_\nu(z) = \sum_{k=0}^\infty \frac{\left(\frac{z}{2}\right)^{\nu+2k}}{\Gamma(k+1)\Gamma(k+\nu+1)}, \qquad (\text{see (B.1)})$$

then integrating term by term on the left hand side of (7.122) and using the formula:

$$\int_0^1 x^{a-1}(1-x)^{b-1}dx = B(a,b) = \frac{\Gamma(a)\Gamma(b)}{\Gamma(a+b)} \qquad (a,b>0).$$

Concerning (7.117) a proof close to the preceding one seems harder since $G_1^{(\nu+\theta)}$ is not a stopping time. Nonetheless, it may be possible to use the enlarged filtration $(\mathscr{G}_t^{(\nu+\theta)}, t \geq 0)$, i.e. the smallest filtration which contains the natural filtration $(\mathscr{R}_t^{(\nu+\theta)}, t \geq 0)$ of $(R_t^{(\nu+\theta)}, t \geq 0)$ and which makes $G_1^{(\nu+\theta)}$ a $(\mathscr{G}_t^{(\nu+\theta)}, t \geq 0)$ stopping time. Explicit computations in this new filtration are made possible from the knowledge of the Azéma supermartingale $\big(\mathbb{P}(G_1^{(\nu+\theta)} \geq t \,|\, \mathscr{R}_t^{(\nu+\theta)}), t \geq 0\big)$ $= \big((R_t^{(\nu+\theta)})^{-2(\nu+\theta)} \wedge 1, t \geq 0\big)$. But the enlargement formulae allowing to express the process $(R_t^{(\nu+\theta)}, t \geq 0)$ in this new filtration lead to complicated formulae which do not seem to provide us with the desired explicit result.

## 7.5.3 Existence and Properties of $(Y_{x,y}^{(\nu,\alpha)}; x < y, \nu \geq 0, \alpha \in [0,1])$

Here is now the main result of this Section 7.5, i.e. the existence and the properties of the r.v.'s $(Y_{x,y}^{(\nu,\alpha)}; x < y, \nu \geq 0, \alpha \in [0,1])$.

**Theorem 7.7.**
*i) For every $\nu \geq 0, x < y$ and $\alpha \in [0,1]$, there exists a positive r.v. $Y_{x,y}^{(\nu,\alpha)}$ such that:*

$$\mathbb{E}\left[e^{-\lambda Y_{x,y}^{(\nu,\alpha)}}\right] = \frac{2^{1-\alpha}\Gamma(\nu+1)}{\Gamma(\nu+\alpha)} \frac{I_\nu(x\sqrt{2\lambda})}{(x\sqrt{2\lambda})^\nu}(y\sqrt{2\lambda})^{\nu+\alpha}K_{\nu+\alpha}(y\sqrt{2\lambda}). \quad (7.123)$$

*Moreover:*
*ii) The r.v.'s $Y_{x,y}^{(\nu,\alpha)}$ interpolate between $Y_{x,y}^{(\nu)}$ and $G_{x,y}^{(\nu)}$, i.e.:*

$$Y_{x,y}^{(\nu,1)} \overset{\text{(law)}}{=} Y_{x,y}^{(\nu)}, \qquad Y_{x,y}^{(\nu,0)} \overset{\text{(law)}}{=} G_{x,y}^{(\nu)}. \quad (7.124)$$

*iii)* $\qquad Y_{x,y}^{(\nu,\alpha)} \overset{\text{(law)}}{=} Y_{x,y}^{(\nu)} + G_y^{(\nu+1,1-\alpha)} \qquad\qquad\qquad (7.125)$

*and, for $(1-\nu)^+ \leq \alpha \leq 1$:*

$$Y_{x,y}^{(\nu,\alpha)} \overset{\text{(law)}}{=} Y_{x,y}^{(\nu+\alpha-1)} + T_x^{(\nu,1-\alpha)}. \quad (7.126)$$

*iv)* $\quad T_x^{(\nu)} + Y_{x,y}^{(\nu,\alpha)} \overset{\text{(law)}}{=} G_y^{(\nu+\alpha)} \overset{\text{(law)}}{=} T_x^{(\nu+\alpha-1)} + Y_{x,y}^{(\nu+\alpha-1)}. \qquad (7.127)$
*v)* $\qquad G_{x,y}^{(\nu)} = G_y^{(\nu+\alpha,\alpha)} + Y_{x,y}^{(\nu,\alpha)}. \qquad\qquad\qquad\qquad (7.128)$

*In particular, for $x = 0$, we recover (7.117):*

$$G_y^{(\nu)} \overset{\text{(law)}}{=} G_y^{(\nu+\alpha,\alpha)} + G_y^{(\nu+\alpha)}. \quad (7.129)$$

*vi)* $\quad Y_{x,y}^{(\nu)} + G_y^{(\nu+1,1-\alpha)} \overset{\text{(law)}}{=} Y_{x,y}^{(\nu+\alpha-1)} + T_x^{(\nu,1-\alpha)} \quad \left(\overset{\text{(law)}}{=} Y_{x,y}^{(\nu,\alpha)}\right). \quad (7.130)$

Observe that, from (7.125) and (7.129), the r.v.'s $Y_{x,y}^{(\nu,\alpha)}$ are stochastically decreasing in $\alpha$.

*Proof.*
*i*) We show (7.123)
We have:

$$
\begin{aligned}
\mathbb{E}\left[e^{-\lambda Y_{x,y}^{(\nu,\alpha)}}\right] &= \frac{2^{1-\alpha}\Gamma(\nu+1)}{\Gamma(\nu+\alpha)}\frac{I_\nu(x\sqrt{2\lambda})}{(x\sqrt{2\lambda})^\nu}(y\sqrt{2\lambda})^{\nu+\alpha}K_{\nu+\alpha}(y\sqrt{2\lambda}) \\
&= \left(\frac{I_\nu(x\sqrt{2\lambda})}{(x\sqrt{2\lambda})^\nu}(y\sqrt{2\lambda})^{\nu+1}K_{\nu+1}(y\sqrt{2\lambda})\right) \\
&\quad \times \left(\frac{2^{1-\alpha}\Gamma(\nu+1)}{\Gamma(\nu+\alpha)}(y\sqrt{2\lambda})^{\alpha-1}\frac{K_{\nu+\alpha}(y\sqrt{2\lambda})}{K_{\nu+1}(y\sqrt{2\lambda})}\right) \\
&= \mathbb{E}\left[e^{-\lambda Y_{x,y}^{(\nu)}}\right]\mathbb{E}\left[e^{-\lambda G_y^{(\nu+1,1-\alpha)}}\right] \qquad \big(\text{from (7.110) and (7.114)}\big).
\end{aligned}
$$

This proves (7.123) and (7.125).

*ii*) We now show (7.126)
We have:

$$
\begin{aligned}
\mathbb{E}\left[e^{-\lambda Y_{x,y}^{(\nu,\alpha)}}\right] &= \frac{2^{1-\alpha}\Gamma(\nu+1)}{\Gamma(\nu+\alpha)}\frac{I_\nu(x\sqrt{2\lambda})}{(x\sqrt{2\lambda})^\nu}(y\sqrt{2\lambda})^{\nu+\alpha}K_{\nu+\alpha}(y\sqrt{2\lambda}) \\
&= \left(\frac{I_{\nu+\alpha-1}(x\sqrt{2\lambda})}{(x\sqrt{2\lambda})^{\nu+\alpha-1}}(y\sqrt{2\lambda})^{\nu+\alpha}K_{\nu+\alpha}(y\sqrt{2\lambda})\right) \\
&\quad \times \left(\frac{2^{1-\alpha}\Gamma(\nu+1)}{\Gamma(\nu+\alpha)(x\sqrt{2\lambda})^{1-\alpha}}\frac{I_\nu(x\sqrt{2\lambda})}{I_{\nu+\alpha-1}(x\sqrt{2\lambda})}\right) \\
&= \mathbb{E}\left[e^{-\lambda Y_{x,y}^{(\nu+\alpha-1)}}\right]\mathbb{E}\left[e^{-\lambda T_x^{(\nu,1-\alpha)}}\right] \qquad \big(\text{from (7.110) and (7.115)}\big)
\end{aligned}
$$

This shows (7.126).

*iii*) The relation $T_x^{(\nu)}+Y_{x,y}^{(\nu,\alpha)}\overset{\text{(law)}}{=}G_y^{(\nu+\alpha)}$ follows immediately from (7.107), (7.123) and (7.109). The relation $G_y^{(\nu+\alpha)}\overset{\text{(law)}}{=}T_x^{(\nu+\alpha-1)}+Y_{x,y}^{(\nu+\alpha-1)}$ follows from (7.111) $\big($or from the preceding relation where we replace $\nu$ by $\nu+\alpha-1$ and we observe, as is obvious, that $Y_{x,y}^{(\nu+\alpha-1)}\overset{\text{(law)}}{=}Y_{x,y}^{(\nu+\alpha-1,1)}\big)$.

*iv*) We now show (7.128)
We deduce, from (7.113), (7.129) and (7.127):

$$
\begin{aligned}
T_x^{(\nu)}+G_{x,y}^{(\nu)}\overset{\text{(law)}}{=}G_y^{(\nu)} &\overset{\text{(law)}}{=} G_y^{(\nu+\alpha,\alpha)}+G_y^{(\nu+\alpha)} \\
&\overset{\text{(law)}}{=} G_y^{(\nu+\alpha,\alpha)}+T_x^{(\nu)}+Y_{x,y}^{(\nu,\alpha)}.
\end{aligned}
$$

Hence:

$$G_{x,y}^{(\nu)} \stackrel{\text{(law)}}{=} G_y^{(\nu+\alpha,\,\alpha)} + Y_{x,y}^{(\nu,\alpha)}.$$

We might also have proven this last relation by using (7.108), (7.114) and (7.123).

*v*) The relation (7.124) obviously holds.

*vi*) Finally, relation (7.130) follows from (7.126) and (7.125). Indeed, from (7.126) and (7.125):

$$Y_{x,y}^{(\nu,\alpha)} \stackrel{\text{(law)}}{=} Y_{x,y}^{(\nu+\alpha-1)} + T_x^{(\nu,\,1-\alpha)}$$
$$\stackrel{\text{(law)}}{=} Y_{x,y}^{(\nu)} + G_y^{(\nu+1,\,1-\alpha)} \qquad \left(x < y, \ (1-\nu)^+ \le \alpha \le 1\right).$$

$\square$

## 7.6 Notes and Comments

The notion of a pseudo-inverse for a process, and in fact the notion of its existence, appears naturally in [49] for the construction of a probability $\gamma$ on $[0,1] \times [0,+\infty[$ (cf. Chapter 6). Most of this Chapter 7 is taken up from B. Roynette and M. Yor [75]. The results found in this Chapter hinge upon:

- formula (7.19) relative to the Bessel bridges, a formula obtained in [94] and [65].
- formula (7.24) which is due to F. Hirsch and S. Song ([30])

The Markov process studied in Subsection 7.4.1 whose infinitesimal generator is given by (7.98) has been introduced by S. Watanabe ([89]). The formulae in Theorem 7.7 extend, in some sense, those obtained by J. Pitman and M. Yor ([65], Th. 4.3, p.312).

# Chapter 8
# Existence of Pseudo-Inverses for Diffusions

**Abstract** In this chapter, we continue the study of pseudo-inverses, extending the previous results of Chapter 7 to the general framework of linear diffusions. We shall focus here on increasing pseudo-inverses, and we shall deal with two cases:

- first, a diffusion taking values in $\mathbb{R}$, and solution of a particular SDE,
- and then, a general diffusion on $\mathbb{R}^+$ starting from 0.

More precisely, we shall prove that, to a positive diffusion $X$ starting from 0, we can associate another diffusion $\overline{X}$ such that the tail probabilities of $X$ are the distribution functions of the last passage times of $\overline{X}$.

## 8.1 Introduction

We consider a regular linear diffusion $X$ taking values in $\mathbb{R}$ or $\mathbb{R}^+$. We denote by $\mathbb{P}_x$ and $\mathbb{E}_x$, respectively, the probability measure and the expectation associated with $X$ when started from $x$. The notion of pseudo-inverse being defined in Chapter 7, Definition 7.1 and 7.2, let us start by making the following remark: if only Point $(ii)$ of Definition 7.1 is satisfied, then, since $t \mapsto \mathbb{P}_x(X_t \geq y)$ is bounded by 1, for every $y > x$ the limit:

$$\lim_{t \to \infty} \mathbb{P}_x(X_t \geq y) =: Z(x,y)$$

exists. Therefore, there exists a family of positive random variables $(Y_{x,y}, y > x)$ such that:

$$\frac{1}{Z(x,y)} \mathbb{P}_x(X_t \geq y) = \mathbb{P}(Y_{x,y} \leq t). \tag{8.1}$$

In this case, we call the family of positive r.v.'s $(Y_{x,y}, y > x)$ the *increasing quasi pseudo-inverse* of the process $(X_t, t \geq 0)$.

We now give the plan and the main results of this chapter:
**1)** We first consider, in Section 8.2, the case of a diffusion on $\mathbb{R}$ which is solution of the stochastic differential equation:

C. Profeta et al., *Option Prices as Probabilities*, Springer Finance,
DOI 10.1007/978-3-642-10395-7_8, © Springer-Verlag Berlin Heidelberg 2010

$$X_t = x + \beta_t + \int_0^t c(X_s)ds.$$

We show that if $c$ is a $\mathscr{C}^2$ positive function which is decreasing and convex, then $X$ admits an increasing quasi pseudo-inverse (see Theorem 8.1). The existence of an increasing pseudo-inverse requires an additional assumption on $c$, which can be given for instance using Feller's criterion.

**2)** In the remaining part of the Chapter, we shall focus on diffusions taking values in $\mathbb{R}^+$. We introduce, in Section 8.3, a family of diffusions $\left(X^{(\alpha)}\right)_{\alpha \geq 0}$ defined as the solutions of the SDEs:

$$X_t^{(\alpha)} = B_t + \int_0^t \left( c - \frac{\alpha e^{2C}}{1 + \alpha \int_0^\cdot e^{2C(y)}dy} \right) (X_s^{(\alpha)})ds$$

where $C(x) := \int_1^x c(y)dy$ with $c :]0, +\infty[ \to \mathbb{R}^+$ a $\mathscr{C}^1$ function such that $c(0) = +\infty$ and $C(0) = -\infty$. We prove that there exists a transient diffusion $\overline{X}$ independent from $\alpha$ such that:

$$\mathbb{P}_0^{(\alpha)}(X_t^{(\alpha)} \geq y) = \frac{1}{1 + \alpha \int_0^y e^{2C(z)}dz} \overline{\mathbb{P}}_0\left(\overline{G}_y \leq t\right).$$

where $\overline{G}_y := \sup\{t \geq 0; \overline{X}_t = y\}$. Therefore, we obtain that:

*i)* if $\alpha > 0$, $X^{(\alpha)}$ admits an increasing quasi pseudo-inverse,
*ii)* if $\alpha = 0$, $X := X^{(0)}$ admits an increasing pseudo-inverse, and the following equality in law holds:

$$X_t \overset{(d)}{=} \inf_{s \geq t} \overline{X}_s.$$

This is Theorem 8.2.

**3)** Then, in Section 8.4, we extend the previous results of Section 8.3 to the general case of a diffusion $X$ taking values on $\mathbb{R}^+$ and started from 0. Denoting by $m$ the speed measure of $X$, we show the existence of a transient process $\overline{X}$ such that:

$$\mathbb{P}_0(X_t \geq y) = \left( 1 - \frac{m[0, y]}{m[0, +\infty[} \right) \overline{\mathbb{P}}_0\left(\overline{G}_y \leq t\right).$$

In particular, we prove in Theorem 8.3, that in our framework, a diffusion admits an increasing pseudo-inverse if and only if it is transient, or null recurrent.

**4)** Finally, in Section 8.5, we derive new relations between the diffusions $X$ and $\overline{X}$ (see Proposition 8.6). We then introduce 3 new processes (see diagram (8.92)) which are linked to $X$ via the notion of pseudo-inverse. One of them, $\widehat{X}$, is obtained by time-reversing $\overline{X}$. Finally, we prove some Zolotarev-like identities, (also known as Kendall's identities, [13], see Proposition 8.8) and some new results between the laws of these processes (see Proposition 8.10).

## 8.2 Pseudo-Inverse for a Brownian Motion with a Convex, Decreasing, Positive Drift

In this section, we establish the existence of pseudo-inverses in the set-up of $\mathbb{R}$-valued diffusions $(X_t, t \geq 0; \mathbb{P}_x, x \in \mathbb{R})$ which solve:

$$X_t = x + \beta_t + \int_0^t c(X_s)ds \tag{8.2}$$

where $(\beta_t, t \geq 0)$ denotes a one-dimensional Brownian motion starting from 0.

**Theorem 8.1.** *Let $c : \mathbb{R} \to \mathbb{R}^+$ a $\mathscr{C}^2$ function which is decreasing and convex. Then:*

*i)  equation (8.2) admits a unique solution, which is strong;*
*ii) for every $y > x$, the function $t \mapsto \mathbb{P}_x(X_t \geq y)$ increases and is continuous;*
*iii) if, for every $y > x$:*

$$\mathbb{P}_x(X_t \geq y) \xrightarrow[t \to \infty]{} 1, \tag{8.3}$$

*then, the process $(X_t, t \geq 0; \mathbb{P}_x, x \in \mathbb{R})$ admits an increasing pseudo-inverse. In other terms, for every $y > x$, there exists a positive r.v. $Y_{x,y}$ such that, for every $t \geq 0$, (7.4) is satisfied:*

$$\mathbb{P}_x(X_t \geq y) = \mathbb{P}(Y_{x,y} \leq t) \qquad (t \geq 0).$$

*Remark 8.1.* Of course, it is desirable to give some conditions which ensure the validity of condition (8.3). For example, if $c(x) > \frac{k}{x}$, for $x \geq A$ for some $A > 0$ and $k > 1/2$, the relation (8.3) is satisfied, since:

$$C(y) := \int_A^y c(x)dx \geq \int_A^y \frac{k}{x}dx \geq k\log(y/A).$$

Thus:

$$\int_A^\infty e^{-2C(y)}dy \leq \int_A^\infty \left(\frac{A}{y}\right)^{2k} dy < \infty,$$

and $\mathbb{P}_x(X_t \xrightarrow[t \to \infty]{} +\infty) = 1$ from Feller's criterion, (see [40], p.345), which implies $\mathbb{P}_x(X_t > y) \xrightarrow[t \to \infty]{} 1$ for every $y > x$.

*Proof.*
**1)** Existence and uniqueness of a strong solution of equation (8.2) is classical. It follows from the fact that $c$ is locally Lipschitz, and from the absence of explosion since:

- $xc(x) \leq 0$   if $x \leq 0$,
- $c$ is bounded on $\mathbb{R}^+$.

**2)** We introduce some notation:
- $p(t,x,y)$ denotes the density with respect to the Lebesgue measure of the r.v. $X_t$ under $\mathbb{P}_x$.

- $h(t,x,y) := \dfrac{1}{\sqrt{2\pi t}} e^{-\frac{(x-y)^2}{2t}}$ denotes the heat semi-group density,
- for any Borel function $g : \mathbb{R} \to \mathbb{R}$, we denote:

$$E_{x,y}^{(g)} := \mathbb{E}_x \left[ \exp \int_0^t g(X_s) ds \Big| X_t = y \right],$$

- if $\gamma : \mathbb{R}^+ \times \mathbb{R} \times \mathbb{R} \to \mathbb{R}$ is a function of $t, x$ and $y$, we write $\dfrac{\partial \gamma}{\partial a}$ (resp. $\dfrac{\partial \gamma}{\partial b}$) for the derivative of $\gamma$ with respect to the second variable $x$ (resp. with respect to the third variable $y$). This will be useful for instance as we shall encounter the quantities:

$$\frac{\partial}{\partial a} p(t,y,x) = \frac{\partial}{\partial a} p(t,a,b)|_{a=y,b=x}$$

- we denote by $C$ the primitive of $c$, such that $C(0) = 0$:

$$C(x) = \int_0^x c(y) dy.$$

### 3) A first Lemma

**Lemma 8.1.**

i)  *The function $t \to \mathbb{P}_x(X_t \geq y)$ is increasing for every $y > x$ if and only if*
    $$\frac{\partial}{\partial a} p(t,x,y) \leq 0 \text{ for every } x > y.$$
ii) *There is a general Hirsch-Song formula, which extends formula (7.24):*

$$\frac{\partial}{\partial a} p(t,x,y) = -\frac{\partial}{\partial b} \left( p(t,x,y) E_{x,y}^{(c')} \right). \tag{8.4}$$

*Proof.*
**a)** We note that, for $y > x$:

$$\frac{\partial}{\partial t} \mathbb{P}_x(X_t \geq y) = \frac{\partial}{\partial t} \int_y^\infty p(t,x,z) dz$$

$$= \int_y^\infty \frac{\partial}{\partial t} p(t,x,z) dz$$

$$= \int_y^\infty \left[ \frac{1}{2} \frac{\partial^2}{\partial b^2} p(t,x,z) - \frac{\partial}{\partial b}(c(z)p(t,x,z)) \right] dz \quad \text{(from Fokker-Planck)}$$

$$= -\frac{1}{2} \frac{\partial}{\partial b} p(t,x,y) + c(y)p(t,x,y). \tag{8.5}$$

On the other hand, since $m(dz) = 2e^{2C(z)} dz$ is the speed measure of the diffusion $(X_t, t \geq 0)$, it is classical (see [12]) that, for every $x, y$:

$$e^{2C(x)} p(t,x,y) = e^{2C(y)} p(t,y,x). \tag{8.6}$$

We shall now use (8.6) in order to establish another expression for the relation (8.5). Differentiating (8.6) with respect to $y$, we get:

$$e^{2C(x)}\frac{\partial}{\partial b}p(t,x,y) = 2c(y)e^{2C(y)}p(t,y,x) + e^{2C(y)}\frac{\partial}{\partial a}p(t,y,x). \tag{8.7}$$

We then plug back (8.7) in (8.5) and we obtain, for every $x,y$:

$$\begin{aligned}\frac{\partial}{\partial t}\mathbb{P}_x(X_t \geq y) &= -\frac{1}{2}\left[2c(y)e^{2C(y)-2C(x)}p(t,y,x) + e^{2C(y)-2C(x)}\frac{\partial}{\partial a}p(t,y,x)\right] \\ &\quad + c(y)e^{2C(y)-2C(x)}p(t,y,x) \\ &= -\frac{1}{2}e^{2C(y)-2C(x)}\frac{\partial}{\partial a}p(t,y,x).\end{aligned} \tag{8.8}$$

This proves Point $(i)$ of Lemma 8.1, after exchanging the roles of $x$ and $y$.

**b) We now prove the general Hirsch-Song formula, i.e. Point $(ii)$ of Lemma 8.1**

For any generic regular function $f$, with compact support, we have, on one hand:

$$\frac{\partial}{\partial x}\mathbb{E}_x[f(X_t)] = \int_{-\infty}^{\infty}\frac{\partial}{\partial a}p(t,x,y)f(y)dy \tag{8.9}$$

whereas, on the other hand, with obvious notation:

$$\frac{\partial}{\partial x}\mathbb{E}_x[f(X_t)] = \frac{\partial}{\partial x}\mathbb{E}[f(X_t^x)] = \mathbb{E}\left[f'(X_t^x)\frac{\partial X_t^x}{\partial x}\right]. \tag{8.10}$$

We deduce from (8.2) that:

$$\frac{\partial X_t^x}{\partial x} = 1 + \int_0^t c'(X_s^x)\frac{\partial X_s^x}{\partial x}ds. \tag{8.11}$$

Integrating the linear equation (8.11), we obtain:

$$\frac{\partial X_t^x}{\partial x} = \exp\left(\int_0^t c'(X_s^x)ds\right). \tag{8.12}$$

Plugging (8.12) in (8.10), we obtain:

$$\begin{aligned}\frac{\partial}{\partial x}\mathbb{E}_x[f(X_t)] &= \mathbb{E}_x\left[f'(X_t)\exp\left(\int_0^t c'(X_s)ds\right)\right] \\ &= \int_{-\infty}^{\infty}\mathbb{E}_x\left[\exp\left(\int_0^t c'(X_s)ds\right)\Big|X_t = y\right]f'(y)p(t,x,y)dy \\ &= -\int_{-\infty}^{\infty}\frac{\partial}{\partial b}\left(E_{x,y}^{(c')}p(t,x,y)\right)f(y)dy\end{aligned} \tag{8.13}$$

(after integrating by parts). The comparison of (8.9) and (8.13) implies Point $(ii)$ of Lemma 8.1.

$\square$

**4)** Lemma 8.1 implies that the function $t \to \mathbb{P}_x(X_t \geq y)$ is increasing for every $y > x$ if and only if $\frac{\partial}{\partial b}\left(p(t,x,y)E_{x,y}^{(c')}\right) \geq 0$ for every $x > y$. To evaluate this quantity, our approach now consists in expressing $E_{x,y}^{(c')}p(t,x,y)$ with the help of the Brownian motion $\beta$.

**Lemma 8.2.** *For every $x, y$ and every $g : \mathbb{R} \to \mathbb{R}$ Borel, one has:*

$$E_{x,y}^{(g)}p(t,x,y) = h(t,x,y)e^{C(y)-C(x)}W_x\left[\exp\int_0^t\left(g - \frac{1}{2}(c' + c^2)\right)(\beta_s)ds\Big|\beta_t = y\right],$$

*where $(W_x, x \in \mathbb{R})$ denotes the family of Wiener measures. In particular, for $g = c'$:*

$$E_{x,y}^{(c')}p(t,x,y) = h(t,x,y)e^{C(y)-C(x)}W_x\left[\exp\frac{1}{2}\int_0^t(c' - c^2)(\beta_s)ds\Big|\beta_t = y\right].$$

*Proof.* For any Borel, positive function $f$, Girsanov's theorem yields:

$$\mathbb{E}_x\left[f(X_t)\exp\left(\int_0^t g(X_s)ds\right)\right]$$
$$= W_x\left[f(\beta_t)\exp\left(\int_0^t g(\beta_s)ds + \int_0^t c(\beta_s)d\beta_s - \frac{1}{2}\int_0^t c^2(\beta_s)ds\right)\right]. \quad (8.14)$$

Actually, a little care is needed in the application of Girsanov's Theorem here, since $c$ is not bounded. However, from Mc Kean's extension (see [55]) involving the explosion time, and since the explosion time in this set-up is infinite a.s., formula (8.14) holds. From Itô's formula:

$$C(\beta_t) = C(x) + \int_0^t c(\beta_s)d\beta_s + \frac{1}{2}\int_0^t c'(\beta_s)ds, \quad (8.15)$$

we deduce, by plugging (8.15) in (8.14):

$$\mathbb{E}_x\left[f(X_t)\exp\left(\int_0^t g(X_s)ds\right)\right]$$
$$= W_x\left[f(\beta_t)\exp\left(C(\beta_t) - C(x) + \int_0^t\left(g - \frac{1}{2}(c' + c^2)\right)(\beta_s)ds\right)\right]. \quad (8.16)$$

However, (8.16) may now be written, after conditioning by $X_t = y$ and $\beta_t = y$:

$$\int_{-\infty}^{\infty}\mathbb{E}_x\left[\exp\left(\int_0^t g(X_s)ds\right)\Big|X_t = y\right]f(y)p(t,x,y)dy$$
$$= \int_{-\infty}^{\infty}W_x\left[\exp\int_0^t\left(g - \frac{1}{2}(c' + c^2)\right)(\beta_s)ds\Big|\beta_t = y\right]e^{C(y)-C(x)}f(y)h(t,x,y)dy.$$
$$(8.17)$$

Letting $f$ vary in (8.17), Point *(ii)* of Lemma 8.2 has been proven.

$\square$

**5)** We are now able to end the proof of Point $(ii)$ in Theorem 8.1

From Lemmas 8.1 and 8.2, it suffices to see that, for $x > y$:

$$\frac{\partial}{\partial b}\left(E_{x,y}^{(c')} p(t,x,y)\right)$$

$$= \frac{\partial}{\partial b}\left(e^{C(y)-C(x)} h(t,x,y) W_x\left[\exp\frac{1}{2}\int_0^t (c'-c^2)(\beta_s)ds\Big|\beta_t = y\right]\right) \quad (8.18)$$

is positive. Now, denoting $\widetilde{E}_{x,y} = W_x\left[\exp\frac{1}{2}\int_0^t (c'-c^2)(\beta_s)ds\Big|\beta_t = y\right]$, we have from (8.18):

$$\frac{\partial}{\partial b}\left(E_{x,y}^{(c')} p(t,x,y)\right) = E_{x,y}^{(c')} p(t,x,y)\left(c(y) + \frac{1}{h(t,x,y)}\frac{\partial}{\partial b}h(t,x,y) + \frac{1}{\widetilde{E}_{x,y}}\frac{\partial}{\partial b}\widetilde{E}_{x,y}\right)$$

$$= E_{x,y}^{(c')} p(t,x,y)\left(c(y) + \frac{x-y}{t} + \frac{1}{\widetilde{E}_{x,y}}\frac{\partial}{\partial b}\widetilde{E}_{x,y}\right).$$

Hence,

$$\frac{\frac{\partial}{\partial b}\left(E_{x,y}^{(c')} p(t,x,y)\right)}{E_{x,y}^{(c')} p(t,x,y)} = c(y) + \frac{x-y}{t} + \frac{\frac{\partial}{\partial b}\widetilde{E}_{x,y}}{\widetilde{E}_{x,y}}.$$

Thus, since $c(y) \geq 0$ and $\frac{x-y}{t} \geq 0$ for $x \geq y$, we will have proven the increase of $t \to \mathbb{P}_x(X_t \geq y)$ once we show:

$$\frac{\partial}{\partial b}\widetilde{E}_{x,y} \geq 0. \quad (8.19)$$

We now show (8.19). The Brownian bridge of length $t$, $(\beta_s^{x,y,t}, s \leq t)$, going from $x$ to $y$, satisfies:

$$\beta_s^{x,y,t} = x + (y-x)\frac{s}{t} + \beta_s^{0,0,t}.$$

Hence:

$$\frac{\partial}{\partial b}\beta_s^{x,y,t} = \frac{s}{t},$$

and consequently,

$$\frac{\partial}{\partial b}\widetilde{E}_{x,y} = W_x\left[\left(\frac{1}{2}\int_0^t \frac{s}{t}(c''-2cc')(\beta_s)ds\right)\exp\frac{1}{2}\int_0^t (c'-c^2)(\beta_s)ds\Big|\beta_t = y\right] \geq 0$$

since, by hypothesis: $c'' \geq 0, c' \leq 0$ and $c \geq 0$. This ends the proof of Theorem 8.1, since Point $(iii)$ is just an immediate consequence of the definition of a pseudo-inverse. Note that, for the proof of Point $(ii)$ of this Theorem, we only needed the relation:

$$c'' - 2cc' \geq 0.$$

$\square$

*Remark 8.2.* The existence of an increasing (quasi) pseudo-inverse for the Bessel process with index $\nu \geq -\frac{1}{2}$ (i.e. Point $(i)$ of Theorem 7.1) may be deduced from Theorem 8.1. Indeed, define, for $\varepsilon > 0$, the function $c^{(\varepsilon)} : \mathbb{R} \to \mathbb{R}^+$ by:

$$
c^{(\varepsilon)}(x) = \begin{cases} \dfrac{2\nu+1}{2x} & \text{if } x \geq \varepsilon, \\[2mm] \dfrac{2\nu+1}{2\varepsilon} - \dfrac{2\nu+1}{2\varepsilon^2}(x-\varepsilon) & \text{if } x \leq \varepsilon. \end{cases}
$$

Let $(X_t^{(\varepsilon)}, t \geq 0)$ denote the solution of:

$$
X_t^{(\varepsilon)} = x + \beta_t + \int_0^t c^{(\varepsilon)}(X_s^{(\varepsilon)}) ds.
$$

Then, the function $t \mapsto \mathbb{P}_x(X_t^{(\varepsilon)} \geq y)$ is increasing for $y > x$ from Theorem 8.1, since $c^{(\varepsilon)}$ satisfies the hypotheses of that Theorem. It then remains to let $\varepsilon \to 0$ to obtain that $t \mapsto \mathbb{P}_x^{(\nu)}(R_t > y)$ is an increasing function of $t$, for $y > x$.

*Remark 8.3.* Note that the density function $f_{Y_{x,y}}$ of the random variable $Y_{x,y}$ is given by:

$$
\begin{aligned}
f_{Y_{x,y}}(t) &= -\frac{e^{2C(y)-2C(x)}}{2Z(x,y)} \frac{\partial}{\partial a} p(t,y,x) \qquad \text{(from (8.8))} \\[2mm]
&= \frac{1}{Z(x,y)} \left( -\frac{1}{2} \frac{\partial}{\partial b} p(t,x,y) + c(y) p(t,x,y) \right) \qquad \text{(from (8.5))}
\end{aligned}
$$

where $Z(x,y)$ is given by (8.1). Likewise, from (7.70), the Laplace transform of $Y_{x,y}$ takes the form:

$$
\mathbb{E}\left[ e^{-\lambda Y_{x,y}} \right] = \frac{\lambda}{Z(x,y)} \int_y^\infty u_\lambda(x,z) dz,
$$

where $u_\lambda(x,z) = \int_0^\infty e^{-\lambda t} p(t,x,z) dt$ is the resolvent kernel of $(X_t, t \geq 0)$.

## 8.3  Study of a Family of $\mathbb{R}^+$-Valued Diffusions

### 8.3.1  Definition of the Operator $T$

Let us introduce the following functional spaces:

$$
\mathscr{H} = \left\{ F : ]0, +\infty[ \mapsto \mathbb{R}, F \text{ of class } \mathscr{C}^1, \int_{0^+} e^{2F(y)} dy < \infty \right\}, \tag{8.20}
$$

$$
\mathscr{H}^\infty = \left\{ F \in \mathscr{H}, \int_0^\infty e^{2F(y)} dy = \infty \right\}. \tag{8.21}
$$

On $\mathcal{H}$, we define the non-linear operator $T$ as follows:

$$TF(x) = \frac{e^{2F(x)}}{\int_0^x e^{2F(y)}dy} - F'(x) \qquad (x > 0, F \in \mathcal{H}). \tag{8.22}$$

This operator was first introduced by Matsumoto and Yor in [53] and [54] in a discussion of extensions of Pitman's theorem about the Bessel process of dimension 3. It was taken up by Roynette, Vallois and Yor in [73] to find a class of max-diffusions (i.e. processes $X$ such that the two-dimensional process $(X_t, S_t^X, t \geq 0)$ where $S_t^X := \sup_{s \leq t} X_s$, is Markov) which enjoy Pitman's property (i.e. $(2S_t^X - X_t, t \geq 0)$ is Markov). We gather here some of their results. Let $F \in \mathcal{H}$. We study the diffusion $(\overline{X}_t, t \geq 0)$ which is solution of:

$$\overline{X}_t = x + B_t + \int_0^t TF(\overline{X}_s)ds \qquad (t, x > 0)$$

where $B$ is a standard Brownian motion started at 0.

**Proposition 8.1 ([73], Section 5).**

i) *The process $(\overline{X}_t, t \geq 0)$ takes its values in $\mathbb{R}^+$, 0 being a not-exit boundary, and $\overline{X}_t$ goes to $+\infty$ as $t \to \infty$.*
ii) *The random variable $\overline{J}_0 = \inf_{t \geq 0} \overline{X}_t$ is finite.*

    *(a) If $F \in \mathcal{H}^\infty$, then, the density function of $\overline{J}_0$ under $\mathbb{P}_x$ is*

$$y \longmapsto \frac{1}{\int_0^x e^{2F(z)}dz}e^{2F(y)}1_{[0,x]}(y). \tag{8.23}$$

    *(b) If $F$ does not belong to $\mathcal{H}^\infty$, then $\mathbb{P}_x(\overline{J}_0 < a) = h(a)/h(x)$ for any $a \in ]0, x[$ where:*

$$h(a) = \frac{\int_0^a e^{2F(y)}dy}{\int_a^\infty e^{2F(y)}dy}, \qquad (0 < a \leq x).$$

We now give some details about the non injectivity of the operator $T$. Indeed, let $U_\alpha(F)$ be the function:

$$U_\alpha(F)(x) := F(x) - \log\left(1 + \alpha \int_0^x e^{2F(y)}dy\right) \tag{8.24}$$

where $F \in \mathcal{H}$ and $\alpha \in \mathbb{R}$ is such that:

$$\alpha \geq -\frac{1}{\int_0^\infty e^{2F(y)}dy}. \tag{8.25}$$

Then, the following Proposition shows precisely how $T$ is not one-to-one:

**Proposition 8.2 ([73], Section 5).**
*Assume $F \in \mathcal{H}$.*

i) *Let $\alpha$ satisfy (8.25), then $U_\alpha(F) \in \mathcal{H}$.*

ii) *Let $G \in \mathcal{H}$. Then $TG = TF$ if and only if $G = \theta + U_\alpha(F)$ for some constant $\theta \in \mathbb{R}$ and $\alpha$ satisfying (8.25).*

### 8.3.2 Study of the Family $(X^{(\alpha)})_{\alpha \geq 0}$

**1) Definition.**

Let $c : ]0, +\infty[ \to \mathbb{R}^+$ a $\mathscr{C}^1$ function such that $c(0) = +\infty$. We denote:

$$C(x) = \int_1^x c(y)dy,$$

and assume that $C(0) = -\infty$. Since $c$ is $\mathbb{R}^+$-valued, $C \in \mathscr{H}^\infty$.
For $\alpha \geq 0$, let us introduce the following family of diffusions:

$$X_t^{(\alpha)} = B_t + \int_0^t \left( c - \frac{\alpha e^{2C}}{1 + \alpha \int_0^s e^{2C(y)}dy} \right) (X_s^{(\alpha)})ds. \tag{8.26}$$

When $\alpha = 0$, we get:

$$X_t = B_t + \int_0^t c(X_s)ds. \tag{8.27}$$

We define:

$$c^{(\alpha)}(x) := c(x) - \frac{\alpha e^{2C(x)}}{1 + \alpha \int_0^x e^{2C(y)}dy}, \tag{8.28}$$

and

$$C^{(\alpha)}(x) := \int_1^x c^{(\alpha)}(y)dy$$
$$= C(x) - \log\left( 1 + \alpha \int_0^x e^{2C(y)}dy \right) + \log\left( 1 + \alpha \int_0^1 e^{2C(y)}dy \right). \tag{8.29}$$

From Proposition 8.2, we have: $TC = TC^{(\alpha)}$ where $T$ is the operator defined by (8.22). Therefore, the process $\overline{X}^{(\alpha)}$ which is solution of:

$$\overline{X}_t^{(\alpha)} = x + B_t + \int_0^t TC^{(\alpha)}(\overline{X}_s^{(\alpha)})ds, \tag{8.30}$$

does not depend on $\alpha$. We will denote it by $\overline{X}$ from now on:

$$\overline{X}_t = x + B_t + \int_0^t \left( \frac{e^{2C(\overline{X}_s)}}{\int_0^{\overline{X}_s} e^{2C(y)}dy} - c(\overline{X}_s) \right) ds. \tag{8.31}$$

Its speed measure $\overline{m}(dx)$ and its scale function $\overline{s}$ are given by:

$$\begin{cases} \overline{s}(x) = -\dfrac{1}{2\int_0^x e^{2C(u)}du}, \\[4mm] \overline{m}(dx) = \left(\displaystyle\int_0^x 2e^{2C(u)}du\right)^2 e^{-2C(x)}dx. \end{cases} \tag{8.32}$$

*Remark 8.4.* Let us note that, since $c$ is positive, $C$ is an increasing function, and 0 is always an entrance endpoint. Indeed, $\forall z > 0$, (See Section 8.4):

$$\int_0^z e^{-2C(x)}\left(\int_0^x e^{2C(y)}dy\right)dx \le \int_0^z e^{-2C(x)}xe^{2C(x)}dx \le z^2 < \infty. \tag{8.33}$$

**2) Intertwining relation.**
Let $(P_t^\alpha, t \ge 0)$ be the semi-group associated to $X^{(\alpha)}$, and $H_\alpha$ defined by:

$$H_\alpha f(x) = \frac{1}{\int_0^x e^{2C^{(\alpha)}(y)}dy}\int_0^x f(y)e^{2C^{(\alpha)}(y)}dy. \tag{8.34}$$

**Proposition 8.3.** *The following intertwining relation:*

$$H_\alpha P_t^\alpha = \overline{P}_t H_\alpha \tag{8.35}$$

*holds.*

*Proof.* It is sufficient to check the identity on the infinitesimal generators. Let $f$ be a $\mathscr{C}^2$ function on $\mathbb{R}^+$. On the one hand, we have:

$$(H_\alpha f)'(x) = -\frac{e^{2C^{(\alpha)}(x)}}{\left(\int_0^x e^{2C^{(\alpha)}(y)}dy\right)^2}\int_0^x f(y)e^{2C^{(\alpha)}(y)}dy + \frac{e^{2C^{(\alpha)}(x)}}{\int_0^x e^{2C^{(\alpha)}(y)}dy}f(x), \tag{8.36}$$

and,

$$\begin{aligned}(H_\alpha f)''(x) = {}& -\frac{2c^{(\alpha)}(x)e^{2C^{(\alpha)}(x)}}{\left(\int_0^x e^{2C^{(\alpha)}(y)}dy\right)^2}\int_0^x f(y)e^{2C^{(\alpha)}(y)}dy - \frac{2e^{4C^{(\alpha)}(x)}}{\left(\int_0^x e^{2C^{(\alpha)}(y)}dy\right)^2}f(x) \\[2mm] & +\frac{2e^{4C^{(\alpha)}(x)}}{\left(\int_0^x e^{2C^{(\alpha)}(y)}dy\right)^3}\int_0^x f(y)e^{2C^{(\alpha)}(y)}dy \\[2mm] & +\frac{2c^{(\alpha)}(x)e^{2C^{(\alpha)}(x)}}{\int_0^x e^{2C^{(\alpha)}(y)}dy}f(x) + \frac{e^{2C^{(\alpha)}(x)}}{\int_0^x e^{2C^{(\alpha)}(y)}dy}f'(x).\end{aligned} \tag{8.37}$$

Gathering (8.36) and (8.37), we obtain after simplifications:

$$\overline{L}H_\alpha f(x) = \frac{1}{2}(H_\alpha f)''(x) + \left( \frac{e^{2C(x)}}{\int_0^x e^{2C(u)}du} - c(x) \right) (H_\alpha f)'(x)$$

$$= \frac{1}{2} \frac{e^{2C^{(\alpha)}(x)}}{\int_0^x e^{2C^{(\alpha)}(y)}dy} f'(x)$$

$$= \frac{1}{2} \frac{e^{2C(x)}}{\left( \int_0^x e^{2C(y)}dy \right) \left( 1 + \alpha \int_0^x e^{2C(y)}dy \right)} f'(x). \qquad (8.38)$$

On the other hand:

$$H_\alpha L^\alpha f(x) = \frac{1}{\int_0^x e^{2C^{(\alpha)}(y)}dy} \int_0^x \left( \frac{1}{2} f''(y) + c^{(\alpha)}(y)f'(y) \right) e^{2C^{(\alpha)}(y)}dy$$

$$= \frac{1}{\int_0^x e^{2C^{(\alpha)}(y)}dy} \left( \int_0^x c^{(\alpha)}(y)f'(y)e^{2C^{(\alpha)}(y)}dy + \frac{1}{2} \left[ e^{2C^{(\alpha)}(y)} f'(y) \right]_0^x \right.$$

$$\left. - \frac{1}{2} \int_0^x f'(y) 2c^{(\alpha)}(y)e^{2C^{(\alpha)}(y)}dy \right) \qquad \text{(after integrating by parts)}$$

$$= \frac{1}{2} \frac{e^{2C^{(\alpha)}(x)}}{\int_0^x e^{2C^{(\alpha)}(y)}dy} f'(x) \qquad \text{(since } e^{2C^{(\alpha)}(0)} = 0)$$

$$= \frac{1}{2} \frac{e^{2C(x)}}{\left( \int_0^x e^{2C(y)}dy \right) \left( 1 + \alpha \int_0^x e^{2C(y)}dy \right)} f'(x). \qquad (8.39)$$

The comparison of (8.38) and (8.39) ends the proof of Proposition 8.3.

□

**3) We now establish a link between $X$ and $X^{(\alpha)}$.**

**Proposition 8.4.** *We have:*

$$\mathbb{P}_0^{(\alpha)} \left( X_t^{(\alpha)} \geq z \right) = \frac{1}{1 + \alpha \int_0^z e^{2C(y)}dy} \mathbb{P}_0(X_t \geq z). \qquad (8.40)$$

*Remark 8.5.* Proposition 8.4 shows that $X^{(\alpha)}$ can admit a pseudo-inverse only if $\alpha = 0$ since, for $\alpha > 0$,

$$\lim_{t \to \infty} \mathbb{P}_0^{(\alpha)}(X_t^{(\alpha)} \geq z) \leq \frac{1}{1 + \alpha \int_0^z e^{2C(y)}dy} < 1.$$

*Proof.* From Proposition 8.3, we have, taking $\alpha = 0$ in (8.35):

$$HP_t = \overline{P}_t H.$$

Hence, from (8.35),

$$P_t^\alpha = H_\alpha^{-1} HP_t H^{-1} H_\alpha.$$

Let us evaluate these operators. For $f$ such that $f(0)=0$, we have:

$$H_\alpha^{-1}f(x) = f(x) + e^{-2C(x)}f'(x)\left(\int_0^x e^{2C(y)}dy\right)\left(1 + \alpha\int_0^x e^{2C(y)}dy\right),$$

$$H^{-1}f(x) = f(x) + e^{-2C(x)}f'(x)\left(\int_0^x e^{2C(y)}dy\right),$$

$$H_\alpha^{-1}Hf(x) = -\alpha\int_0^x f(y)e^{2C(y)}dy + \left(1 + \alpha\int_0^x e^{2C(y)}dy\right)f(x),$$

$$H^{-1}H_\alpha f(x) = \frac{f(x)}{1 + \alpha\int_0^x e^{2C(y)}dy} + \int_0^x \frac{\alpha e^{2C(y)}f(y)}{\left(1 + \alpha\int_0^y e^{2C(z)}dz\right)^2}dy$$

$$= \int_0^x \frac{f'(y)}{1 + \alpha\int_0^y e^{2C(z)}dz}dy.$$

Therefore, we obtain:

$$P_t H^{-1}H_\alpha f(x) = \mathbb{E}_x\left[\int_0^{X_t} \frac{f'(y)}{1 + \alpha\int_0^y e^{2C(z)}dz}dy\right]$$

and:

$$H_\alpha^{-1}HP_t H^{-1}H_\alpha f(x) = -\alpha\int_0^x e^{2C(y)}\mathbb{E}_y\left[\int_0^{X_t} \frac{f'(u)}{1 + \alpha\int_0^u e^{2C(z)}dz}du\right]dy \qquad (8.41)$$

$$+ \left(1 + \alpha\int_0^x e^{2C(y)}dy\right)\mathbb{E}_x\left[\int_0^{X_t} \frac{f'(y)}{1 + \alpha\int_0^y e^{2C(z)}dz}dy\right].$$

Letting $x$ tend toward 0 in (8.41), we obtain:

$$P_t^\alpha f(0) = \mathbb{E}_0\left[\int_0^{X_t} \frac{f'(y)}{1 + \alpha\int_0^y e^{2C(z)}dz}dy\right]. \qquad (8.42)$$

We take $f(x) = \int_0^x h(y)dy$ with $h$ positive. (8.42) becomes:

$$P_t^\alpha f(0) = \int_0^\infty\left(\int_0^y h(z)dz\right)p^{(\alpha)}(t,0,y)dy \qquad (8.43)$$

$$= \mathbb{E}_0\left[\int_0^{X_t} \frac{h(y)}{1 + \alpha\int_0^y e^{2C(z)}dz}dy\right]$$

$$= \int_0^\infty\left(\int_0^y \frac{h(z)}{1 + \alpha\int_0^z e^{2C(u)}du}dz\right)p(t,0,y)dy. \qquad (8.44)$$

Applying Fubini in (8.43) and (8.44), we obtain:

$$\int_0^\infty h(z)\left(\int_z^\infty p^{(\alpha)}(t,0,y)dy\right)dz = \int_0^\infty \frac{h(z)dz}{1 + \alpha\int_0^z e^{2C(u)}du}\int_z^\infty p(t,0,y)dy,$$

which ends the proof of Proposition 8.4.

$\square$

*Remark 8.6.* Let us introduce the $\left(\mathbb{P}_x^{(\alpha)}, (\mathscr{F}_t, t \geq 0)\right)$ martingale:

$$M_t^{(\alpha)} := \left(\frac{e^{-2C(X_t)} \int_0^{X_t} e^{2C(u)} du}{e^{-2C(x)} \int_0^x e^{2C(u)} du}\right) \left(\frac{1 + \alpha \int_0^{X_t} e^{2C(u)} du}{1 + \alpha \int_0^x e^{2C(u)} du}\right) \exp\left(\int_0^t \left(c^{(\alpha)}\right)'(X_u) du\right).$$

Then, Girsanov's formula gives:

$$\overline{\mathbb{P}}_{x|\mathscr{F}_t} = M_t^{(\alpha)} \cdot \mathbb{P}_{x|\mathscr{F}_t}^{(\alpha)}.$$

This leads to the relation:

$$\frac{\overline{p}(t,x,y)}{p^{(\alpha)}(t,x,y)}$$

$$= \left(\frac{e^{-2C(y)} \int_0^y e^{2C(u)} du}{e^{-2C(x)} \int_0^x e^{2C(u)} du}\right) \left(\frac{1 + \alpha \int_0^y e^{2C(u)} du}{1 + \alpha \int_0^x e^{2C(u)} du}\right) \mathbb{E}_x^{(\alpha)}\left[e^{\int_0^t \left(c^{(\alpha)}\right)'(X_s) ds} | X_t = y\right].$$

$$(8.45)$$

Therefore, plugging (8.45) in the general Hirsch-Song formula (8.4), one obtains in this case:

$$\frac{e^{2C(x)}}{\left(\int_0^x e^{2C(u)} du\right) \left(1 + \alpha \int_0^x e^{2C(u)} du\right)} \frac{\partial p^{(\alpha)}}{\partial x}(t,x,y)$$

$$= -\frac{\partial}{\partial y}\left(\frac{e^{2C(y)} \overline{p}(t,x,y)}{\left(\int_0^y e^{2C(u)} du\right) \left(1 + \alpha \int_0^y e^{2C(u)} du\right)}\right). \quad (8.46)$$

Introducing the differential operator:

$$D_\alpha f(x) = \frac{e^{2C(x)}}{\left(\int_0^x e^{2C(u)} du\right) \left(1 + \alpha \int_0^x e^{2C(u)} du\right)} f'(x),$$

it follows easily from (8.46) that:

$$D_\alpha P_t^\alpha = \overline{P}_t D_\alpha.$$

We can also note that, as a transient diffusion whose endpoint 0 is entrance-not-exit, $\overline{X}$ is not "too far" from a Bessel process of dimension 3. Indeed, let us introduce $m(dx) = 2e^{2C(x)} dx$ the speed measure of the diffusion $(X_t, t \geq 0)$. We define the time change $\left(\overline{A}_t := \int_0^t 4e^{4C(\overline{X}_s)} ds, t \geq 0\right)$ and $(\overline{\tau}_u, u \geq 0)$ its right-continuous inverse. Then:

**Proposition 8.5.** $\left(m[0, \overline{X}_{\overline{\tau}_u}], u \geq 0\right)$ *is a Bessel process of dimension 3 starting from* $m[0, x]$.

*Proof.* Let us apply Itô formula to $m[0,\overline{X}_t]$, using (8.31):

$$m[0,\overline{X}_t] = m[0,x] + \int_0^t 2e^{2C(\overline{X}_s)}dB_s + \int_0^t 2\frac{e^{4C(\overline{X}_s)}}{\int_0^{\overline{X}_s} e^{2C(y)}dy}ds.$$

Denoting by $\beta$ the Dambis, Dubins, Schwarz's Brownian motion associated to the local martingale $\left(N_t := \int_0^t 2e^{2C(\overline{X}_s)}dB_s, t \geq 0\right)$, we obtain $N_t = \beta_{A_t}$ and:

$$m[0,\overline{X}_t] = m[0,x] + \beta_{A_t} + \int_0^t \frac{d\overline{A}_s}{m[0,\overline{X}_s]}.$$

Finally, the time change $t = \overline{\tau}_u$ gives:

$$m[0,\overline{X}_{\overline{\tau}_u}] = m[0,x] + \beta_u + \int_0^u \frac{ds}{m[0,\overline{X}_{\overline{\tau}_s}]},$$

and we recognize the SDE satisfied by the Bessel process of dimension 3 starting from $m[0,x]$. This ends the proof of Proposition 8.5.

$\square$

## 8.3.3 Existence of a Pseudo-Inverse when $\alpha = 0$

We now state the main result of this subsection.

**Theorem 8.2.** *Let $\alpha \geq 0$ and $X^{(\alpha)}$ the diffusion solution of (8.26). Then:*

*i) The function $t \mapsto \mathbb{P}_0^{(\alpha)}(X_t^{(\alpha)} \geq y)$ increases and equals:*

$$\mathbb{P}_0^{(\alpha)}(X_t^{(\alpha)} \geq y) = \frac{1}{1 + \alpha \int_0^y e^{2C(z)}dz}\mathbb{P}_0\left(\overline{G}_y \leq t\right) \tag{8.47}$$

*where $\overline{G}_y := \sup\{t \geq 0; \overline{X}_t = y\}$.*

*ii) $\lim_{t\to\infty} \mathbb{P}_0^{(\alpha)}(X_t^{(\alpha)} \geq y) = \frac{1}{1 + \alpha \int_0^y e^{2C(z)}dz}.$*

*iii) Therefore:*

*1) if $\alpha = 0$, $X$ admits an increasing pseudo-inverse $(Y_{0,y}, y > 0)$ and we have the following equalities in law:*

*(a) $X_t \overset{(law)}{=} \inf_{s \geq t}\overline{X}_s$,*

*(b) $Y_{0,y} \overset{(law)}{=} \overline{G}_y$.*

*2) if $\alpha > 0$, $X^{(\alpha)}$ admits an increasing quasi pseudo-inverse $(Y_{0,y}^{(\alpha)}, y > 0)$.*

*Proof.*
We start by a useful Lemma

Let $u(t,y)$ denote the density of $\inf\limits_{s\geq t}\overline{X}_s$ under $\overline{\mathbb{P}}_0$. We have:

**Lemma 8.3.**

$$u(t,y) = e^{2C(y)} \int_y^\infty \frac{\overline{p}(t,0,z)}{\int_0^z e^{2C(u)}du} dz. \tag{8.48}$$

*Proof.* Using (8.23) (since $C \in \mathscr{H}^\infty$) and the Markov property for the diffusion $\overline{X}$ we have:

$$\overline{\mathbb{P}}_0\left(\inf_{s\geq t}\overline{X}_s > y\right) = \int_y^\infty \overline{\mathbb{P}}_0\left(\inf_{s\geq t}\overline{X}_s > y | \overline{X}_t = z\right) \overline{p}(t,0,z)dz$$

$$= \int_y^\infty \overline{\mathbb{P}}_z\left(\inf_{s\geq 0}\overline{X}_s > y\right) \overline{p}(t,0,z)dz$$

$$= \int_y^\infty \frac{\int_y^z e^{2C(u)}du}{\int_0^z e^{2C(a)}da}\overline{p}(t,0,z)dz \qquad \text{(from (8.23))}$$

$$= \int_y^\infty du\, e^{2C(u)} \int_u^\infty \frac{\overline{p}(t,0,z)}{\int_0^z e^{2C(a)}da}dz \quad \text{(by Fubini).} \tag{8.49}$$

Differentiating (8.49) with respect to $y$ ends the proof of Lemma 8.3.

$\square$

We go back to the proof of Theorem 8.2

We start by showing Point (*iii*) item (*a*). We need to show that $u(t,y) = p(t,0,y)$. We will prove that $u$ satisfies the same parabolic equation as $p$, namely:

$$\frac{\partial p}{\partial t}(t,0,y) = L^* p(t,0,y) \tag{8.50}$$

where

$$L^* f = \frac{1}{2}f'' - (cf)'. \tag{8.51}$$

1) On the one hand, let us calculate $\dfrac{\partial u}{\partial t}$ using (8.48):

$$\frac{\partial u}{\partial t}(t,y) = e^{2C(y)} \int_y^\infty \frac{\partial \overline{p}}{\partial t}(t,0,z)\frac{dz}{\int_0^z e^{2C(u)}du}$$

$$= -2e^{2C(y)} \int_y^\infty \overline{L}^*\overline{p}(t,0,z)\overline{s}(z)dz \tag{8.52}$$

from Fokker-Planck and (8.32). We integrate (8.52) by parts. $\overline{s}$ being a scale function for the process $\overline{X}$, we have for all $z$, $\overline{L}\overline{s}(z) = 0$. Consequently, only the boundary terms coming from the integration by parts are not null. We obtain:

$$\frac{\partial u}{\partial t}(t,y)$$

$$= -2e^{2C(y)} \int_y^\infty \left( \frac{1}{2}\overline{p}''(t,0,z) - \left( \left( \frac{e^{2C(z)}}{\int_0^z e^{2C(u)}du} - c(z) \right) \overline{p}(t,0,z) \right)' \right) \overline{s}(z)dz$$

$$= e^{2C(y)} \left( -\frac{1}{2}\frac{\overline{p}'(t,0,y)}{\int_0^y e^{2C(u)}du} + \left( \frac{e^{2C(y)}}{\int_0^y e^{2C(u)}du} - c(y) \right) \frac{\overline{p}(t,0,y)}{\int_0^y e^{2C(u)}du} - \frac{1}{2}\frac{e^{2C(y)}\overline{p}(t,0,y)}{\left(\int_0^y e^{2C(u)}du\right)^2} \right)$$

$$= e^{2C(y)} \left( -\frac{1}{2}\frac{\overline{p}'(t,0,y)}{\int_0^y e^{2C(u)}du} - c(y)\frac{\overline{p}(t,0,y)}{\int_0^y e^{2C(u)}du} + \frac{1}{2}\frac{e^{2C(y)}\overline{p}(t,0,y)}{\left(\int_0^y e^{2C(u)}du\right)^2} \right). \tag{8.53}$$

2) On the other hand, we calculate $L^*u$ still using (8.48).We have:

$$\frac{\partial u}{\partial y}(t,y) = 2c(y)u(t,y) - e^{2C(y)}\frac{\overline{p}(t,0,y)}{\int_0^y e^{2C(u)}du}, \tag{8.54}$$

and

$$\frac{\partial^2 u}{\partial y^2}(t,y) = 2c'(y)u(t,y) + 2c(y)\left(2c(y)u(t,y) - e^{2C(y)}\frac{\overline{p}(t,0,y)}{\int_0^y e^{2C(u)}du}\right) \tag{8.55}$$

$$-2c(y)e^{2C(y)}\frac{\overline{p}(t,0,y)}{\int_0^y e^{2C(u)}du} - e^{2C(y)}\frac{\overline{p}'(t,0,y)}{\int_0^y e^{2C(u)}du} + \frac{e^{4C(y)}\overline{p}(t,0,y)}{\left(\int_0^y e^{2C(u)}du\right)^2}.$$

Hence, gathering (8.54) and (8.55), we obtain:

$$L^*u(\cdot,t)(y) = e^{2C(y)}\left( -\frac{1}{2}\frac{\overline{p}'(t,0,y)}{\int_0^y e^{2C(u)}du} - c(y)\frac{\overline{p}(t,0,y)}{\int_0^y e^{2C(u)}du} + \frac{1}{2}\frac{e^{2C(y)}\overline{p}(t,0,y)}{\left(\int_0^y e^{2C(u)}du\right)^2} \right). \tag{8.56}$$

Comparing (8.53) and (8.56), we see that:

$$\frac{\partial u}{\partial t}(t,y) = L^*u(\cdot,t)(y).$$

We must now check that the 2 functions $(t,y) \mapsto p(t,0,y)$ and $(t,y) \mapsto u(t,y)$ satisfy the same initial conditions. First, it is clear that $p(t,0,\cdot)$ and $u(t,\cdot)$ are 2 density functions on $\mathbb{R}^+$ satisfying: $u(0,\cdot) = p(0,0,\cdot) = \delta_0$. Then, for every $t > 0$, $u(t,0) = 0$, since $\overline{X}$ is a transient diffusion whose endpoint 0 is entrance-not-exit (and therefore 0 does not belong to the state space of $\overline{X}$). The same is true for $X$: if 0 is entrance-not-exit for the diffusion $X$, then for every $t > 0$, $p(t,0,0) = 0$. Otherwise, 0 is reflecting. In this case, let us introduce $q(t,x,y)$ the transition density of $X$ with respect to its speed measure $m(dy) = 2e^{2C(y)}dy$. $q$ is well-defined on $[0,+\infty[^3$ (see [12], p.13) and we have $p(t,x,y)dy = q(t,x,y)m(dy)$. Thus, for $x = y = 0$, we get $p(t,0,0) = q(t,0,0)2e^{2C(0)} = 0$ since $C(0) = -\infty$. Consequently, by uniqueness of the solution of equation (8.50) (when proper initial conditions are given, see [43], p.145), we obtain that:

$$\forall t \geq 0, \ \forall y \geq 0, \quad u(t,y) = p(t,0,y). \tag{8.57}$$

This ends the proof of $(iii)$ item $(a)$.

We now end the proof of Theorem 8.2

From Lemma 8.3, we have, with $\overline{G}_y := \sup\{s \geq 0; \ \overline{X}_s = y\}$:

$$\mathbb{P}_0(X_t \geq y) = \overline{\mathbb{P}}_0 \left( \inf_{s \geq t} \overline{X}_s \geq y \right) = \overline{\mathbb{P}}_0(\overline{G}_y \leq t). \tag{8.58}$$

Plugging (8.58) in (8.40) gives (8.47) (i.e. Point $(i)$).

Point $(ii)$ follows from the fact that $\overline{X}$ is transient from Proposition 8.1. Therefore, if $\alpha = 0$, $X$ admits an increasing pseudo-inverse, and we have:

$$\overline{\mathbb{P}}_0(\overline{G}_y \leq t) = \mathbb{P}(Y_{0,y} \leq t).$$

This ends the proof of Theorem 8.2.

$\square$

*Remark 8.7.* If $c$ is a decreasing function, then, for all $x \geq 0$, $TC(x) \geq c(x)$. Indeed, we have:

$$
\begin{aligned}
TC(x) &= \frac{e^{2C(x)}}{\int_0^x e^{2C(y)}dy} - c(x) \geq c(x) \\
&\iff e^{2C(x)} - 2c(x) \int_0^x e^{2C(y)}dy \geq 0 \\
&\iff e^{2C(x)} - c(x) \left( \left[ \frac{e^{2C(y)}}{c(y)} \right]_0^x + \int_0^x \frac{e^{2C(y)}c'(y)}{c^2(y)}dy \right) \geq 0 \\
&\iff -c(x) \int_0^x \frac{e^{2C(y)}c'(y)}{c^2(y)}dy \geq 0.
\end{aligned}
$$

Therefore, applying the stochastic comparison theorem, if we realize the 2 processes $X$ and $\overline{X}$ on the same space (with respect to the same Brownian motion), then,

$$\mathbb{P}\left(\overline{X}_t \geq X_t \text{ for all } t \geq 0\right) = 1.$$

## 8.4 Existence of Pseudo-Inverses for a $\mathbb{R}^+$-Valued Diffusion Started at 0

### 8.4.1 Notations

Our aim now is to extend the results proven previously for the family $(X^{(\alpha)})_{\alpha \geq 0}$ to the general framework of a linear regular diffusion $X$ taking values in $\mathbb{R}^+$. Let us denote by $m$ its speed measure, and $s$ its scale function. We assume that $m$ is absolutely continuous with respect to the Lebesgue measure:

$$m(dx) = \rho(x)dx, \qquad (\rho > 0)$$

and that $s$ is a strictly increasing function. We also introduce $q(t,x,y)$ its transition density with respect to the speed measure,

$$u_\lambda(x,y) := \int_0^\infty e^{-\lambda t} q(t,x,y)dt$$

its resolvent kernel and

$$L := \frac{\partial^2}{\partial m \partial s}$$

its infinitesimal generator. Let us recall the following classification of boundaries, for the left hand endpoint 0 (See [12], p.14):

i)   0 is called exit if, for $z > 0$:

$$\int_0^z m[a,z]s'(a)da < \infty, \tag{8.59}$$

ii) 0 is called entrance if, for $z > 0$:

$$\int_0^z (s(z) - s(a))m(da) < \infty. \tag{8.60}$$

Let us note that integrating (8.60) by parts gives an equivalent condition:

$$\lim_{a \to 0}(s(a)m[0,a]) + \int_0^z m[0,a]s'(a)da < \infty. \tag{8.61}$$

In our study, we assume that the diffusion is started at 0. Therefore, inequality (8.60) (or equivalently (8.61)) must be satisfied, and 0 is an entrance endpoint. Furthermore:

• If 0 is also an exit endpoint, it is called a non-singular boundary. A diffusion reaches its non-singular boundaries with positive probability. In this case, it is necessary, in order to describe the diffusion process, to add a boundary condition. Here, since $m$ is assumed to be absolutely continuous with respect to the Lebesgue measure, we have $m(\{0\}) = 0$, and 0 will be a reflecting boundary.

• If 0 is not an exit endpoint (i.e. equation (8.59) is not satisfied), then, the diffusion cannot reach it from an interior point of its state space. Therefore, since the diffusion is assumed to be regular, 0 does not belong to the state space. But, as an entrance endpoint, it is nevertheless possible to start the diffusion from 0.

We also assume that $+\infty$ is a natural boundary (i.e. neither entrance, nor exit).

## 8.4.2 Biane's Transformation

### 1) Associating $\overline{X}$ to $X$

**Definition 8.1.** To the diffusion $X$, we now associate another diffusion, $\overline{X}$, whose speed measure $\overline{m}(dx) = \overline{\rho}(x)dx$ and scale function $\overline{s}$ are defined by:

$$\begin{cases} \overline{\rho}(x) = (m[0,x])^2 s'(x), \\[2mm] \overline{s}(x) = \dfrac{1}{m[0,+\infty[} - \dfrac{1}{m[0,x]}. \end{cases} \tag{8.62}$$

$\overline{X}$ is a transient diffusion, since $s$ is increasing and:

$$\overline{s}(0) = -\infty \qquad \text{and} \qquad \overline{s}(+\infty) = 0.$$

We assume furthermore that $\lim_{a \to 0} m[0,a]s(a) > -\infty$. From (8.61), this implies that the function $a \mapsto m[0,a]s'(a)$ is integrable at 0. Therefore, from (8.62), the endpoint 0 is entrance-not-exit for the diffusion $\overline{X}$.

For example, if $X$ is a Bessel process with index $\nu$, then $\overline{X}$ is a Bessel process with index $\nu + 1$. This transformation was first introduced by Biane [9] in order to generalize a celebrated identity from Ciesielski and Taylor ([17]), which was originally obtained for Bessel processes.

*Remark 8.8.*
**a)** We must stress the fact that this transformation is not injective. For instance, as for the operator $T$ (see (8.22)), if $X^{(\alpha)}$ is the solution of (8.26), then $\overline{X}^{(\alpha)}$ defined by Biane's transformation does not depend on $\alpha$. Indeed, from (8.26):

$$\left( s^{(\alpha)} \right)'(x) = \left( 1 + \alpha \int_0^x e^{2C(u)} du \right)^2 e^{-2C(x)}, \tag{8.63}$$

and

$$m^{(\alpha)}(dx) = \frac{2 e^{2C(x)}}{\left( 1 + \alpha \int_0^x e^{2C(u)} du \right)^2} dx. \tag{8.64}$$

Integrating (8.64) gives:

$$\begin{aligned} m^{(\alpha)}[0,x] &= 2 \int_0^x \frac{e^{2C(y)}}{\left( 1 + \alpha \int_0^y e^{2C(u)} du \right)^2} dy \\[2mm] &= 2 \left[ -\frac{1}{\alpha} \frac{1}{1 + \alpha \int_0^y e^{2C(u)} du} \right]_0^x \\[2mm] &= \frac{2 \int_0^x e^{2C(u)} du}{1 + \alpha \int_0^x e^{2C(u)} du}. \end{aligned} \tag{8.65}$$

Therefore:

$$\overline{s}^{(\alpha)}(x) := \frac{\alpha}{2} - \frac{1}{m^{(\alpha)}[0,x]} = -\frac{1}{2\int_0^x e^{2C(u)}du},$$

and

$$\overline{\rho}^{(\alpha)}(x) := (m^{(\alpha)}[0,x])^2 \left(\overline{s}^{(\alpha)}\right)'(x) = \left(2\int_0^x e^{2C(u)}du\right)^2 e^{-2C(x)}.$$

We thus see that the characteristics of the diffusion $\overline{X}^{(\alpha)}$ do not depend on $\alpha$. In fact, $\overline{X}^{(\alpha)} = \overline{X}$ where $\overline{X}$ is the diffusion defined by (8.31).

**b)** More generally, if $X$ is a linear diffusion with speed measure $m(dx) = \rho(x)dx$ and scale function $s$ which we suppose to be strictly increasing, then, the 2 processes $s(X)$ and $s(\overline{X})$ have the same law. This can easily be shown by computing the 2 infinitesimal generators.

**2) We now study some links between $X$ and $\overline{X}$.**
Let $H$ be the functional:

$$Hf(x) := \frac{1}{m[0,x]} \int_0^x f(y)m(dy) \tag{8.66}$$

defined on the space of continuous functions with compact support in $]0, +\infty[$.

**Lemma 8.4.** *The following intertwining relation:*

$$HP_t = \overline{P}_t H$$

*holds.*

*Proof.* It is sufficient to check the identity on the infinitesimal generators. Let $f$ be a $\mathscr{C}^2$ function with compact support in $]0, +\infty[$. On the one hand, we have:

$$
\begin{aligned}
HLf(x) &= \frac{1}{m[0,x]} \int_0^x Lf(y)m(dy) \\
&= \frac{1}{m[0,x]} \int_0^x \frac{\partial}{\partial m}\frac{\partial}{\partial s}f(y)m(dy) \\
&= \frac{1}{m[0,x]} \int_0^x \frac{1}{\rho(y)}\frac{\partial}{\partial y}\left(\frac{f'(y)}{s'(y)}\right)\rho(y)dy \\
&= \frac{1}{m[0,x]} \left(\frac{f'(x)}{s'(x)} - \frac{f'(0)}{s'(0)}\right) \\
&= \frac{f'(x)}{m[0,x]s'(x)} \quad \text{(since $f$ has a compact support in $]0, +\infty[$).} \tag{8.67}
\end{aligned}
$$

On the other hand, since from (8.62) $\overline{s}'(x) = \dfrac{\rho(x)}{(m[0,x])^2}$, we have:

$$\overline{L}Hf(x) = \frac{\partial^2}{\overline{\rho}(x)\partial x\overline{s}'(x)\partial x}\left(\frac{1}{m[0,x]}\int_0^x f(y)m(dy)\right)$$

$$= \frac{\partial}{\overline{\rho}(x)\partial x}\frac{(m[0,x])^2}{\rho(x)}\left(-\frac{\rho(x)}{(m[0,x])^2}\int_0^x f(y)m(dy) + \frac{\rho(x)}{(m[0,x])}f(x)\right)$$

$$= \frac{\partial}{\overline{\rho}(x)\partial x}\left(-\int_0^x f(y)m(dy) + m[0,x]f(x)\right)$$

$$= \frac{m[0,x]}{\overline{\rho}(x)}f'(x)$$

$$= \frac{f'(x)}{m[0,x]s'(x)}. \tag{8.68}$$

The comparison of (8.67) and (8.68) ends the proof of Lemma 8.4.                                    $\square$

Lemma 8.4 will allow us to deduce the following relation between the transition densities of $X$ and $\overline{X}$:

**Lemma 8.5.**

$$\frac{1}{m[0,x]}\int_0^x q(t,y,z)m(dy) = \int_z^\infty \overline{q}(t,x,y)m[0,y]s'(y)dy. \tag{8.69}$$

*In particular, letting x tend to 0, we obtain:*

$$q(t,0,z) = \int_z^\infty \overline{q}(t,0,y)m[0,y]s'(y)dy. \tag{8.70}$$

*Proof.* We use the intertwining relation between the semi-groups of $X$ and $\overline{X}$. Let $f$ be a Borel function on $]0,+\infty[$ with compact support. On the one hand, we have, applying Fubini:

$$HP_tf(x) = \frac{1}{m[0,x]}\int_0^x\left(\int_0^\infty f(z)q(t,y,z)m(dz)\right)m(dy)$$

$$= \int_0^\infty f(z)\left(\frac{1}{m[0,x]}\int_0^x q(t,y,z)m(dy)\right)m(dz). \tag{8.71}$$

On the other hand, using (8.62):

$$\overline{P}_tHf(x) = \int_0^\infty\left(\frac{1}{m[0,y]}\int_0^y f(z)m(dz)\right)\overline{q}(t,x,y)\overline{m}(dy)$$

$$= \int_0^\infty f(z)\left(\int_z^\infty\frac{1}{m[0,y]}\overline{q}(t,x,y)\overline{m}(dy)\right)m(dz)$$

$$= \int_0^\infty f(z)\left(\int_z^\infty \overline{q}(t,x,y)m[0,y]s'(y)dy\right)m(dz). \tag{8.72}$$

The comparison of (8.71) and (8.72) ends the proof of Lemma 8.5.                                    $\square$

*Remark 8.9.* Lemma 8.4 and Lemma 8.5 have differential counterparts. Indeed, if we define:

$$\widetilde{D}f(x) = \frac{1}{m[0,x]s'(x)}f'(x),$$

on the space of $\mathscr{C}^1$ functions, with derivative equal to 0 at 0, and which are bounded as well as their derivative, we obtain, following the same pattern of proof as for Lemma 8.4:

$$\widetilde{D}P_t = \overline{P}_t\widetilde{D}. \tag{8.73}$$

This easily implies the generalized Hirsch-Song formula:

$$\frac{1}{m[0,x]s'(x)}\frac{\partial p}{\partial x}(t,x,y) = -\frac{\partial}{\partial y}\left(\frac{\overline{p}(t,x,y)}{m[0,y]s'(y)}\right)$$

where $p$ (resp. $\overline{p}$) is the transition density function of $X$ (resp. $\overline{X}$) with respect to the Lebesgue measure. Furthermore, denoting by $Z$ the Markov process:

$$Z_t := \int_0^{X_t} m[0,a]s'(a)da,$$

and $(Q_t, t \geq 0)$ its semi-group, we have:

$$DQ_t = \overline{Q}_t D$$

where $D$ denotes the differentiation operator: $Df(x) = f'(x)$ defined on the space of $\mathscr{C}^1$ functions bounded as well as their first derivative.

### 8.4.3 Existence of Pseudo-Inverses

We now state our main result:

**Theorem 8.3.** *Let $X$ be a regular linear diffusion on $\mathbb{R}^+$ with speed measure $m(dx) = \rho(x)dx$ and scale function $s$. We assume that $s$ is a strictly increasing $\mathscr{C}^2$ function such that:*

$$\lim_{a \to 0} m[0,a]s(a) > -\infty,$$

*and that 0 is an entrance endpoint and $+\infty$ a natural boundary. Let $y > 0$. Then:*

*i)   The function $t \mapsto \mathbb{P}_0(X_t \geq y)$ increases and satisfies:*

$$\mathbb{P}_0(X_t \geq y) = \left(1 - \frac{m[0,y]}{m[0,+\infty[}\right)\overline{\mathbb{P}}_0\left(\overline{G}_y \leq t\right) \tag{8.74}$$

*where $\overline{G}_y := \sup\{s \geq 0; \overline{X}_s = y\}$.*

*ii)* $\lim_{t\to\infty} \mathbb{P}_0(X_t \geq y) = 1 - \dfrac{m[0, y]}{m[0, +\infty[}$.

*iii) Therefore:*

1) *If* $m[0, +\infty[= +\infty$, *X admits an increasing pseudo-inverse* $(Y_{0,y}, y > 0)$ *and we have the following equalities in law:*

   (a) $X_t \overset{(law)}{=} \inf_{s\geq t} \overline{X}_s$,

   (b) $Y_{0,y} \overset{(law)}{=} \overline{G}_y$.

2) *If* $m[0, +\infty[< +\infty$, *X admits an increasing quasi pseudo-inverse* $(Y_{0,y}, y > 0)$.

*Then, in our framework, a diffusion admits an increasing pseudo-inverse if and only if it is transient, or null recurrent.*

**Remark 8.10.**
**a)** Observe that, unlike $(X_t, t \geq 0)$, the process $(\inf_{s\geq t} \overline{X}_s, t \geq 0)$ is increasing, see Remark 7.2 and [39].
**b)** It is clear that a positively recurrent diffusion cannot admit an increasing pseudo-inverse since, denoting $\pi(dz) := \dfrac{\rho(z)}{m[0, +\infty[}dz$ the stationary probability measure of $X$, we have, for $y > 0$ (see [12], p.35):

$$\mathbb{P}_0(X_t \geq y) \xrightarrow[t\to\infty]{} \int_0^\infty 1_{\{z\geq y\}} \pi(dz) = \frac{m[y, +\infty[}{m[0, +\infty[} < 1.$$

**Remark 8.11.** Let us study, using this theorem, the case of the diffusions $X^{(\alpha)}$ which are solutions of (8.26). From (8.63), we see that $s^{(\alpha)}$ is strictly increasing and:

$$\left|m^{(\alpha)}[0, a]s^{(\alpha)}(a)\right| = \frac{2\int_0^a e^{2C(u)}du}{1+\alpha\int_0^a e^{2C(u)}du} \int_a^1 \left(1+\alpha\int_0^x e^{2C(u)}du\right)^2 e^{-2C(x)}dx$$

$$\leq (1-a)\left(1+\alpha\int_0^1 e^{2C(u)}du\right)^2 e^{-2C(a)}2ae^{2C(a)}$$

$$\leq 2a(1-a)\left(1+\alpha\int_0^1 e^{2C(u)}du\right)^2 \xrightarrow[a\to 0]{} 0.$$

Thus Theorem 8.3 applies, and the application $t \mapsto \mathbb{P}_0^{(\alpha)}(X_t^{(\alpha)} \geq y)$ increases. Furthermore, since $C \in \mathscr{H}^\infty$ (cf. (8.21)), we obtain, letting $x$ tend to $+\infty$ in (8.65):

$$m^{(\alpha)}[0, +\infty[= \frac{2}{\alpha}.$$

Hence, from Point *(ii)* of Theorem 8.3,

$$\lim_{t\to\infty} \mathbb{P}_0^{(\alpha)}(X_t^{(\alpha)} \geq y) = 1 - \frac{\alpha}{2}\left(\frac{\int_0^y 2e^{2C(u)}du}{1+\alpha\int_0^y e^{2C(u)}du}\right) = \frac{1}{1+\alpha\int_0^y e^{2C(u)}du}.$$

This is Point $(ii)$ of Theorem 8.2. Hence, Theorem 8.2 is a particular case of Theorem 8.3.

*Proof.*
**a)** It is known from [35], p.149, that $(t,z) \mapsto q(t,0,z)$ solves the parabolic equation:

$$\frac{\partial u}{\partial t}(t,z) = Lu(t,z).$$

Hence:

$$\frac{\partial}{\partial t}\mathbb{P}_0(X_t \geq y) = \int_y^\infty \frac{\partial q}{\partial t}(t,0,z)m(dz)$$

$$= \int_y^\infty Lq(t,0,z)m(dz)$$

$$= \int_y^\infty \frac{\partial}{\partial m}\frac{\partial q}{\partial s}(t,0,z)m(dz)$$

$$= -\frac{\partial q}{\partial s}(t,0,y). \tag{8.75}$$

Now, (8.75) can be rewritten, thanks to (8.70):

$$\frac{\partial}{\partial t}\mathbb{P}_0(X_t \geq y) = -\frac{1}{s'(y)}\frac{\partial}{\partial y}\left(\int_y^\infty \bar{q}(t,0,z)m[0,z]s'(z)dz\right)$$

$$= \bar{q}(t,0,y)m[0,y]$$

$$= -\left(1 - \frac{m[0,y]}{m[0,+\infty[}\right)\frac{1}{\bar{s}(y)}\bar{q}(t,0,y) \quad \text{(from (8.62))}. \tag{8.76}$$

Then, $\overline{X}$ being transient, from Theorem 2.4, $t \mapsto -\frac{1}{\bar{s}(y)}\bar{q}(t,0,y)$ is the density function of the last passage time of $\overline{X}$ at level y, starting from 0:

$$\mathbb{P}_0(\overline{G}_y \in dt) = -\frac{1}{\bar{s}(y)}\bar{q}(t,0,y)dt. \tag{8.77}$$

Thus, integrating (8.76) with respect to $t$ yields Point $(i)$ of Theorem 8.3.

**b)** Point $(ii)$ is immediate since $\overline{X}$ is transient, and finally, items $(a)$ and $(b)$ follow easily from (8.74) and the identity:

$$\mathbb{P}_0\left(\inf_{s \geq t} \overline{X}_s \geq y\right) = \mathbb{P}_0\left(\overline{G}_y \leq t\right) = \mathbb{P}_0(X_t \geq y).$$

This ends the proof of Theorem 8.3.

$\square$

### 8.4.4 A Second Proof of Theorem 8.3

We now give another proof of Theorem 8.3, by showing that the Laplace transforms of both sides of (8.74) coincide. Let $\overline{X}$ be the process associated to $X$ by Biane's transformation. From (8.77) we have for $\lambda \geq 0$:

$$\overline{\mathbb{E}}_0 \left[ e^{-\lambda \overline{G}_y} \right] = - \int_0^\infty e^{-\lambda t} \frac{\overline{q}(t,0,y)}{\overline{s}(y)} dt = \frac{\overline{u}_\lambda(0,y)}{-\overline{s}(y)}. \tag{8.78}$$

Let us remark that $y \mapsto \overline{u}_\lambda(0,y)$ is the unique (up to a multiplicative constant) eigenfunction of the operator $\overline{L} = \dfrac{\partial^2}{\partial \overline{m} \partial \overline{s}}$, associated to the eigenvalue $\lambda$, which is decreasing and satisfies $\overline{u}_\lambda(0,+\infty) = 0$ (see [12], p. 18). Let $\phi$ be the function defined by:

$$\phi(y,\lambda) := \lambda \frac{1}{m[0,y]} \int_y^\infty u_\lambda(0,x) m(dx).$$

We show that $y \mapsto \phi(y,\lambda)$ satisfies the same conditions as $y \mapsto \overline{u}_\lambda(0,y)$. First, it is clear that $\phi(\cdot,\lambda)$ is a decreasing function such that $\phi(+\infty,\lambda) = 0$. Furthermore, we have:

$$\begin{aligned}
\overline{L}\phi(y,\lambda) &= \frac{\partial^2}{\partial \overline{m} \partial \overline{s}} \phi(y,\lambda) \\
&= \lambda \frac{\partial}{\overline{\rho}(y)\partial y} \frac{(m[0,y])^2}{\rho(y)} \left( -\frac{\rho(y)}{(m[0,y])^2} \int_y^\infty u_\lambda(0,x) m(dx) - \frac{\rho(y)}{m[0,y]} u_\lambda(0,y) \right) \\
&= \lambda \frac{\partial}{\overline{\rho}(y)\partial y} \left( -\int_y^\infty u_\lambda(0,x) m(dx) - m[0,y] u_\lambda(0,y) \right) \\
&= \frac{\lambda}{(m[0,y])^2 s'(y)} \left( u_\lambda(0,y)\rho(y) - \rho(y) u_\lambda(0,y) - m[0,y] \frac{\partial}{\partial y} u_\lambda(0,y) \right) \\
&= -\frac{\lambda}{(m[0,y])s'(y)} \frac{\partial}{\partial y} u_\lambda(0,y). \tag{8.79}
\end{aligned}$$

But, since $y \mapsto u_\lambda(0,y)$ is an eigenfunction of the operator $L = \dfrac{\partial^2}{\partial m \partial s}$ associated to the eigenvalue $\lambda$, we have:

$$\frac{\partial^2}{\rho(y)\partial y s'(y)\partial y} u_\lambda(0,y) = \lambda u_\lambda(0,y). \tag{8.80}$$

Integrating (8.80) gives:

$$\frac{1}{s'(+\infty)} \frac{\partial}{\partial y} u_\lambda(0,+\infty) - \frac{1}{s'(y)} \frac{\partial}{\partial y} u_\lambda(0,y) = \lambda \int_y^\infty u_\lambda(0,x)\rho(x) dx. \tag{8.81}$$

But the endpoint $+\infty$ is assumed to be natural, and therefore, from ([12], p.19):

$$\frac{1}{s'(+\infty)}\frac{\partial}{\partial y}u_\lambda(0,+\infty) = \frac{\partial u_\lambda}{\partial s}(0,+\infty) = 0.$$

Then, plugging (8.81) into (8.79), we obtain:

$$\overline{L}\phi(y,\lambda) = \lambda\left(\lambda\frac{1}{m[0,y]}\int_y^\infty u_\lambda(0,x)m(dx)\right) = \lambda\phi(y,\lambda),$$

which means that $\phi$ is an eigenfunction of $\overline{L}$ associated to the eigenvalue $\lambda$. Therefore, there is a constant $\gamma > 0$ such that:

$$\overline{u}_\lambda(0,y) = \gamma\frac{\lambda}{m[0,y]}\int_y^\infty u_\lambda(0,x)m(dx). \tag{8.82}$$

Plugging (8.82) into (8.78) we finally get:

$$
\begin{aligned}
\overline{\mathbb{E}}_0\left[e^{-\lambda\overline{G}_y}\right] &= -\frac{\gamma\lambda}{\overline{s}(y)m[0,y]}\int_y^\infty u_\lambda(0,x)m(dx) \\
&= \frac{\gamma\lambda}{1-\frac{m[0,y]}{m[0,+\infty[}}\int_y^\infty\left(\int_0^\infty e^{-\lambda t}q(t,0,x)dt\right)\rho(x)dx \quad \text{(from (8.62))} \\
&= \frac{\gamma\lambda}{1-\frac{m[0,y]}{m[0,+\infty[}}\int_0^\infty e^{-\lambda t}\mathbb{P}_0(X_t \geq y)dt \quad \text{(applying Fubini)} \\
&= \frac{\gamma}{1-\frac{m[0,y]}{m[0,+\infty[}}\int_0^\infty e^{-\lambda t}\left(\frac{\partial}{\partial t}\mathbb{P}_0(X_t \geq y)\right)dt
\end{aligned}
$$

after an integration by parts. Therefore, from the injectivity of the Laplace transform, we deduce that $t \mapsto \dfrac{\gamma}{1-\frac{m[0,y]}{m[0,+\infty[}}\dfrac{\partial}{\partial t}\mathbb{P}_0(X_t \geq y)$ is the density function of a random variable $Y_{0,y}$ (having the same law as $\overline{G}_y$ under $\overline{\mathbb{P}}_0$) which satisfies:

$$
\begin{aligned}
\mathbb{P}(Y_{0,y} \leq t) = \overline{\mathbb{P}}_0(\overline{G}_y \leq t) &= \int_0^t \frac{\gamma}{1-\frac{m[0,y]}{m[0,+\infty[}}\left(\frac{\partial}{\partial s}\mathbb{P}_0(X_s \geq y)\right)ds \\
&= \frac{\gamma}{1-\frac{m[0,y]}{m[0,+\infty[}}\mathbb{P}_0(X_t \geq y).
\end{aligned}
$$

Then, letting $y$ tend to 0 in this last identity, we obtain:

$$\overline{\mathbb{P}}_0(\overline{G}_0 \leq t) = \gamma\mathbb{P}_0(X_t \geq 0) = \gamma \quad \text{(X being a } \mathbb{R}^+\text{-valued diffusion).} \tag{8.83}$$

But, since $\overline{X}$ is transient:

$$\overline{\mathbb{P}}_0(\overline{G}_0 \leq t) \xrightarrow[t\to\infty]{} 1,$$

and, from (8.83), $\gamma = 1$. The fact that $\overline{\mathbb{P}}_0(\overline{G}_0 \le t)$ does not depend on $t$ is not a surprise since $0$ is an entrance-not-exit endpoint for the diffusion $\overline{X}$, and thus $\overline{G}_0 = 0$ $\overline{\mathbb{P}}_0$-a.s. This ends the second proof of Theorem 8.3.

□

## Exercise 8.1 (Existence of pseudo-inverses for nearest neighbor random walk on $\mathbb{Z}$).

Let $(X_n, n \ge 0)$ denote the nearest neighbor random walk on $\mathbb{Z}$, started from $0$, and defined by:

$$X_0 = 0, \qquad X_n = Y_1 + \ldots + Y_n$$

where the r.v.'s $(Y_i)_{i \in \mathbb{N}^*}$ are i.i.d. and such that $\mathbb{P}(Y_1 = 1) = p$ and $\mathbb{P}(Y_1 = -1) = q$ with $p + q = 1$.

**1)** Let $k \in \mathbb{N}^*$. Prove that, for all $n \ge 0$:

$$\mathbb{P}(X_{2n+2} \ge 2k) = \mathbb{P}(X_{2n} \ge 2k) + p^2 \mathbb{P}(X_{2n} = 2k-2) - q^2 \mathbb{P}(X_{2n} = 2k). \quad (1)$$

Let $\varphi_k(n) := \mathbb{P}(X_{2n} \ge 2k)$. It follows from (1) that, for $k \ge 1$ fixed, $\varphi_k(n)$ is an increasing function of $n$ if and only if

$$p^2 \mathbb{P}(X_{2n} \ge 2k-2) \ge q^2 \mathbb{P}(X_{2n} \ge 2k).$$

**2)** Prove that, for all $n \ge k$:

$$\mathbb{P}(X_{2n} = 2k) = C_{2n}^{n+k} p^{n+k} q^{n-k}.$$

Deduce then that the function $n \mapsto \varphi_k(n)$ (defined on $\mathbb{N}$) is increasing if and only if $p \ge q$. Check that $\varphi_k(0) = 0$ and $\varphi_k(n) \xrightarrow[n \to \infty]{} 1$ if and only if $p > q$.

**3)** Prove that, for $p > q$, there exists a family of r.v.'s $(Y_{2k}, k \ge 1)$ taking values in $\{2k, 2k+2, \ldots\}$ such that:

$$\mathbb{P}(Y_{2k} \le 2n) = \mathbb{P}(X_{2n} \ge 2k), \qquad (n \in \mathbb{N}).$$

Consequently, the family $(Y_{2k}, k \ge 1)$ is the increasing pseudo-inverse of the random walk $(X_{2n}, n \ge 0)$.

Give an expression of $\mathbb{P}(Y_{2k} = 2n)$ $(n \ge k)$ and compute the expectation of $Y_{2k}$ in terms of the potential kernel of the random walk $(X_{2n}, n \ge 0)$.

## Exercise 8.2 (Existence of pseudo-inverses for a birth-death process).

Let $(X_n, n \ge 0)$ denote the Markov chain started from $0$, whose transition matrix is given by:

$$\Pi(0,1) = 1$$

$$\Pi(k, k+1) = \frac{k+2}{2k+2}$$

$$\Pi(k, k-1) = \frac{k}{2k+2} \qquad (k \ge 1).$$

**1)** Let $k \geq 1$. Prove that, for all $n \in \mathbb{N}$:

$$\mathbb{P}(X_{2n+2} \geq 2k) = \mathbb{P}(X_{2n} \geq 2k) + \mathbb{P}(X_{2n} = 2k - 2)b_k - \mathbb{P}(X_{2n} = 2k)a_k$$

where:

$$\begin{cases} a_k &= \mathbb{P}(X_{2n+2} = 2k - 2 | X_{2n} = 2k) = \dfrac{2k - 1}{2(4k + 2)}, \\ b_k &= \mathbb{P}(X_{2n+2} = 2k | X_{2n} = 2k - 2) = \dfrac{2k + 1}{2(4k - 2)}. \end{cases}$$

Let $\varphi_k(n) := \mathbb{P}(X_{2n} \geq 2k)$. Then, the function $n \mapsto \varphi_k(n)$, for $k \geq 1$ fixed, is an increasing function of $n$ if and only if

$$\mathbb{P}(X_{2n} = 2k - 2)b_k \geq \mathbb{P}(X_{2n} = 2k)a_k.$$

**2)** It is shown in [50, F], p.266, that:

$$\mathbb{P}(X_{2n} = 2k) = \frac{2}{\pi}(2k + 1) \int_0^{\pi} (\cos(\theta))^{2n} \sin(\theta) \sin((2k + 1)\theta)d\theta.$$

Prove, using the previous formula, that:

$$\mathbb{P}(X_{2n} = 2k) = 4\frac{(2k + 1)^2 C_n}{B\left(\frac{2n+1+2k+1}{2} + 1, \frac{2n+1-2k-1}{2} + 1\right)} \qquad (k \leq n)$$

where $C_n$ is a constant which depends only on $n$, and $B(s,t) = \int_0^1 u^{s-1}(1 - u)^{t-1}du$.

$\Bigg($ [27, F], p.375, formula 3.633:

$$\int_0^{\pi/2} (\cos(\theta))^{p-1} \sin(a\theta) \sin(\theta)d\theta = \frac{a\pi}{2^{p+1}p(p + 1)B\left(\frac{p+a}{2} + 1, \frac{p-a}{2} + 1\right)}\Bigg).$$

**3)** Deduce from the previous questions that $\varphi_k(n)$ is an increasing function of $n$ for $n \geq k$. The chain $(X_n, n \geq 0)$ being transient (cf. [50, F], p.268), show that: $\lim_{n \to \infty} \varphi_k(n) = 1$. Consequently, the chain $(X_{2n}, n \geq 0)$ admits an increasing pseudo-inverse $(Y_{2k}, k \geq 1)$ which satisfies:

$$\mathbb{P}(X_{2n} \geq 2k) = \mathbb{P}(Y_{2k} \leq 2n).$$

**4)** Express the expectation of $Y_{2k}$ in terms of the potential kernel of the chain $(X_n, n \geq 0)$. (cf. [50, F] p.268 and the followings).

*Comments and References: These exercises rely mainly on the lectures by B. Roynette at the Ecole d'été de St Flour in summer 1977 [50, F].*

*Exercise 8.2 may be generalized by replacing the chain $(X_n, n \geq 0)$ by the chain "associated to the ultraspherical polynomials", see [26, F].*
*For Table of integral computations, see [27, F].*

## 8.5 Some Consequences of the Existence of Pseudo-Inverses

### 8.5.1 Another Relation Between the Processes $X$ and $\overline{X}$ Started from 0

**Proposition 8.6.** *Let $t \geq 0$ and $U$ a uniform r.v. on $[0,1]$ independent from $\overline{X}_t$. Then, under the hypotheses of Theorem 8.3, the law of the r.v. $m[0, X_t]$ under $\mathbb{P}_0$ is the same as the law of $m[0, \overline{X}_t]U$ under $\overline{\mathbb{P}}_0$:*

$$m[0, X_t] \overset{(law)}{=} m[0, \overline{X}_t]U.$$

We provide two proofs.
**1) An analytic proof of Proposition 8.6.**
We write for every positive Borel function $f$:

$$\begin{aligned}
\mathbb{E}_0\left[f(m[0, X_t])\right] &= \int_0^\infty f(m[0, y])q(t, 0, y)\rho(y)dy \\
&= \int_0^\infty f(m[0, y])\left(\int_y^\infty \overline{q}(t, 0, z)m[0, z]s'(z)dz\right)\rho(y)dy \text{ (from (8.70))} \\
&= \int_0^\infty \left(\frac{1}{m[0, z]}\int_0^z f(m[0, y])\rho(y)dy\right)\overline{q}(t, 0, z)\overline{\rho}(z)dz \\
&\quad \text{(applying Fubini and (8.62))} \\
&= \int_0^\infty \left(\frac{1}{m[0, z]}\int_0^{m[0, z]} f(u)du\right)\overline{q}(t, 0, z)\overline{\rho}(z)dz \\
&\quad \text{(with the change of variable } u = m[0, y]) \\
&= \overline{\mathbb{E}}_0\left[\frac{1}{m[0, \overline{X}_t]}\int_0^{m[0, \overline{X}_t]} f(u)du\right] \\
&= \overline{\mathbb{E}}_0\left[f(m[0, \overline{X}_t]U)\right] \quad \text{(after the change of variable } u = m[0, \overline{X}_t]v).
\end{aligned}$$

This ends the proof of Proposition 8.6.

$\square$

**2) A more probabilistic proof of Proposition 8.6 under the assumption $m[0, +\infty[= +\infty$.**
Since $\overline{s}(x) = -\dfrac{1}{m[0, x]}$ is a scale function for $\overline{X}$, the process $\left(\dfrac{1}{m[0, \overline{X}_{t+s}]}, s < \overline{T}_0^{(t)}\right)$
where $\overline{T}_0^{(t)} := \inf\{s \geq 0; \overline{X}_{t+s} = 0\}$, is a positive continuous $(\overline{\mathscr{F}}_{t+s}, s \geq 0)$-local

martingale. But, 0 being entrance-not-exit for $\overline{X}$, we have $\overline{T}_0^{(t)} = +\infty$ $\mathbb{P}_0$–a.s., and $\overline{X}$ being transient, $\left( \dfrac{1}{m[0,\overline{X}_{t+s}]}, s \geq 0 \right)$ is in fact a positive continuous local martingale which converges towards 0. We apply Doob's maximal identity (Lemma 2.1):

$$\sup_{s \geq t} \frac{1}{m[0,\overline{X}_s]} \overset{(law)}{=} \frac{1}{m[0,\overline{X}_t]U}. \tag{8.84}$$

Then, since $x \mapsto m[0,x]$ is increasing, applying Point $(iii)$ of Theorem 8.3, we obtain:

$$\frac{1}{m[0,\overline{X}_t]U} \overset{(law)}{=} \sup_{s \geq t} \frac{1}{m[0,\overline{X}_s]} = \frac{1}{\inf_{s \geq t} m[0,\overline{X}_s]} = \frac{1}{m[0,\inf_{s \geq t}\overline{X}_s]} \overset{(law)}{=} \frac{1}{m[0,X_t]}. \tag{8.85}$$

$\square$

## 8.5.2 A Time Reversal Relationship

Assume that $X$ is a diffusion satisfying the hypotheses of Theorem 8.3. Then the process $\widehat{X} := \left( \overline{X}_{\overline{G}_y - t}, t < \overline{G}_y \right)$ is a diffusion started at $y$ with semi-group (see [70], Exercise 4.18, p.322):

$$\widehat{P}_t f(x) := \frac{1}{\overline{s}(x)} \overline{P}_t(f\overline{s})(x).$$

Using (8.62), its speed measure $\widehat{m}(dx) = \widehat{\rho}(x)dx$ and scale function $\widehat{s}(x)$ are given by:

$$\begin{cases} \widehat{s}(x) = -\dfrac{1}{\overline{s}(x)} = \dfrac{m[0,x]m[0,+\infty[}{m[0,+\infty[-m[0,x]}, \\[2ex] \widehat{\rho}(x) = \overline{s}^2(x)\overline{\rho}(x) = \left(1 - \dfrac{m[0,x]}{m[0,+\infty[}\right)^2 s'(x). \end{cases} \tag{8.86}$$

As a result, the law of $\overline{G}_y$ under $\mathbb{P}_0$ is the same as the law of $\widehat{T}_0$ for the process $\widehat{X}$. This leads to a new formulation of identity (8.74):

**Proposition 8.7.** *Let $X$ be a diffusion satisfying the hypotheses of Theorem 8.3, and $\widehat{X}$ defined by (8.86). Then:*

$$\mathbb{P}_0 \left( X_t \geq y \right) = \left( 1 - \frac{m[0,y]}{m[0,+\infty[} \right) \widehat{\mathbb{P}}_y \left( \widehat{T}_0 \leq t \right). \tag{8.87}$$

In fact, (see [79], Section 2.2), the diffusion $\overline{X}$ can also be obtained by conditioning $\widehat{X}$ not to hit 0, and we have:

$$\overline{q}(t,x,y) = \frac{\widehat{q}(t,x,y)}{\widehat{s}(x)\widehat{s}(y)} \xrightarrow[x \to 0]{} \frac{\widehat{f}_{y0}(t)}{\widehat{s}(y)} = \left( \frac{1}{m[0,y]} - \frac{1}{m[0,+\infty[} \right) \widehat{f}_{y0}(t), \tag{8.88}$$

where $\widehat{f}_{y0}$ is the density function of the first hitting time of 0 of $\widehat{X}$ under $\widehat{\mathbb{P}}_y$:

$$\widehat{\mathbb{P}}_y(\widehat{T}_0 \in dt) = \widehat{f}_{y0}(t)dt.$$

Therefore, plugging (8.88) in (8.70), we get:

**Proposition 8.8.**

$$q(t,0,y) = \int_y^\infty \widehat{f}_{z0}(t) \left(1 - \frac{m[0,z]}{m[0,+\infty[}\right) s'(z)dz.$$

Let us note that (8.86) reduces significantly when $m[0,+\infty[= +\infty$. Indeed, we obtain in this case:

$$\begin{cases} \widehat{s}(x) = m[0,x] \\ \widehat{\rho}(x) = s'(x) \end{cases}$$

and differentiating formula (8.87) with respect to $y$ and $t$ gives:

**Proposition 8.9 (Zolotarev-like identity, or Kendall's identity, [13]).**
*If $m[0,+\infty[= +\infty$:*

$$-\frac{\partial}{\partial t}q(t,0,y)\rho(y) = \frac{\partial}{\partial y}\widehat{f}_{y0}(t).$$

*Remark 8.12.*
**a)** Proposition 8.9 can also be proven using Krein's spectral representations of $q(t,0,y)$ and $f_{y0}(t)$, see [78] and [42].

**b)** If $X$ is a diffusion such that $m[0,+\infty[= +\infty$ and $s(x) = m[0,x]$, then, $X$ is a null recurrent diffusion, and, with our notations, the 2 processes $(X_t, t < T_0)$ and $(\widehat{X}_t, t < \widehat{T}_0)$ started from $y$ have the same law. Proposition 8.8 implies then:

$$\mathbb{P}_y(T_0 \in dt) = -\frac{1}{s'(y)}\frac{\partial q}{\partial y}(t,0,y)dt.$$

Note that this formula is very similar to (8.77) for transient diffusions:

$$\overline{\mathbb{P}}_0(\overline{G}_y \in dt) = -\frac{1}{\overline{s}(y)}\overline{q}(t,0,y)dt.$$

*Example 8.1.* If $X$ is a Brownian motion reflected at 0, then $\overline{X}$ is a Bessel process of dimension 3, and $\widehat{X}$ is a Brownian motion killed at 0. From (8.87), we thus get the well-known formula:

$$W_0(X_t \geq y) = W_y(T_0 \leq t).$$

## 8.5.3  Back to the Family $(X^{(\alpha)})_{\alpha \geq 0}$

All the results of Section 8.4 apply to the family $(X^{(\alpha)})_{\alpha \geq 0}$ defined by (8.26). In particular, we can define the diffusion $(\widehat{X}_t, t < \widehat{T}_0)$ on $\mathbb{R}^+$ as the solution of:

$$X_t = y + B_t - \int_0^t c(X_s)ds.$$

We note that:

$$\begin{cases} \widehat{s}(x) = 2 \int_0^x e^{2C(u)}du, \\ \widehat{\rho}(x) = e^{-2C(x)}. \end{cases}$$

Note that, since $\overline{X}$ does not depend on $\alpha$, neither does $\widehat{X}$. Let us also introduce the max-diffusion:

$$Z_t = B_t + \int_0^t c(2S_u^Z - Z_u)du \qquad \text{where } S_u^Z := \sup_{t \leq u} Z_t. \tag{8.89}$$

Then it is known from ([73], Theorem 5.4) that the process $(2S_t^Z - Z_t, t \geq 0)$ is distributed as $(\overline{X}_t, t \geq 0)$ under $\overline{\mathbb{P}}_0$. Hence, since $2S_t^Z - Z_t \geq S_t$, we have:

$$\overline{G}_y \stackrel{(law)}{=} T_y^Z \qquad \text{where } T_y^Z := \inf\{t \geq 0, Z_t > y\},$$

and formula (8.74) can be rewritten as follows:

$$\mathbb{P}_0 (X_t \geq y) = \left(1 - \frac{m[0,y]}{m[0,+\infty[}\right) \mathbb{P}_0^Z \left(T_y^Z \leq t\right).$$

Note that we can also construct, in the general case, the process $(Z_t, t \leq T_y^Z)$ from $(\overline{X}_t, t \leq \overline{G}_y)$ by the relation:

$$Z_t = 2J_t^{\overline{X}} - \overline{X}_t \qquad \text{where } J_t^{\overline{X}} := \inf_{\overline{G}_y \geq s \geq t} \overline{X}_s. \tag{8.90}$$

Now, introduce the process:

$$\widehat{Z}_t = 2I_t^{\widehat{X}} - \widehat{X}_t \qquad \text{where } I_t^{\widehat{X}} := \inf_{s \leq t} \widehat{X}_s. \tag{8.91}$$

Like $Z$, $\widehat{Z}$ is a priori not Markov, but, since $\widehat{Z}_t \leq \widehat{X}_t$, we have $G_0^{\widehat{Z}} = \widehat{T}_0$ $\widehat{\mathbb{P}}_y$-a.s.

**Proposition 8.10.** *The following time reversal relationship holds:*

$$\left(Z_{T_y^Z - t}, t \leq T_y^Z\right) \stackrel{(law)}{=} \left(\widehat{Z}_t, t \leq G_0^{\widehat{Z}}\right).$$

*Proof.*

$$\left(Z_{T_y^Z-t}, t \le T_y^Z\right) \stackrel{(law)}{=} \left(2J_{\overline{G}_y-t}^{\overline{X}} - \overline{X}_{\overline{G}_y-t}, t \le \overline{G}_y\right) \qquad \text{(from (8.90))}$$

$$\stackrel{(law)}{=} \left(2 \inf_{\overline{G}_y \ge s \ge \overline{G}_y-t} \overline{X}_s - \overline{X}_{\overline{G}_y-t}, t \le \overline{G}_y\right)$$

$$\stackrel{(law)}{=} \left(2 \inf_{s \le t} \overline{X}_{\overline{G}_y-s} - \overline{X}_{\overline{G}_y-t}, t \le \overline{G}_y\right)$$

$$\stackrel{(law)}{=} \left(2 \inf_{s \le t} \widehat{X}_s - \widehat{X}_t, t \le \widehat{T}_0\right)$$

$$\stackrel{(law)}{=} \left(\widehat{Z}_t, t \le G_0^{\widehat{Z}}\right) \qquad \text{(from (8.91))}.$$

$\square$

The links between these 5 processes can be summed up in the following diagrams:

$$X \xrightarrow{\quad \text{Biane} \quad} \overline{X}$$

and,

$$\left(Z_t, t \le T_y^Z\right) \xleftarrow{\quad \text{Time Reversal} \quad} \left(\widehat{Z}_t, t \le G_0^{\widehat{Z}}\right) \qquad (8.92)$$

$$2J^{\overline{X}}-\overline{X} \Big\updownarrow 2S^Z-Z \qquad\qquad\qquad 2I^{\widehat{X}}-\widehat{X} \Big\updownarrow 2V^{\widehat{Z}}-\widehat{Z}$$

$$\left(\overline{X}_t, t \le \overline{G}_y\right) \xleftarrow{\quad \text{Time Reversal} \quad} \left(\widehat{X}_t, t \le \widehat{T}_0\right)$$

where, in this commutative diagram, $V_t^{\widehat{Z}} := \sup_{s \ge t} \widehat{Z}_s$. (See relations (8.89), (8.90) and (8.91), for the definition of $S, J$ and $I$.) In particular, we have:

$$\overline{\mathbb{P}}_0 \left(\overline{G}_y \le t\right) = \widehat{\mathbb{P}}_y \left(\widehat{T}_0 \le t\right) = \mathbb{P}_y^{\widehat{Z}} \left(G_0^{\widehat{Z}} \le t\right) = \mathbb{P}_0^Z \left(T_y^Z \le t\right).$$

*Example 8.2.* If $X$ is a Bessel process of index $\nu > -1$, $\nu \ne 0$, then:

- $\overline{X}$ is a Bessel process of index $\nu + 1$ started at 0.

- $\widehat{X}$ is a Bessel process of index $-(\nu+1)$ started at $y$ and killed when it first hits 0 if $-(\nu+1) > -1$. In the case $-(\nu+1) < -1$, $\widehat{X}$ can be obtained as the root of a "square" Bessel process of negative dimension, see ([70], Exercise 1.33, p. 453).

**Exercise 8.3.** (Around Zolotarev's formula)
Let $(X_t, t \ge 0)$ be a $\mathbb{R}^+$-valued diffusion satisfying the hypotheses of this section. We assume furthermore that $(X_t, t \ge 0)$ satisfies a scaling property, namely, there

exist $\alpha, \beta \in \mathbb{R}^*$ such that

$$\forall t \geq 0, \forall c > 0, \ X_t \stackrel{(law)}{=} c^\alpha X_{t/c^\beta}.$$

**1)** Prove that, $\forall c > 0, \forall y \geq 0$:

$$c^\alpha q(t, 0, y)\rho(y) = q\left(\frac{t}{c^\beta}, 0, \frac{y}{c^\alpha}\right)\rho\left(\frac{y}{c^\alpha}\right).$$

**2)** Apply Proposition 8.7 to deduce that:

$$\left(1 - \frac{m[0, y]}{m[0, +\infty[}\right)\beta t \widehat{\mathbb{P}}_y\left(\widehat{T}_0 \in dt\right) dy = \alpha y \mathbb{P}_0\left(X_t \in dy\right) dt. \tag{1}$$

**3)** Let $(R_t^{(\nu)}, t \geq 0)$ be a Bessel process of index $\nu \in\ ]-1, 0[$. Using (1), prove that:

$$\mathbb{P}_y^{(\nu)}\left(T_0^{(\nu)} \in dt\right) = \frac{1}{\Gamma(-\nu)}\left(\frac{y^2}{2t}\right)^{-\nu}\exp\left(-\frac{y^2}{2t}\right) dt$$

and recover Getoor's result: under $\mathbb{P}_y^{(\nu)}$,

$$T_0^{(\nu)} \stackrel{(law)}{=} \frac{y^2}{2\gamma_{-\nu}}$$

where $\gamma_{-\nu}$ is a gamma r.v. of parameter $-\nu$.

**4)** Similarly, prove that, if $(Y_t^{(\nu)}, t \geq 0)$ is a square Bessel process of index $\nu \in\ ]-1, 0[$, then:

$$\mathbb{Q}_y^{(\nu)}\left(T_0^{(\nu)} \in dt\right) = \frac{2^\nu t^{\nu+1}}{y^\nu \Gamma(-\nu)}\exp\left(-\frac{y}{2t}\right) dt.$$

## 8.6 Notes and Comments

This Chapter is, essentially, taken from [68]. Proposition 8.1 is due to B. Roynette, P. Vallois and M. Yor ([73]) where it has been established in order to extend in a non-Markovian framework the celebrated Theorem of J. Pitman expressing $2S - X$ with $X$ a Brownian motion, and $S$ its supremum, as a Bessel process of dimension 3. The transformation of Definition 8.1 has been introduced by P. Biane to generalize to arbitrary diffusions the famous result of Ciesielski-Taylor ([17]) obtained in the set-up of Bessel processes (see for instance [52], Chapter 4, for a description and extensions of this result). Formulae such as these of Remark 8.9 (intertwining with the first derivative operator) may be found in Hirsch-Yor [31]. Zolotarev's formula – of which Proposition 8.9 is a slightly different version – is classical for Lévy processes without positive jumps (see [13] where the term Kendall's identity is also used).

# Appendix A
# Complements

## A.1 Study of the Call Associated to a Strict Local Martingale (see Yen-Yor [93])

### A.1.1 Introduction

Let $(M_t, t \geq 0)$ be a "true" positive, continuous martingale converging towards 0 when $t \to \infty$. We showed in Chapter 2, Theorem 2.2 that for all $\mathscr{F}_t$-measurable and bounded r.v. $F_t$, and all $K \geq 0$:

$$\mathbb{E}\left[F_t(M_t - K)^+\right] = \mathbb{E}^{(M)}\left[F_t 1_{\{\mathscr{G}_K^{(M)} \leq t\}}\right] \tag{A.1}$$

where $\mathscr{G}_K^{(M)} := \sup\{t \geq 0; \ M_t = K\}$ and $\mathbb{P}^{(M)}$ is the probability defined by:

$$\mathbb{P}_{|\mathscr{F}_t}^{(M)} = M_t \cdot \mathbb{P}_{|\mathscr{F}_t}. \tag{A.2}$$

In particular, taking $F_t = 1$ in (A.1), we see that $t \mapsto \mathbb{E}\left[(M_t - K)^+\right]$ is an increasing function. What happens if $(M_t, t \geq 0)$ is not a martingale, but only a local martingale ? As an example, we examine the case of the strict local martingale $\left(M_t = \frac{1}{R_t}, t \geq 0\right)$, where $(R_t, t \geq 0)$ is a Bessel process of dimension 3 starting at 1. (The study for a Bessel process of dimension 3 starting at $r > 0$ can be reduced to this case by scaling.) We shall see that in this set-up Theorem 2.2 is no longer true; in particular, the function $t \mapsto \mathbb{E}\left[(M_t - K)^+\right]$ is not increasing (not even monotonous). We refer the reader to S. Pal and P. Protter ([62]) and to Ju-Yi Yen and M. Yor ([93]) from which the following results are taken.

### A.1.2 Main Results

We denote by $(X_t, t \geq 0)$ the canonical process on $\mathscr{C}(\mathbb{R}^+, \mathbb{R})$; $W_x$ is the Wiener measure such that $W_x(X_0 = x) = 1$ and $\mathbb{P}_1^{(1/2)}$ the law of the Bessel process of di-

C. Profeta et al., *Option Prices as Probabilities*, Springer Finance,
DOI 10.1007/978-3-642-10395-7, © Springer-Verlag Berlin Heidelberg 2010

mension 3 (i.e. index $\nu = 1/2$) starting at 1. We define:

- $C_K(t) := \mathbb{E}_1^{(1/2)}\left[\left(\dfrac{1}{X_t} - K\right)^+\right]$,
- $\mathcal{G}_1 := \sup\{t \geq 0;\ X_t = 1\}$,
- $T_k := \inf\{t \geq 0;\ X_t = k\}$, with $k = \dfrac{1}{K}$,

  (these two r.v.'s $g_1$ and $T_k$ being considered under $\mathbb{P}_1^{1/2}$),
- $\widetilde{T}_k$ the size-biased sampling of $T_k$, i.e., for every Borel function $f$,

$$\mathbb{E}\left[f(\widetilde{T}_k)\right] = 3K^2\mathbb{E}_0^{(1/2)}\left[f(T_k)T_k\right].$$

**Theorem A.1 (Yen-Yor, [93]).**

i)   *The function* $(3K^2C_K(t), t \geq 0)$ *is a probability density on* $\mathbb{R}^+$.
ii)  *It is the density function of the r.v.* $\Lambda_K$, *with:*

$$\Lambda_K \overset{(law)}{=} (\mathcal{G}_1 - \widetilde{T}_k) + \widetilde{T}_k U \tag{A.3}$$

   *where in (A.3), U is uniform on* $[0,1]$, *and* $\mathcal{G}_1, \widetilde{T}_k$ *and U are independent.*
iii) *The Laplace transform of* $\Lambda_K$ *is given by:*

$$\mathbb{E}\left[e^{-\lambda\Lambda_K}\right] = \int_0^\infty 3K^2C_K(t)e^{-\lambda t}dt = \frac{3K^2}{\lambda}e^{-\sqrt{2\lambda}}\left(\frac{\sinh(k\sqrt{2\lambda})}{k\sqrt{2\lambda}} - 1\right).$$

iv)  *There are the asymptotic formulae at* $\infty$:

$$C_K(t) \underset{t\to\infty}{\sim} \frac{1}{3\sqrt{2\pi}K^2}\frac{1}{t^{3/2}} \qquad (K > 0), \tag{A.4}$$

$$C_0(t) \underset{t\to\infty}{\sim} \sqrt{\frac{2}{\pi t}} \qquad (K = 0). \tag{A.5}$$

v)   *There are the equivalents at 0:*

$$C_K(t) \underset{t\to 0}{\longrightarrow} (1 - K)^+ \qquad (K \neq 1), \tag{A.6}$$

$$C_1(t) \underset{t\to 0}{\sim} \sqrt{\frac{t}{2\pi}} \qquad (K = 1). \tag{A.7}$$

*It is then clear from (iv) and (v), and from the first statement (i) of the Theorem, that* $C_K(t)$ *is not an increasing function in t.*

This Theorem relies mainly on Theorem 2.1, on the well-known Doob $h$-process relationship between Brownian motion and the Bessel process of dimension 3:

$$\mathbb{P}_{1|\mathscr{F}_t}^{(1/2)} = (X_{t\wedge T_0}) \cdot W_{1|\mathscr{F}_t},$$

and on the classical time reversal result by D. Williams, namely: the law of $(X_{T_0-t}, t \leq T_0)$ under $W_1$ is the same as that of $(X_t, t \leq \mathscr{G}_1)$ under $\mathbb{P}_0^{(1/2)}$.

## A.1.3 An Extension

Theorem A.1 may be extended to the case of the local martingale $\left(M_t := \frac{1}{R_t^{2\nu}}, t \geq 0\right)$ where $(R_t, t \geq 0)$ is a Bessel process of index $\nu$, i.e. dimension $\delta = 2\nu + 2$, with $2 < \delta < 4$. We now state this result, see [93]:

**Theorem A.2 (Yen-Yor, [93]).**
*Let, for $K \geq 0$, $C_K^{(\nu)}(t) := \mathbb{E}_1^{(\nu)}\left[\left(\frac{1}{R_t^{2\nu}} - K\right)^+\right]$, where $\mathbb{E}_1^{(\nu)}$ denotes the expectation relative to the Bessel process of index $\nu$ started from 1. Then:*

i) *The function $(2(\nu+1)K^{\frac{1}{\nu}}C_K^{(\nu)}(t), t \geq 0)$ is a probability density on $\mathbb{R}^+$.*
ii) *It is the density function of the r.v. $\Lambda_K^{(\nu)}$, with:*

$$\Lambda_K^{(\nu)} \stackrel{(law)}{=} (\mathscr{G}_1 - \tilde{T}_k) + \tilde{T}_k U \tag{A.8}$$

*where $k = \frac{1}{K^{\frac{1}{\delta-2}}}$, and $\tilde{T}_k$ is the sized-biased sampling of $T_k$, with $\mathscr{G}_1$ and $T_k$ defined with respect to $\mathbb{P}_0^{(\nu)}$.*
iii) *The Laplace transform of $\Lambda_K^{(\nu)}$ is given by:*

$$\mathbb{E}\left[e^{-\lambda\Lambda_K^{(\nu)}}\right] = \frac{4(\nu+1)K^{\frac{1}{\nu}}}{\lambda}K_\nu\left(\sqrt{2\lambda}\right)\left(\nu\frac{I_\nu(k\sqrt{2\lambda})}{k^\nu} - \frac{1}{\Gamma(\nu)}\left(\frac{\sqrt{2\lambda}}{2}\right)^\nu\right)$$

*where $I_\nu$ and $K_\nu$ denote the usual modified Bessel functions with parameter $\nu$.*
iv) *There are the asymptotic formulae at $\infty$:*

$$C_K^{(\nu)}(t) \underset{t \to \infty}{\sim} \frac{\alpha_K^{(\nu)}}{t^{3/2}} \quad (K > 0) \quad \text{with } \alpha_K^{(\nu)} := \frac{1}{2^{\nu+1}(\nu+1)\Gamma(\nu)K^{1/\nu}}, \tag{A.9}$$

$$C_0^{(\nu)}(t) \underset{t \to \infty}{\sim} \frac{1}{2^\nu\Gamma(\nu+1)}\frac{1}{t^\nu} \quad (K = 0). \tag{A.10}$$

v) *There are the equivalents at 0:*

$$C_K^{(\nu)}(t) \xrightarrow[t \to 0]{} (1-K)^+ \quad (K \neq 1), \tag{A.11}$$

$$C_1^{(\nu)}(t) \underset{t \to 0}{\sim} \nu\sqrt{\frac{2t}{\pi}} \quad (K = 1). \tag{A.12}$$

## A.2  Measuring the "Non-Stopping Timeness" of Ends of Previsible Sets (see Yen-Yor, [92])

### A.2.1  About Ends of Previsible Sets

In this section, we take up the framework of Chapter 3. We are interested in random times $L$ defined on a filtered probability space $(\Omega, \mathscr{F}, (\mathscr{F}_t)_{t\geq 0}, \mathbb{P})$ as ends of $(\mathscr{F}_t)$-previsible sets $\Gamma$, that is:

$$L = \sup\{t \geq 0; \, ]t, +\infty[ \subset \Gamma\}. \tag{A.13}$$

We suppose that assumptions (A) and (C) of Subsection 3.1.2 hold, namely:

(C)              All $(\mathscr{F}_t)$-martingales are continuous.

(A)              For any stopping time $T$, $\mathbb{P}(L = T) = 0$.

We also define the Azéma supermartingale:

$$\left( Z_t^{(L)} = \mathbb{P}(L > t | \mathscr{F}_t), t \geq 0 \right)$$

which, under (CA), admits a continuous version.

We have seen in Chapters 1 and 2 that the Black-Scholes type formulae are deeply linked with certain last passage times. For example, if $(M_t, t \geq 0)$ is a positive continuous martingale converging a.s. towards 0 when $t \to \infty$, then:

$$\mathbb{E}\left[ (K - M_t)^+ \right] = K\mathbb{P} \left( \mathscr{G}_K^{(M)} \leq t \right) \tag{A.14}$$

$$\mathbb{E}\left[ (M_t - K)^+ \right] = \mathbb{P}^{(M)} \left( \mathscr{G}_K^{(M)} \leq t \right) \tag{A.15}$$

with $\mathscr{G}_K^{(M)} := \sup\{t \geq 0; \, M_t \in [K, +\infty[\}$. In the present section, we would like to measure "how much $L$ differs from a $(\mathscr{F}_t, t \geq 0)$ stopping time". For that purpose, we shall first give some criterions to measure the NST (=Non-Stopping Timeness) of $L$, and then make explicit computations on some examples.

### A.2.2  Some Criterions to Measure the NST

#### A.2.2.1  A Fundamental Function: $m_L(t)$

Let $(m_L(t), t \geq 0)$ be the function defined by:

$$m_L(t) := \mathbb{E}\left[ \left( 1_{\{L \geq t\}} - \mathbb{P}(L \geq t | \mathscr{F}_t) \right)^2 \right]. \tag{A.16}$$

If $L$ were a $(\mathscr{F}_t)$-stopping time, then the process $1_{\{L \geq t\}}$ would be identically equal to $Z_t^{(L)} := \mathbb{P}(L > t | \mathscr{F}_t)$ and $m_L$ would equal 0. Thus the function $(m_L(t), t \geq 0)$ tells us about the NST of $L$. Note furthermore that:

$$m_L(t) = \mathbb{E}\left[Z_t^{(L)}(1 - Z_t^{(L)})\right]. \tag{A.17}$$

### A.2.2.2 Some Other Criterions

Instead of considering the "full function" $(m_L(t), t \geq 0)$, we may consider only:

$$m_L^* := \sup_{t \geq 0} m_L(t) \tag{A.18}$$

as a global measurement of the NST of $L$.

Here are also two other (a priori natural) measurements of the NST of $L$:

$$m_L^{**} = \mathbb{E}\left[\sup_{t \geq 0}\left(Z_t^{(L)}(1 - Z_t^{(L)})\right)\right] \tag{A.19}$$

and

$$\widetilde{m}_L = \sup_{T \geq 0} \mathbb{E}\left[Z_T^{(L)}\left(1 - Z_T^{(L)}\right)\right] \tag{A.20}$$

where $T$ runs over all $(\mathscr{F}_t)$-stopping times. However, we cannot expect to learn very much from $m_L^{**}$ and $\widetilde{m}_L$, since it is easily shown that:

$$m_L^{**} = 1/4 = \widetilde{m}_L. \tag{A.21}$$

Proof of (A.21)

i) The fact that $m_L^{**} = 1/4$ follows immediately from $\sup\limits_{x \in [0,1]}(x(1-x)) = 1/4$ and the fact that, a.s., the range of the process $(Z_t^{(L)}, t \geq 0)$ is $[0,1]$ since $Z_0^{(L)} = 1$, $Z_\infty^{(L)} = 0$ and $(Z_t^{(L)}, t \geq 0)$ is continuous.

ii) Let us consider $T_a = \inf\{t \geq 0; Z_t^{(L)} = a\}$, for $0 < a < 1$. Then, $Z_t^{(L)}(1 - Z_t^{(L)})|_{t=T_a} = a(1-a)$; hence,

$$\sup_{a \in ]0,1[} \mathbb{E}\left[Z_{T_a}^{(L)}(1 - Z_{T_a}^{(L)})\right] = \sup_{a \in ]0,1[}(a(1-a)) = 1/4.$$

$\square$

### A.2.2.3 A Distance from Stopping Times

As a natural criterion, one could think of the distance:

$$\nu_L := \inf_{T \geq 0} \mathbb{E}\left[|L - T|\right]$$

where $T$ runs over all $(\mathscr{F}_t, t \geq 0)$ stopping times. However, this quantity may be infinite as $L$ may have infinite expectation. A more adequate distance may be given by:

$$\widetilde{\nu}_L := \inf_{T \geq 0} \mathbb{E}\left[\frac{|L-T|}{1+|L-T|}\right].$$

Note that this distance was precisely computed by du Toit-Peskir-Shiryaev in the example they consider [86]. Other criterions are proposed and discussed in [92], from which this section is taken. We shall now focus on $m_L$ and compute this function in a few examples.

### A.2.3 Computations of Several Examples of Functions $m_L(t)$

#### A.2.3.1 A General Formula

We shall compute $(m_L(t), t \geq 0)$ in the case where:

$$L = \mathscr{G}_K := \sup\{t \geq 0;\ M_t = K\} \qquad (K \leq 1)$$

with $(M_t, t \geq 0) \in \mathscr{M}_+^{0,c}$, $M_0 = 1$. From (2.6),

$$\mathbb{P}(\mathscr{G}_K \leq t | \mathscr{F}_t) = \left(1 - \frac{M_t}{K}\right)^+,$$

thus:

$$Z_t = 1 - \left(1 - \frac{M_t}{K}\right)^+ = \left(\frac{M_t}{K}\right) \wedge 1$$

and,

$$m_L(t) = \mathbb{E}\left[Z_t(1 - Z_t)\right] = \frac{1}{K^2}\mathbb{E}\left[M_t(K - M_t)^+\right]. \qquad (A.22)$$

We now particularize this formula when $M_t := \mathscr{E}_t = \exp(B_t - t/2)$ with $(B_t, t \geq 0)$ a standard Brownian motion and $\mathscr{G}_K := \sup\{t \geq 0;\ \mathscr{E}_t = K\}$ $(K \leq 1)$. From formula (A.22), we deduce:

$$
\begin{aligned}
m_{\mathscr{G}_K}(t) &= \frac{1}{K^2}\mathbb{E}\left[\mathscr{E}_t(K - \mathscr{E}_t)^+\right] \\
&= \frac{1}{K^2}\mathbb{E}\left[\left(K - \exp\left(B_t + \frac{t}{2}\right)\right)^+\right] \qquad \text{(from the Cameron-Martin formula)} \\
&= \frac{1}{K^2}\left\{K\mathbb{P}\left(\exp\left(B_t + \frac{t}{2}\right) < K\right) - \mathbb{E}\left[1_{\{\exp(B_t + t/2) < K\}}\exp\left(B_t + \frac{t}{2}\right)\right]\right\} \\
&= e^{-l}\mathbb{P}\left(B_t + \frac{t}{2} < l\right) - e^t e^{-2l}\mathbb{P}\left(B_t + \frac{3t}{2} < l\right) \qquad \text{(with } K = e^l\text{)} \\
&= \mathbb{P}\left(B_1 < -\frac{3\sqrt{t}}{2} + \frac{l}{\sqrt{t}}\right)(e^{-l} - e^t e^{-2l}) \\
&\quad + e^{-l}\mathbb{P}\left(-\frac{3\sqrt{t}}{2} + \frac{l}{\sqrt{t}} < B_1 < -\frac{\sqrt{t}}{2} + \frac{l}{\sqrt{t}}\right).
\end{aligned}
$$

For a comparative study of the graphs of $(m_{\mathscr{G}_K}(t), t \geq 0)$ when $K$ varies, we refer the reader to [92], from which this Section A.2 is taken.

### A.2.3.2 The Case $L = \mathscr{G}_{T_a} = \sup\{t < T_a; B_t = 0\}$

We apply Theorem 2.1 to the positive martingale $(M_t := a - B_{t \wedge T_a}, t \geq 0)$, with $\mathscr{G}_a^{(M)} := \sup\{t \geq 0; M_t = a\}$. This gives:

$$\mathbb{P}\left(\mathscr{G}_a^{(M)} \leq t \,|\, \mathscr{F}_t\right) = \left(1 - \frac{a - B_{t \wedge T_a}}{a}\right)^+ = \frac{B_{t \wedge T_a}^+}{a}.$$

Hence:

$$Z_t = \mathbb{P}\left(\mathscr{G}_{T_a} \geq t \,|\, \mathscr{F}_t\right) = \mathbb{P}\left(\mathscr{G}_a^{(M)} \geq t \,|\, \mathscr{F}_t\right) = 1 - \frac{B_{t \wedge T_a}^+}{a}.$$

Thus, we obtain:

$$
\begin{aligned}
m_{\mathscr{G}_{T_a}}(t) = \mathbb{E}\left[\frac{B_{t \wedge T_a}^+}{a}\left(1 - \frac{B_{t \wedge T_a}^+}{a}\right)\right] &= \frac{1}{a^2}\mathbb{E}\left[1_{\{t < T_a\}} 1_{\{B_t > 0\}} B_t (a - B_t)\right] \\
&= \frac{1}{a^2}\mathbb{E}\left[1_{\{S_t < a\}} 1_{\{B_t > 0\}} B_t (a - B_t)\right] \\
&= \frac{1}{a^2}\mathbb{E}\left[1_{\{S_1 < \frac{a}{\sqrt{t}}\}} 1_{\{B_1 > 0\}} t B_1 \left(\frac{a}{\sqrt{t}} - B_1\right)\right] \\
&= \frac{1}{x^2}\varphi(x),
\end{aligned}
$$

where $x = \dfrac{a}{\sqrt{t}}$, and:

$$\varphi(x) = \mathbb{E}\left[1_{\{S_1 < x\}} 1_{\{B_1 > 0\}} B_1 (x - B_1)\right].$$

It remains to compute the function $\varphi$. We note that:

$$\varphi(x) = \mathbb{E}\left[B_1^+ (x - B_1)^+\right] - \mathbb{E}\left[1_{\{S_1 > x\}} B_1^+ (x - B_1)^+\right].$$

Recall the useful formula:

$$\mathbb{P}(S_1 > x \,|\, B_1 = a) = \exp(-2x(x - a)), \qquad (x \geq a > 0)$$

(see Chapter 1 (1.50), or Chapter 5, (5.14)). Then, we find:

$$
\begin{aligned}
\varphi(x) &= \frac{1}{\sqrt{2\pi}} \int_0^x y(x - y)\left(e^{-\frac{y^2}{2}} - e^{\frac{1}{2}(2x - y)^2} dy\right) \\
&= \frac{x^3}{\sqrt{2\pi}} \int_0^1 u(1 - u)\left(e^{-\frac{x^2 u^2}{2}} - e^{\frac{x^2}{2}(2 - u)^2} du\right).
\end{aligned}
$$

Thus:

$$\frac{\varphi(x)}{x^2} = \frac{x}{\sqrt{2\pi}} \int_0^1 u(1-u) \left( e^{-\frac{x^2 u^2}{2}} - e^{\frac{x^2}{2}(2-u)^2} \right) du.$$

(See [92] for a study and the graph of this function.)

## A.3 Some Connexions with Dupire's Formula

### A.3.1 Dupire's Formula (see [20, F])

Let $(M_t, t \geq 0)$ be the solution of the SDE:

$$M_t = x + \int_0^t \sigma(s, M_s) dB_s \tag{A.23}$$

where $(B_s, s \geq 0)$ is a Brownian motion started at 0. Let, for every $K$ and $t$:

$$\Pi(t, K) := \mathbb{E}\left[ (K - M_t)^+ \right]. \tag{A.24}$$

Under some regularity and slow growth conditions on the function $\Pi$ and its derivatives, B. Dupire shows that the knowledge of $\Pi$ makes it possible to determine $\sigma$. More precisely, the following formulae hold:

$$\frac{1}{2}\sigma^2(t, K)\frac{\partial^2 \Pi}{\partial K^2}(t, K) = \frac{\partial \Pi}{\partial t}(t, K) \tag{A.25}$$

i.e.

$$\sigma(t, K) = \left( \frac{2\frac{\partial \Pi}{\partial t}(t, K)}{\frac{\partial^2 \Pi}{\partial K^2}(t, K)} \right)^{\frac{1}{2}}. \tag{A.26}$$

### A.3.2 Extension of Dupire's Formula to a General Martingale in $\mathcal{M}_+^{0,c}$

Let $(M_t, t \geq 0)$ denote a positive, continuous martingale converging towards 0 when $t \to \infty$ and such that $M_0$ is a.s. constant. We assume that the hypotheses of Subsection 2.4.1 are satisfied:

- $(m_t(x), x \geq 0)$ denotes the density of the r.v. $M_t$,
- $d\langle M \rangle_t = \sigma_t^2 dt$ and $\theta_t(x) := \mathbb{E}\left[ \sigma_t^2 | M_t = x \right]$,

the functions $m$ and $\theta$ being continuous in both variables. Let us define:

$$\Pi(t, K) := \mathbb{E}\left[ (K - M_t)^+ \right] \quad \text{and} \quad C(t, K) := \mathbb{E}\left[ (M_t - K)^+ \right]. \tag{A.27}$$

**Theorem A.3.** *The following quantities are equal:*

$$i) \qquad \frac{\partial \Pi}{\partial t}(t,K) = \frac{1}{2}\theta_t(K)\frac{\partial^2 \Pi}{\partial K^2}(t,K) = Kf_{\mathscr{G}_K}(t) \qquad (A.28)$$

$$ii) \qquad \frac{\partial C}{\partial t}(t,K) = \frac{1}{2}\theta_t(K)\frac{\partial^2 C}{\partial K^2}(t,K) = Kf_{\mathscr{G}_K}(t) \qquad (A.29)$$

*where $f_{\mathscr{G}_K}$ denotes the density function of the r.v. $\mathscr{G}_K := \sup\{t \geq 0; M_t = K\}$.*

*Proof.* We first prove (A.28)
We have:

$$\Pi(t,K) = \mathbb{E}\left[(K - M_t)^+\right] = \int_0^K \mathbb{P}(M_t \leq x)dx.$$

Thus:

$$\frac{\partial^2 \Pi}{\partial K^2}(t,K) = m_t(K). \qquad (A.30)$$

On the other hand, from Tanaka's formula:

$$\mathbb{E}\left[(K - M_t)^+\right] = (K - M_0)^+ + \frac{1}{2}\mathbb{E}\left[L_t^K\right] \qquad (A.31)$$

where $(L_t^K, t \geq 0)$ denotes the local time of $(M_t, t \geq 0)$ at level $K$. Thus:

$$\frac{\partial \Pi}{\partial t}(t,K) = \frac{1}{2}\frac{\partial}{\partial t}\mathbb{E}\left[L_t^K\right] = \frac{1}{2}\theta_t(K)m_t(K) \qquad \text{(from (2.44))}. \qquad (A.32)$$

Comparing (A.30) and (A.32), we obtain the first part of (A.28).
   The second part of this equality relies on Theorem 2.1, formula (2.7):

$$\Pi(t,K) = K\mathbb{P}(\mathscr{G}_K \leq t). \qquad (A.33)$$

Hence: $\dfrac{\partial \Pi}{\partial t}(t,K) = Kf_{\mathscr{G}_K}(t)$. This proves (A.28).
   Then, (A.29) follows immediately from (A.28) since:

$$\mathbb{E}\left[M_t - K\right] = \mathbb{E}\left[M_0 - K\right] = \mathbb{E}\left[(M_t - K)^+\right] - \mathbb{E}\left[(M_t - K)^-\right]$$
$$= C(t,K) - \Pi(t,K), \qquad (A.34)$$

thus:

$$\frac{\partial C}{\partial t} = \frac{\partial \Pi}{\partial t} \qquad \text{and} \qquad \frac{\partial^2 C}{\partial K^2} = \frac{\partial^2 \Pi}{\partial K^2}.$$

Observe that the hypothesis: $(M_t, t \geq 0)$ *is a true martingale* is necessary to obtain (A.34). Moreover, the analogue of formula (A.31) for the call:

$$\mathbb{E}\left[(M_t - K)^+\right] = (M_0 - K)^+ + \frac{1}{2}\mathbb{E}\left[L_t^K\right]$$

is only correct if $(M_t, t \geq 0)$ is a true martingale. When it is a strict local martingale, a correction term must be added on the right hand side of the preceding formula. This is the content of the paper [39, F].

□

### A.3.3  A Formula Relative to Lévy Processes Without Positive Jumps

We shall now use the formulae of Theorem A.3 to obtain an interesting formula relative to Lévy processes without positive jumps.

Let $(X_t, t \geq 0)$ be a Lévy process without positive jumps; then it admits exponential moments, and we have:

$$\mathbb{E}\left[e^{\lambda X_t}\right] := \exp(t\psi(\lambda)). \tag{A.35}$$

Let $(M_t, t \geq 0)$ be the martingale defined by:

$$(M_t := \exp(X_t - t\psi(1)), t \geq 0). \tag{A.36}$$

We assume that $M_t \xrightarrow[t \to \infty]{} 0$ a.s., which is satisfied for instance if $\psi'(0) - \psi'(1) < 0$. This martingale having no positive jumps, we can apply the results of Chapter 2 and of Theorem A.3. We now slightly modify our notation, and assume that all the functions (of two variables) that appear below are continuous. We denote:

- $(m(t,x), x \geq 0)$ the density of the r.v. $M_t$,
- $(l(t,x), t \geq 0)$ the density of $\mathscr{G}_x^{(M)} := \sup\{t \geq 0;\ M_t = x\}$,
  (i.e. $l(t,x) = f_{\mathscr{G}_x}(t)$ with the notation of the previous subsection),
- $(f(t,x), x \geq 0)$ the density of $X_t - t\psi(1)$,
- $(\tau(t,x), t \geq 0)$ the density of $T_x := \inf\{t \geq 0;\ X_t - t\psi(1) \geq x\}$
  $= \inf\{t \geq 0;\ M_t \geq e^x\}$,
- $(\gamma(t,x), t \geq 0)$ the density of $G_x := \sup\{t \geq 0;\ X_t - t\psi(1) \geq x\}$
  $= \sup\{t \geq 0;\ M_t \geq e^x\}$.

**Theorem A.4.** *We have, for all $t, x \geq 0$:*

$$\frac{\partial f}{\partial t}(t,x) = \frac{\partial^2 \gamma}{\partial x^2}(t,x) + \frac{\partial \gamma}{\partial x}(t,x) = \frac{1}{x}\left(\tau(t,x) + t\frac{\partial \tau}{\partial t}(t,x)\right). \tag{A.37}$$

*Proof.*
*i)* It follows from (A.30) and (A.33) that:

$$\frac{\partial^2 \Pi}{\partial x^2}(t,x) = m(t,x) = \frac{\partial^2}{\partial x^2}\left(x\int_0^t l(u,x)du\right),$$

hence, on differentiating with respect to $t$:

$$\frac{\partial m}{\partial t}(t,x) = \frac{\partial^2}{\partial x^2}(xl(t,x)). \tag{A.38}$$

*ii)* Since $\mathscr{G}_x^{(M)} = G_{\log(x)}$, we have:

$$l(t,x) = \gamma(t,\log(x)), \tag{A.39}$$

and, since $M_t = \exp(X_t - t\psi(1))$:

$$m(t,x) = \frac{1}{x} f(t, \log(x)). \tag{A.40}$$

Furthermore, it follows from Zolotarev's formula, since $(X_t - t\psi(1), t \geq 0)$ is a Lévy process without positive jumps (see [13], or [8] p.190) that:

$$t\tau(t,x) = xf(t,x) \qquad (t \geq 0, \ x > 0). \tag{A.41}$$

Hence, from (A.38), (A.40) and (A.41):

$$\frac{\partial m}{\partial t}(t,x) = \frac{\partial}{\partial t}\left(\frac{1}{x}f(t, \log(x))\right) = \frac{\partial}{\partial t}\left(\frac{t}{x\log(x)}\tau(t, \log(x))\right)$$

$$= \frac{\partial^2}{\partial x^2}\{x\gamma(t, \log(x))\} \tag{A.42}$$

which reduces to:

$$\frac{1}{\log(x)}\tau(t, \log(x)) + \frac{t}{\log(x)}\frac{\partial \tau}{\partial t}(t, \log(x)) = \frac{\partial \gamma}{\partial x}(t, \log(x)) + \frac{\partial^2 \gamma}{\partial x^2}(t, \log(x)),$$

i.e., replacing $\log(x)$ by $x$:

$$\tau(t,x) + t\frac{\partial \tau}{\partial t}(t,x) = x\left(\frac{\partial^2 \gamma}{\partial x^2}(t,x) + \frac{\partial \gamma}{\partial x}(t,x)\right). \tag{A.43}$$

This is the second part of (A.37).

*iii*) From (A.42), we have:

$$\frac{1}{x}\frac{\partial f}{\partial t}(t, \log(x)) = \frac{\partial}{\partial x}\left(\frac{\partial \gamma}{\partial x}(t, \log(x)) + \gamma(t, \log(x))\right)$$

$$= \frac{1}{x}\left(\frac{\partial^2 \gamma}{\partial x^2}(t, \log(x)) + \frac{\partial \gamma}{\partial x}(t, \log(x))\right)$$

which, after replacing $\log(x)$ by $x$, leads to:

$$\frac{\partial f}{\partial t}(t,x) = \frac{\partial^2 \gamma}{\partial x^2}(t,x) + \frac{\partial \gamma}{\partial x}(t,x). \tag{A.44}$$

This is the first part of (A.37).

$\square$

Note that (A.37) gives a relationship between the densities of $(Y_t := X_t - t\psi(1), t \geq 0)$, $T_x = \inf\{t \geq 0; \ Y_t \geq x\}$ and $G_x = \sup\{t \geq 0; \ Y_t \geq x\}$, and that this relationship no longer makes use of $\Pi(t, K)$ or $C(t, K)$, although it was obtained (partially) thanks to the formula $\Pi(t, K) = \mathbb{E}[(K - M_t)^+] = K\mathbb{P}\left(\mathcal{G}_K^{(M)} \leq t\right)$.

(A.37) can also be seen, in some way, as a counterpart for Lévy processes of the formulae (8.74) and (8.87) obtained in the framework of positive linear diffusions starting from 0. Indeed, assuming (to simplify) that $m[0,+\infty[=+\infty$ and denoting $f_{\overline{G}_x}$ the density function of $\overline{G}_x := \sup\{t \geq 0;\ \overline{X}_t = x\}$ under $\mathbb{P}_0$, we have proven in Theorem 8.3 and Proposition 8.9 (with the notations of Section 8.5) that:

$$-\frac{\partial}{\partial t}q(t,0,x)\rho(x) = \frac{\partial}{\partial x}f_{\overline{G}_x}(t) = \frac{\partial}{\partial x}\widehat{f}_{x0}(t),$$

a formula which gives a relationship between the densities of $(X_t, t \geq 0)$ under $\mathbb{P}_0$, $(\overline{G}_x, x \geq 0)$ under $\overline{\mathbb{P}}_0$ and $\widehat{T}_0$ under $\widehat{\mathbb{P}}_x$.

# Appendix B
# Bessel Functions and Bessel Processes

## B.1 Bessel Functions (see [46], p. 108-136)

Let $I_\nu$ denote the modified Bessel function defined by:

$$I_\nu(z) = \sum_{k=0}^{\infty} \frac{(z/2)^{\nu+2k}}{\Gamma(k+1)\Gamma(k+\nu+1)} \qquad z \in \mathbb{C}\backslash]-\infty,0[, \qquad \text{(B.1)}$$

and $K_\nu$ the McDonald function defined, for $\nu \notin \mathbb{Z}$, by:

$$K_\nu(z) = \frac{\pi}{2}\frac{I_{-\nu}(z)-I_\nu(z)}{\sin(\nu\pi)} \qquad z \in \mathbb{C}\backslash]-\infty,0[, \qquad \text{(B.2)}$$

and for $\nu = n \in \mathbb{Z}$ by:

$$K_n(z) = \lim_{\substack{\nu\to n \\ \nu\neq n}} K_\nu(z). \qquad \text{(B.3)}$$

These functions are analytic functions in $z$ for $z \in \mathbb{C}\backslash]-\infty,0[$ and entire functions in $\nu$. It is known that $I_\nu$ and $K_\nu$ generate the set of solutions of the linear differential equation:

$$u'' + \frac{1}{x}u' - \left(1 + \frac{\nu^2}{x^2}\right)u = 0,$$

an equation which is often encountered in mathematical physics. They also appear in many computations of probability laws. The derivatives of these functions are seen to satisfy some simple recurrence relations:

$$\frac{\partial}{\partial z}\left[z^\nu K_\nu(z)\right] = -z^\nu K_{\nu-1}(z) \qquad \frac{\partial}{\partial z}\left[z^{-\nu}K_\nu(z)\right] = -z^{-\nu}K_{\nu+1}(z) \qquad \text{(B.4)}$$

$$\frac{\partial}{\partial z}\left[z^\nu I_\nu(z)\right] = z^\nu I_{\nu-1}(z) \qquad \frac{\partial}{\partial z}\left[z^{-\nu}I_\nu(z)\right] = z^{-\nu}I_{\nu+1}(z) \qquad \text{(B.5)}$$

C. Profeta et al., *Option Prices as Probabilities*, Springer Finance,
DOI 10.1007/978-3-642-10395-7, © Springer-Verlag Berlin Heidelberg 2010

Note moreover that, $I_{-n} = I_n$ for $n \in \mathbb{N}$ (since the first $n$ terms of the expansion vanish if $\nu = -n$), $K_\nu = K_{-\nu}$ for $\nu \in \mathbb{R}$, and that these functions simplify when $\nu = \pm 1/2$:

$$I_{-1/2}(z) = \sqrt{\frac{2}{\pi z}}\cosh(z), \qquad I_{1/2}(z) = \sqrt{\frac{2}{\pi z}}\sinh(z) \tag{B.6}$$

and

$$K_{-1/2}(z) = K_{1/2}(z) = \sqrt{\frac{\pi}{2z}}e^{-z}. \tag{B.7}$$

There exist also some useful integral representations for $K_\nu$. For instance, for $x > 0$ and $\nu \in \mathbb{C}$:

$$K_\nu(x) = \int_0^\infty e^{-x\cosh(u)}\cosh(\nu u)du \tag{B.8}$$

or, after a change of variable,

$$K_\nu(x) = \frac{1}{2}\left(\frac{x}{2}\right)^\nu \int_0^\infty t^{-\nu-1}e^{-t-x^2/4t}dt. \tag{B.9}$$

Finally, looking at $I_\nu$ and $K_\nu$ as functions from $\mathbb{R}^+$ to $\mathbb{R}$, it is seen that, for $x > 0$ and $\nu \geq 0$, $I_\nu$ is a positive increasing function while $K_\nu$ is a positive decreasing function. We have the following equivalents when $x \to 0$:

$$I_\nu(x) \underset{x\to 0}{\sim} \frac{x^\nu}{2^\nu \Gamma(\nu+1)} \tag{B.10}$$

$$K_\nu(x) \underset{x\to 0}{\sim} \frac{2^{\nu-1}\Gamma(\nu)}{x^\nu} \tag{B.11}$$

$$K_0(x) \underset{x\to 0}{\sim} \log\left(\frac{2}{x}\right), \tag{B.12}$$

and the asymptotic formulae when $x \to \infty$:

$$I_\nu(x) \underset{x\to\infty}{\sim} \frac{e^x}{\sqrt{2\pi x}} \tag{B.13}$$

$$K_\nu(x) \underset{x\to\infty}{\sim} \sqrt{\frac{\pi}{2x}}e^{-x}. \tag{B.14}$$

Clearly, neither function has any strictly positive zeros.

## B.2 Squared Bessel Processes (see [70] Chapter XI, or [26])

### B.2.1 Definition of Squared Bessel Processes

**Definition B.1.** For every $\delta \geq 0$ and $x \geq 0$, the unique (strong) solution of the equation

$$Z_t = x + 2 \int_0^t \sqrt{Z_s} dB_s + \delta t, \qquad Z_t \geq 0 \tag{B.15}$$

where $(B_t, t \geq 0)$ is a standard Brownian motion, is called the squared Bessel process of dimension $\delta$ started at $x$, and is denoted $\text{BESQ}^\delta(x)$. The real $\nu := \frac{\delta}{2} - 1$ is called the index of the process $\text{BESQ}^\delta(x)$.

For an integer dimension $\delta \in \mathbb{N}$, $(Z_t, t \geq 0)$ may be realized as the square of the Euclidean norm of a $\delta$-dimensional Brownian motion. Note immediately that a straightforward change of variable in equation (B.15) gives a scaling property of BESQ, namely: if $Z$ is a $\text{BESQ}^\delta(x)$, then, for every $c > 0$, the process $(c^{-1} Z_{ct}, t \geq 0)$ is a $\text{BESQ}^\delta(x/c)$.

Denoting by $\mathbb{Q}_x^\delta$ the law of $(Z_t, t \geq 0)$, solution of (B.15) on the canonical space $\mathscr{C}(\mathbb{R}^+, \mathbb{R}^+)$ (where $(X_t, t \geq 0)$ is taken as the coordinate process), there is the convolution property:

$$\mathbb{Q}_x^\delta * \mathbb{Q}_{x'}^{\delta'} = \mathbb{Q}_{x+x'}^{\delta+\delta'} \tag{B.16}$$

which holds for all $x, x', \delta, \delta' \geq 0$ (see Shiga-Watanabe [84]); in other terms, adding two independent BESQ processes yields another BESQ process, whose starting point (resp. dimension) is the sum of the starting points (resp. dimensions). It follows from (B.16) that for any positive measure $\mu(du)$ on $\mathbb{R}^+$ such that $\int (1 + u)\mu(du) < \infty$, then, if $X^{(\mu)} := \int X_u \mu(du)$,

$$\mathbb{Q}_x^\delta \left[ \exp\left( -\frac{1}{2} X^{(\mu)} \right) \right] = (A_\mu)^\delta (B_\mu)^x \tag{B.17}$$

with $A_\mu = \sqrt{\Phi_\mu(\infty)}$ and $B_\mu := \exp(\Phi_\mu'(0^+))$ for $\Phi_\mu$ the unique decreasing solution of the Sturm-Liouville equation: $\Phi'' = \Phi\mu; \Phi(0) = 1$. (see [70], Appendix §8). (B.17) may be considered as the (generalized) Laplace transform (with argument $\mu$) of the probability $\mathbb{Q}_x^\delta$. Furthermore, for any fixed $\delta$ and $x$, $\mathbb{Q}_x^\delta$ is infinitely divisible, and its Lévy Khintchine representation is given by:

$$\mathbb{Q}_x^\delta \left[ \exp\left( -\frac{1}{2} X^{(\mu)} \right) \right] = \exp\left( -\int_{\mathscr{C}(\mathbb{R}^+, \mathbb{R}^+)} M_x^\delta(d\omega) \left( 1 - e^{-\frac{1}{2} X^{(\mu)}(\omega)} \right) \right) \tag{B.18}$$

where $M_x^\delta = xM + \delta N$ for $M$ and $N$ two $\sigma$-finite measures on $\mathscr{C}(\mathbb{R}^+, \mathbb{R}^+)$ which are described in details in Pitman-Yor [66], Revuz-Yor [70], ...

## B.2.2  BESQ as a Diffusion

From (B.15), it is seen that $\mathrm{BESQ}^{\delta}$ is a linear diffusion, whose infinitesimal generator $\Gamma$ equals on $\mathscr{C}_K^2(]0,+\infty[)$ the operator:

$$\Gamma^{(\delta)} := 2x\frac{\partial^2}{\partial x^2} + \delta\frac{\partial}{\partial x}.$$

Its speed measure $m(dx)$ is the measure whose density with respect to the Lebesgue measure is given by:

$$\frac{x^{\nu}}{2\nu} \quad \text{for } \nu > 0, \qquad \frac{1}{2} \quad \text{for } \nu = 0, \qquad -\frac{x^{\nu}}{2\nu} \quad \text{for } \nu < 0,$$

where $\nu := \delta/2 - 1$. Its scale function $s$ is the function:

$$-x^{-\nu} \quad \text{for } \nu > 0, \qquad \log(x) \quad \text{for } \nu = 0, \qquad x^{-\nu} \quad \text{for } \nu < 0.$$

Furthermore, taking $\mu(du) = \lambda\varepsilon_t(du)$ in (B.17), where $\varepsilon_t$ is the Dirac measure at $t$, we get the Laplace transform of the transition density function of $\mathrm{BESQ}^{\delta}$:

$$\mathbb{Q}_x^{\delta}\left[e^{-\lambda X_t}\right] = (1+2\lambda t)^{-\frac{\delta}{2}} \exp\left(-\frac{\lambda x}{1+2\lambda t}\right).$$

By inverting this Laplace transform, it follows that, for $\delta > 0$, the semi-group of $\mathrm{BESQ}^{\delta}$ has a density in $y$ equal to:

$$q_t^{\delta}(x,y) = \frac{1}{2t}\left(\frac{y}{x}\right)^{\nu/2} \exp\left(-\frac{x+y}{2t}\right) I_{\nu}\left(\frac{\sqrt{xy}}{t}\right) \qquad (t>0,\ x,y>0), \qquad \text{(B.19)}$$

where $I_{\nu}$ is the modified Bessel function with index $\nu$.

More generally, replacing $\mu$ by $\lambda\mu$ (for any scalar $\lambda \geq 0$) in (B.17) yields the Laplace transform of $X^{(\mu)}$, provided the function $\Phi_{\lambda\mu}$ is known explicitly, which is the case for instance when $\mu(dt) = at^{\alpha}1_{\{t\leq K\}}dt + b\varepsilon_K(dt)$, and in many other cases. Consequently, several quantities associated to $\mathrm{BESQ}^{\delta}$, such as first hitting times (see Kent [41]) or distributions of last passage times (see Pitman-Yor [65]) may be expressed explicitly in terms of Bessel functions (see Chapter 7, Section 7.2 of the present monograph).

## B.2.3  Brownian Local Times and BESQ Processes

The Ray-Knight theorems for Brownian local times $(L_t^y;\ y \in \mathbb{R}, t \geq 0)$ express the laws of $(L_T^y, y \in \mathbb{R})$ for some very particular stopping times in terms of certain $\mathbb{Q}_x^{\delta}$; namely:

- if $T = T_a$ is the first hitting time of level $a$ of some Brownian motion, and $(L_{T_a}^z, z \in \mathbb{R})$ its local times up to time $T_a$ then: $(Z_y^{(a)} := L_{T_a}^{a-y}, y \geq 0)$ satisfies the SDE:

$$Z_y^{(a)} = 2 \int_0^y \sqrt{Z_z^{(a)}} d\gamma_z + 2(y \wedge a) \tag{B.20}$$

for some Brownian motion $(\gamma_z, z \geq 0)$ indexed by "space". In particular:

i) Conditionally on $L_{T_a}^0$, the processes $(L_{T_a}^{a-x}, 0 \leq x \leq a)$ and $(L_{T_a}^{a-x}, x \geq a)$ are independent,

ii) $(L_{T_a}^{a-x}, 0 \leq x \leq a)$ is distributed as $\mathrm{BESQ}^2(0)$,

iii) Conditionally on $L_{T_a}^0 = l$, $(L_{T_a}^{-x}, x \geq 0)$ is distributed as $\mathrm{BESQ}^0(l)$.

- if $T = \tau_l$ is the right-continuous inverse of the local time at 0: $\tau_l := \inf\{t \geq 0; L_t > l\}$, then, $(L_{\tau_l}^y, y \geq 0)$ and $(L_{\tau_l}^{-y}, y \geq 0)$ are two independent $\mathrm{BESQ}^0(l)$.

## B.3 Bessel Processes (see [70] Chapter XI, or [26])

### B.3.1 Definition

**Definition B.2.** For $\delta \geq 0$, $x \geq 0$, the square root of $\mathrm{BESQ}^\delta(x^2)$ is called the Bessel process of dimension $\delta$ started at $x$, and is denoted $\mathrm{BES}^\delta(x)$. As for $\mathrm{BESQ}^\delta$, we call $\nu := \frac{\delta}{2} - 1$ the index of $\mathrm{BES}^\delta$, and shall denote by $\mathbb{P}_x^{(\nu)}$, the law of the Bessel process of index $\nu$ started at $x$.

The function $x \mapsto \sqrt{x}$ being a homeomorphism, $\mathrm{BES}^\delta$ is still a linear diffusion taking positive values. Its generator is given by:

$$\Lambda^{(\delta)} := \frac{1}{2}\frac{\partial^2}{\partial x^2} + \frac{\delta - 1}{2x}\frac{\partial}{\partial x} \quad \left( = \frac{1}{2}\frac{\partial^2}{\partial x^2} + \frac{2\nu + 1}{2x}\frac{\partial}{\partial x} \right)$$

on the set of $\mathscr{C}^2(\mathbb{R}^+, \mathbb{R}^+)$ functions $f$ such that $\lim_{x \to 0} x^{2\nu+1} f'(x) = 0$. As a result, its scale function may be chosen equal to:

$$-x^{-2\nu} \quad \text{for } \nu > 0, \qquad 2\log(x) \quad \text{for } \nu = 0, \qquad x^{-2\nu} \quad \text{for } \nu < 0,$$

and with this choice, its speed measure is given by the densities (with respect to the Lebesgue measure):

$$\frac{x^{2\nu+1}}{\nu} \quad \text{for } \nu > 0, \qquad x \quad \text{for } \nu = 0, \qquad -\frac{x^{2\nu+1}}{\nu} \quad \text{for } \nu < 0.$$

Moreover, for $\delta > 0$, a straightforward change of variable in relation (B.19) gives the density of the semi-group of $\mathrm{BES}^\delta$:

$$p_t^\delta(x,y) = \frac{1}{t}\left(\frac{y}{x}\right)^\nu y\exp\left(-\frac{x^2+y^2}{2t}\right)I_\nu\left(\frac{xy}{t}\right) \qquad (t>0,\ x,y>0). \qquad \text{(B.21)}$$

Note also that for integer dimension $\delta \in \mathbb{N}$, $\mathrm{BES}^\delta$ can be realized as the modulus of a $\delta$-dimensional Brownian motion. Moreover, the scaling property of $\mathrm{BESQ}^\delta$ makes it possible to derive a similar scaling property for Bessel processes, namely: if $R$ is a $\mathrm{BES}^\delta(x)$, then, for every $c>0$, the process $(cR_{t/c^2}, t\geq 0)$ is a $\mathrm{BES}^\delta(cx)$.

## B.3.2 An Implicit Representation in Terms of Geometric Brownian Motions

J. Lamperti [45] showed a one-to-one correspondence between Lévy processes $(\xi_t, t\geq 0)$ and semi-stable Markov processes (i.e. strong Markov processes on $(0,+\infty)$ which satisfy a scaling property) $(\Sigma_u, u\geq 0)$ via the (implicit) formula:

$$\exp(\xi_t) = \Sigma_{\int_0^t ds\exp(\xi_s)} \qquad (t\geq 0). \qquad \text{(B.22)}$$

In the particular case where $(\xi_t := 2(B_t+\nu t), t\geq 0)$, formula (B.22) becomes:

$$\exp(2(B_t+\nu t)) = X^{(\nu)}\left(\int_0^t ds\exp(2(B_s+\nu s))\right) \qquad (t\geq 0) \qquad \text{(B.23)}$$

where $(X_u^{(\nu)}, u\geq 0)$ denotes a squared Bessel process with index $\nu$ started at 1. Taking the square root of both sides, (B.23) may be rewritten:

$$\exp(B_t+\nu t) = R^{(\nu)}\left(\int_0^t ds\exp(2(B_s+\nu s))\right) \qquad (t\geq 0) \qquad \text{(B.24)}$$

with $(R_u^{(\nu)}, u\geq 0)$ a Bessel process with index $\nu$ started at 1. Note that more generally, thanks to the scaling property of Brownian motion, we have for every $a\in\mathbb{R}$:

$$\exp(aB_t+\nu t) = R^{(\nu/a^2)}\left(a^2\int_0^t ds\exp(2(aB_s+\nu s))\right) \qquad (t\geq 0). \qquad \text{(B.25)}$$

Absolute continuity relationships between the laws of different BES processes may be derived from (B.24), combined with the Cameron-Martin relationships (1.5) between the laws of $(B_t+\nu t, t\geq 0)$ and $(B_t, t\geq 0)$. Precisely, one obtains, for $\nu\geq 0$:

$$\mathbb{P}_{x|\mathscr{R}_u}^{(\nu)} = \left(\frac{R_u}{x}\right)^\nu \exp\left(-\frac{\nu^2}{2}\int_0^u \frac{ds}{R_s^2}\right)\cdot \mathbb{P}_{x|\mathscr{R}_u}^{(0)} \qquad \text{(B.26)}$$

where $\mathscr{R}_u := \sigma\{R_s, s\leq u\}$ is the natural filtration of $R$, and $\nu = \frac{\delta}{2}-1$. The combination of (B.24) and (B.26) may be used to derive results about $(B_t+\nu t, t\geq 0)$ from results about the $\mathrm{BES}^\delta(x)$ process (and vice-versa). In particular, the law of

$$A_{T_\lambda}^{(\nu)} := \int_0^{T_\lambda} ds \exp\left(2(B_s + \nu s)\right)$$

where $T_\lambda$ denotes an independent exponential time, was derived in ([98], Paper 2) from this combination.

Other relations about Bessel processes, such as resolvent kernels, Bessel realizations of Hartman-Watson laws, or the Hirsch-Song formula are given in Chapter 7, Section 7.2.

# References

1. Akahori, J., Imamura, Y., Yano, Y.: On the pricing of options written on the last exit time. Methodol. Comput. Appl. Probab. **11**(4), 661–668 (2009)
2. Azéma, J., Jeulin, T.: Précisions sur la mesure de Föllmer. Ann. Inst. H. Poincaré Sect. B (N.S.) **12**(3), 257–283 (1976)
3. Azéma, J., Yor, M.: En guise d'introduction. In: Temps locaux, *Astérisque*, vol. 52–53, 3–16. Société Mathématique de France, Paris (1978)
4. Azéma, J., Yor, M.: Une solution simple au problème de Skorokhod. In: Séminaire de Probabilités, XIII (Univ. Strasbourg, Strasbourg, 1977/78), *Lecture Notes in Math.*, vol. 721, pp. 90–115. Springer, Berlin (1979)
5. Bentata, A., Yor, M.: Around the last passage times viewpoint on Black-Scholes type options: some new bits and pieces. Research Notes (April 2008)
6. Bentata, A., Yor, M.: Ten notes on three lectures: From Black-Scholes and Dupire formulae to last passage times of local martingales. Part B: The finite time horizon. Preprint LPMA-Paris VI **1232** (July 2008)
7. Bertoin, J.: Excursions of a $BES_0(d)$ and its drift term $(0 < d < 1)$. Probab. Theory and Related Fields **84**(2), 231–250 (1990)
8. Bertoin, J.: Lévy processes, *Cambridge Tracts in Mathematics*, vol. 121. Cambridge University Press, Cambridge (1996)
9. Biane, P.: Comparaison entre temps d'atteinte et temps de séjour de certaines diffusions réelles. In: Séminaire de Probabilités, XIX, 1983/84, *Lecture Notes in Math.*, vol. 1123, pp. 291–296. Springer, Berlin (1985)
10. Biane, P., Yor, M.: Quelques précisions sur le méandre brownien. Bull. Sci. Math. (2) **112**(1), 101–109 (1988)
11. Black, F., Scholes, M.: The pricing of options and corporate liabilities. Journal of Political Economy **81**, 637–654 (1973)
12. Borodin, A.N., Salminen, P.: Handbook of Brownian motion—facts and formulae, second edn. Probability and its Applications. Birkhäuser Verlag, Basel (2002)
13. Borovkov, K., Burq, Z.: Kendall's identity for the first crossing time revisited. Electron. Comm. Probab. **6**, 91–94 (electronic) (2001)
14. Carmona, P., Petit, F., Yor, M.: Beta-gamma random variables and intertwining relations between certain Markov processes. Rev. Mat. Iberoamericana **14**(2), 311–367 (1998)
15. Carr, P.: Put No Touch-Inversion. Unpublished Notes (July 2007)
16. Choquet, G., Meyer, P.A.: Existence et unicité des représentations intégrales dans les convexes compacts quelconques. Ann. Inst. Fourier (Grenoble) **13**, 139–154 (1963)
17. Ciesielski, Z., Taylor, S.J.: First passage times and sojourn times for Brownian motion in space and the exact Hausdorff measure of the sample path. Trans. Amer. Math. Soc. **103**, 434–450 (1962)

C. Profeta et al., *Option Prices as Probabilities*, Springer Finance, 259
DOI 10.1007/978-3-642-10395-7, © Springer-Verlag Berlin Heidelberg 2010

18. De Meyer, B., Roynette, B., Vallois, P., Yor, M.: On independent times and positions for Brownian motions. Rev. Mat. Iberoamericana **18**(3), 541–586 (2002)
19. Donati-Martin, C., Roynette, B., Vallois, P., Yor, M.: On constants related to the choice of the local time at 0, and the corresponding Itô measure for Bessel processes with dimension $d = 2(1 - \alpha), 0 < \alpha < 1$. Studia Sci. Math. Hungar. **45**(2), 207–221 (2008)
20. Dubins, L.E., Émery, M., Yor, M.: On the Lévy transformation of Brownian motions and continuous martingales. In: Séminaire de Probabilités, XXVII, *Lecture Notes in Math.*, vol. 1557, pp. 122–132. Springer, Berlin (1993)
21. Dufresne, D.: The distribution of a perpetuity, with applications to risk theory and pension funding. Scand. Actuar. J. (1-2), 39–79 (1990)
22. Eisenbaum, N.: Integration with respect to local time. Potential Anal. **13**(4), 303–328 (2000)
23. Eisenbaum, N.: Local time-space calculus for reversible semimartingales. In: Séminaire de Probabilités XL, *Lecture Notes in Math.*, vol. 1899, pp. 137–146. Springer, Berlin (2007)
24. Fitzsimmons, P., Pitman, J., Yor, M.: Markovian bridges: construction, Palm interpretation, and splicing. In: Seminar on Stochastic Processes, 1992 (Seattle, WA, 1992), *Progr. Probab.*, vol. 33, pp. 101–134. Birkhäuser Boston, Boston, MA (1993)
25. Getoor, R.K.: The Brownian escape process. Ann. Probab. **7**(5), 864–867 (1979)
26. Göing-Jaeschke, A., Yor, M.: A survey and some generalizations of Bessel processes. Bernoulli **9**(2), 313–349 (2003)
27. Guasoni, P.: Excursions in the martingale hypothesis. In: Stochastic processes and applications to mathematical finance, pp. 73–95. World Sci. Publ., River Edge, NJ (2004)
28. Hartman, P., Watson, G.S.: "Normal" distribution functions on spheres and the modified Bessel functions. Ann. Probability **2**, 593–607 (1974)
29. Hibino, Y., Hitsuda, M., Muraoka, H.: Construction of noncanonical representations of a Brownian motion. Hiroshima Math. J. **27**(3), 439–448 (1997)
30. Hirsch, F., Song, S.: Two-parameter Bessel processes. Stochastic Process. Appl. **83**(1), 187–209 (1999)
31. Hirsch, F., Yor, M.: Fractional Intertwinings between two Markov semigroups. Potential Anal. **31**, 133–146 (2009)
32. Ismail, M.E.H.: Complete monotonicity of modified Bessel functions. Proc. Amer. Math. Soc. **108**(2), 353–361 (1990)
33. Ismail, M.E.H., Kelker, D.H.: Special functions, Stieltjes transforms and infinite divisibility. SIAM J. Math. Anal. **10**(5), 884–901 (1979)
34. Itô, K.: Extension of stochastic integrals. In: Proceedings of the International Symposium on Stochastic Differential Equations (Res. Inst. Math. Sci., Kyoto Univ., Kyoto, 1976), pp. 95–109. Wiley, New York (1978)
35. Itô, K., McKean Jr., H.P.: Diffusion processes and their sample paths. Springer-Verlag, Berlin (1974). Second printing, corrected, Die Grundlehren der mathematischen Wissenschaften, Band 125. Now in: *Classics in Mathematics*
36. Jacod, J., Protter, P.: Time reversal on Lévy processes. Ann. Probab. **16**(2), 620–641 (1988)
37. Jeulin, T.: Semi-martingales et grossissement d'une filtration, *Lecture Notes in Mathematics*, vol. 833. Springer, Berlin (1980)
38. Jeulin, T., Yor, M.: Inégalité de Hardy, semimartingales, et faux-amis. In: Séminaire de Probabilités, XIII (Univ. Strasbourg, Strasbourg, 1977/78), *Lecture Notes in Math.*, vol. 721, pp. 332–359. Springer, Berlin (1979)
39. Kamae, T., Krengel, U.: Stochastic partial ordering. Ann. Probab. **6**(6), 1044–1049 (1978)
40. Karatzas, I., Shreve, S.E.: Brownian motion and stochastic calculus, *Graduate Texts in Mathematics*, vol. 113, second edn. Springer-Verlag, New York (1991)
41. Kent, J.: Some probabilistic properties of Bessel functions. Ann. Probab. **6**(5), 760–770 (1978)
42. Kotani, S., Watanabe, S.: Kreĭn's spectral theory of strings and generalized diffusion processes. In: Functional analysis in Markov processes (Katata/Kyoto, 1981), *Lecture Notes in Math.*, vol. 923, pp. 235–259. Springer, Berlin (1982)
43. Ladyženskaja, O.A., Solonnikov, V.A., Uralceva, N.N.: Linear and Quasilinear Equations of Parabolic Type, vol. 23. American Mathematical Society (1968)

44. Lamberton, D., Lapeyre, B.: Introduction to stochastic calculus applied to finance, second edn. Chapman & Hall/CRC Financial Mathematics Series. Chapman & Hall/CRC, Boca Raton, FL (2008)
45. Lamperti, J.: Semi-stable Markov processes. I. Z. Wahrscheinlichkeitstheorie und Verw. Gebiete **22**, 205–225 (1972)
46. Lebedev, N.N.: Special functions and their applications. Revised English edition. Translated and edited by Richard A. Silverman. Prentice-Hall Inc., Englewood Cliffs, N.J. (1965)
47. Madan, D., Roynette, B., Yor, M.: Option Prices as Probabilities. Finance Research Letters **5**, 79–87 (2008)
48. Madan, D., Roynette, B., Yor, M.: Unifying Black-Scholes type formulae which involves last passage times up to finite horizon. Asia-Pacific Journal of Financial Markets **15**(2), 97–115 (2008)
49. Madan, D., Roynette, B., Yor, M.: Put option prices as joint distribution functions in strike and maturity: the Black-Scholes case. Preprint IEC Nancy **41** (2008). Submitted (January 2009). Forthcoming in Quantitative Finance
50. Madan, D., Roynette, B., Yor, M.: An alternative expression for the Black-Scholes formula in terms of Brownian first and last passage times. Preprint IEC Nancy **8** (February 2008)
51. Mansuy, R., Yor, M.: Random times and enlargements of filtrations in a Brownian setting, *Lecture Notes in Mathematics*, vol. 1873. Springer-Verlag, Berlin (2006)
52. Mansuy, R., Yor, M.: Aspects of Brownian motion. Universitext. Springer-Verlag, Berlin (2008)
53. Matsumoto, H., Yor, M.: An analogue of Pitman's $2M - X$ theorem for exponential Wiener functionals. I: A time-inversion approach. Nagoya Math. J. **159**, 125–166 (2000)
54. Matsumoto, H., Yor, M.: An analogue of Pitman's $2M - X$ theorem for exponential Wiener functionals. II: The role of the generalized inverse Gaussian laws. Nagoya Math. J. **162**, 65–86 (2001)
55. McKean Jr., H.P.: Stochastic integrals. Probability and Mathematical Statistics, No. 5. Academic Press, New York (1969)
56. Meyer, P.A.: Démonstration simplifiée d'un théorème de Knight. In: Séminaire de Probabilités, V (Univ. Strasbourg, année universitaire 1969–1970), pp. 191–195. Lecture Notes in Math., Vol. 191. Springer, Berlin (1971)
57. Najnudel, J., Nikeghbali, A.: On some universal sigma finite measures and some extensions of Doob's optional stopping theorem. arXiv:0906.1782 (June 2009). Submitted to Stochastic Process. Appl.
58. Najnudel, J., Roynette, B., Yor, M.: A remarkable $\sigma$-finite measure on $\mathscr{C}(\mathbf{R}_+, \mathbf{R})$ related to many Brownian penalisations. C. R. Math. Acad. Sci. Paris **345**(8), 459–466 (2007)
59. Najnudel, J., Roynette, B., Yor, M.: A global view of Brownian penalisations, *Monograph*, vol. 19. Memoirs of the Japan Math. Soc. (2009)
60. Nikeghbali, A., Yor, M.: Doob's maximal identity, multiplicative decompositions and enlargements of filtrations. Illinois J. Math. **50**(1-4), 791–814 (electronic) (2006)
61. Pagès, G.: Introduction to Numerical Probability for Finance. Notes from Lectures in LPMA-Paris VI (2008)
62. Pal, S., Protter, P.: Strict local martingales, bubbles and no early exercise. Preprint. Cornell University (November 2007)
63. Pal, S., Protter, P.: Analysis of continuous strict local martingales, via h-transforms. (July 2009). Submitted
64. Pitman, J.: The distribution of local times of a Brownian bridge. In: Séminaire de Probabilités, XXXIII, *Lecture Notes in Math.*, vol. 1709, pp. 388–394. Springer, Berlin (1999)
65. Pitman, J., Yor, M.: Bessel processes and infinitely divisible laws. In: Stochastic integrals (Proc. Sympos., Univ. Durham, Durham, 1980), *Lecture Notes in Math.*, vol. 851, pp. 285–370. Springer, Berlin (1981)
66. Pitman, J., Yor, M.: A decomposition of Bessel bridges. Z. Wahrsch. Verw. Gebiete **59**(4), 425–457 (1982)
67. Profeta, C.: Autour de la pénalisation et des pseudo-inverses de diffusions linéaires. Thèse de l'Université Henri Poincaré, Nancy (2010)

68. Profeta, C., Roynette, B., Yor, M.: Existence of pseudo-inverses for diffusions. Preprint IEC Nancy, **55** (2008)
69. Qian, M.: Personal communication (August 2007)
70. Revuz, D., Yor, M.: Continuous martingales and Brownian motion, *Grundlehren der Mathematischen Wissenschaften [Fundamental Principles of Mathematical Sciences]*, vol. 293, third edn. Springer-Verlag, Berlin (1999)
71. Rogers, L.C.G.: The joint law of the maximum and terminal value of a martingale. Probab. Theory Related Fields **95**(4), 451–466 (1993)
72. Roynette, B., Vallois, P., Yor, M.: Limiting laws associated with Brownian motion perturbed by its maximum, minimum and local time. II. Studia Sci. Math. Hungar. **43**(3), 295–360 (2006)
73. Roynette, B., Vallois, P., Yor, M.: Some extensions of Pitman and Ray-Knight theorems for penalized Brownian motions and their local times. IV. Studia Sci. Math. Hungar. **44**(4), 469–516 (2007)
74. Roynette, B., Yano, Y., Yor, M.: A last passage times view point for an extension of Black-Scholes formula involving a finite-number of continuous local martingales
75. Roynette, B., Yor, M.: Existence and properties of pseudo-inverses for Bessel and related processes. Preprint IEC Nancy, **50** (2008)
76. Roynette, B., Yor, M.: An interesting family of Black-Scholes perpetuities. Preprint IEC Nancy, **54** (2008)
77. Roynette, B., Yor, M.: Penalising Brownian paths, *Lecture Notes in Maths*, vol. 1969. Springer-Berlin (2009)
78. Salminen, P., Vallois, P.: On the subexponentiality of the Lévy measure of the diffusion inverse local time; with applications to penalizations. Electron. J. Probab. **14**, 1963–1991 (2009)
79. Salminen, P., Vallois, P., Yor, M.: On the excursion theory for linear diffusions. Jpn. J. Math. **2**(1), 97–127 (2007)
80. Salminen, P., Yor, M.: On Dufresne's perpetuity, translated and reflected. In: Stochastic processes and applications to mathematical finance, pp. 337–354. World Sci. Publ., River Edge, NJ (2004)
81. Salminen, P., Yor, M.: Perpetual integral functionals as hitting and occupation times. Electron. J. Probab. **10**, no. 11, 371–419 (electronic) (2005)
82. Salminen, P., Yor, M.: Properties of perpetual integral functionals of Brownian motion with drift. Ann. Inst. H. Poincaré Probab. Statist. **41**(3), 335–347 (2005)
83. Seshadri, V.: Exponential models, Brownian motion, and independence. Canad. J. Statist. **16**(3), 209–221 (1988)
84. Shiga, T., Watanabe, S.: Bessel diffusions as a one-parameter family of diffusion processes. Z. Wahrscheinlichkeitstheorie und Verw. Gebiete **27**, 37–46 (1973)
85. Tindel, S.: On the stochastic calculus method for spins systems. Ann. Probab. **33**(2), 561–581 (2005)
86. du Toit, J., Peskir, G., Shiryaev, A.N.: Predicting the last zero of Brownian motion with drift. Stochastics **80**(2-3), 229–245 (2008)
87. Vallois, P.: Sur la loi du maximum et du temps local d'une martingale continue uniformément intégrable. Proc. London Math. Soc. (3) **69**(2), 399–427 (1994)
88. Vostrikova, L.: On regularity properties of Bessel flow. LAREMA Preprint; also on arXiv:0902.4232v1
89. Watanabe, S.: On time inversion of one-dimensional diffusion processes. Z. Wahrscheinlichkeitstheorie und Verw. Gebiete **31**, 115–124 (1974/75)
90. Watson, G.N.: A Treatise on the Theory of Bessel Functions. Cambridge University Press, Cambridge, England (1944)
91. Williams, D.: Decomposing the Brownian path. Bull. Amer. Math. Soc. **76**, 871–873 (1970)
92. Yen, J.Y., Yor, M.: Measuring the "non-stopping timeness" of ends of previsible sets. Submitted (2008)
93. Yen, J.Y., Yor, M.: Call option prices based on Bessel processes. In: Methodology and Computing in Applied Probability. Springer U.S. (Online 08/2009)
94. Yor, M.: Loi de l'indice du lacet brownien, et distribution de Hartman-Watson. Z. Wahrsch. Verw. Gebiete **53**(1), 71–95 (1980)

95. Yor, M.: Some aspects of Brownian motion. Part I: Some special functionals. Lectures in Mathematics ETH Zürich. Birkhäuser Verlag, Basel (1992)
96. Yor, M.: The distribution of Brownian quantiles. J. Appl. Probab. **32**(2), 405–416 (1995)
97. Yor, M.: Some remarks about the joint law of Brownian motion and its supremum. In: Séminaire de Probabilités, XXXI, *Lecture Notes in Math.*, vol. 1655, pp. 306–314. Springer, Berlin (1997)
98. Yor, M.: Exponential functionals of Brownian motion and related processes. Springer Finance. Springer-Verlag, Berlin (2001). With an introductory chapter by Hélyette Geman, Chapters 1, 3, 4, 8 translated from the French by Stephen S. Wilson

# Further Readings

In the main, these references are not cited in our text, but they are closely related to it. Some of them, say number N, appear in Exercises, Footnotes, Remarks ... where they are referenced as [N, F].

1. Andreasen, J., Jensen, B., Poulsen, R.: Eight valuation methods in financial mathematics: the Black-Scholes formula as an example. Math. Scientist **23**(1), 18–40 (1998)
2. Azéma, J., Gundy, R.F., Yor, M.: Sur l'intégrabilité uniforme des martingales continues. In: Séminaire de Probabilités, XIV (Paris, 1978/1979), *Lecture Notes in Math.*, vol. 784, pp. 53–61. Springer, Berlin (1980)
3. Azéma, J., Yor, M.: Etude d'une martingale remarquable. In: Séminaire de Probabilités, XXIII, *Lecture Notes in Math.*, vol. 1372, pp. 88–130. Springer, Berlin (1989)
4. Azéma, J., Yor, M.: Sur les zéros des martingales continues. In: Séminaire de Probabilités, XXVI, *Lecture Notes in Math.*, vol. 1526, pp. 248–306. Springer, Berlin (1992)
5. Bentata, A., Yor, M.: Further developments stemming from the last passage times viewpoint on Black-Scholes formula. Research Notes (April 2008)
6. Bentata, A., Yor, M.: Ten notes on three lectures: From Black-Scholes and Dupire formulae to last passage times of local martingales. Preprint (February 2008)
7. Bentata, A., Yor, M.: From Black-Scholes and Dupire formulae to last passage times of local martingales. Part A: The infinite time horizon. Preprint LPMA-Paris VI **1223** (June 2008)
8. Bertoin, J., Martinelli, F., Peres, Y.: Lectures on probability theory and statistics, *Lecture Notes in Mathematics*, vol. 1717. Springer-Verlag, Berlin (1999). Lectures from the 27th Summer School on Probability Theory held in Saint-Flour, July 7–23, 1997
9. Carr, P., Ewald, C.O., Xiao, Y.: On the qualitative effect of volatility and duration on prices of Asian options. Finance Research Letters **5**(3), 162–171 (September 2008)
10. Carraro, L., El Karoui, A., Obloj, J.: On Azéma-Yor processes, their optimal properties and the Bachelier-Drawdown equation. arXiv:0902.1328, August 2009
11. Chaumont, L., Yor, M.: Exercises in Probability, *Cambridge Series in Statistical and Probabilistic Mathematics*, vol. 13. Cambridge University Press, Cambridge (2003). A guided tour from measure theory to random processes, via conditioning
12. Chaumont, L., Yor, M.: Exercises in Probability, second edn. In preparation. Cambridge University Press, Cambridge (2008)
13. Chung, K.L.: Probabilistic approach in potential theory to the equilibrium problem. Ann. Inst. Fourier (Grenoble) **23**(3), 313–322 (1973)
14. Chybiryakov, O.: Itô's integrated formula for strict local martingales with jumps. In: Séminaire de Probabilités XL, *Lecture Notes in Math.*, vol. 1899, pp. 375–388. Springer, Berlin (2007)
15. Dang Ngoc, N., Yor, M.: Champs markoviens et mesures de Gibbs sur **R**. Ann. Sci. École Norm. Sup. (4) **11**(1), 29–69 (1978)
16. Dassios, A.: The distribution of the quantile of a Brownian motion with drift and the pricing of related path-dependent options. Ann. Appl. Probab. **5**(2), 389–398 (1995)

17. Delbaen, F., Schachermayer, W.: Arbitrage possibilities in Bessel processes and their relations to local martingales. Probab. Theory and Related Fields **102**(3), 357–366 (1995)
18. Donati-Martin, C., Yor, M.: Some Brownian functionals and their laws. Ann. Probab. **25**(3), 1011–1058 (1997)
19. Doob, J.L.: Classical potential theory and its probabilistic counterpart, *Grundlehren der Mathematischen Wissenschaften [Fundamental Principles of Mathematical Sciences]*, vol. 262. Springer-Verlag, New York (1984)
20. Dupire, B.: Pricing with a smile. Risk Magazine **7**, 17–20 (1994)
21. Dupire, B.: Pricing and hedging with smiles. In: Mathematics of derivative securities (Cambridge, 1995), *Publ. Newton Inst.*, vol. 15, pp. 103–111. Cambridge Univ. Press, Cambridge (1997)
22. El Karoui, N., Meziou, A.: Constrained optimization with respect to stochastic dominance: application to portfolio insurance. Math. Finance **16**(1), 103–117 (2006)
23. El Karoui, N., Meziou, A.: Max-plus decomposition of supermartingales and convex order. Application to American option and portfolio insurance. Ann. Prob. **36**(2), 647–697 (2008)
24. Embrechts, P., Rogers, L.C.G., Yor, M.: A proof of Dassios' representation of the $\alpha$-quantile of Brownian motion with drift. Ann. Appl. Probab. **5**(3), 757–767 (1995)
25. Fujita, T., Yor, M.: Some past-future martingales for Symmetric Random Walk and Symmetric Lévy processes. Preprint (April 2008)
26. Guivarch, Y., Keane, M., Roynette, B.: Marches aléatoires sur les groupes de Lie, *Lecture Notes in Math.*, vol. 624. Springer, Berlin (1978)
27. Gradshteyn, I.S., Ryshik, I.M.: Table of Integrals, Series and Products. *A.P. Inc.* (1980)
28. Jeanblanc, M., Yor, M., Chesney, M.: Mathematical Methods for Financial Markets. 2009, Springer-Finance XXVI, 732 p., 9 illustrations
29. Jeulin, T., Yor, M. (eds.): Grossissements de filtrations: exemples et applications, *Lecture Notes in Mathematics*, vol. 1118. Springer-Verlag, Berlin (1985). Papers from the seminar on stochastic calculus held at the Université de Paris VI, Paris, 1982/1983
30. Jeulin, T., Yor, M.: Filtration des ponts browniens et équations différentielles stochastiques linéaires. In: Séminaire de Probabilités, XXIV, 1988/89, *Lecture Notes in Math.*, vol. 1426, pp. 227–265. Springer, Berlin (1990)
31. Kaji, S.: The tail estimation of the quadratic variation of a quasi left continuous local martingale. Osaka J. Math. **44**(4), 893–907 (2007)
32. Kamke, E.: Differentialgleichungen. Lösungmethoden and Lösungen. Becker-Erler Verlag, Leipzig (1942)
33. Klebaner, F.: Option price when the stock is a semimartingale. Electron. Comm. Probab. **7**, 79–83 (electronic) (2002)
34. Knight, F.B.: Characterization of the Lévy measures of inverse local times of gap diffusion. In: Seminar on Stochastic Processes, 1981 (Evanston, Ill., 1981), *Progr. Prob. Statist.*, vol. 1, pp. 53–78. Birkhäuser Boston, Mass. (1981)
35. Knight, F.B., Maisonneuve, B.: A characterization of stopping times. Ann. Probab. **22**(3), 1600–1606 (1994)
36. Kunita, H.: Absolute continuity of Markov processes and generators. Nagoya Math. J. **36**, 1–26 (1969)
37. Kunita, H.: Absolute continuity of Markov processes. Séminaire de Probabilités, X. *Lecture Notes in Math.*, vol. 511, pp. 44–77. Springer (1976)
38. Madan, D., Roynette, B., Yor, M.: From Black-Scholes formula, to local times and last passage times for certain submartingales. Preprint IEC Nancy **14** (March 2008)
39. Madan, D., Yor, M.: Ito's integrated formula for strict local martingales. In: In memoriam Paul-André Meyer: Séminaire de Probabilités XXXIX, *Lecture Notes in Math.*, vol. 1874, pp. 157–170. Springer, Berlin (2006)
40. Mansuy, R., Yor, M.: Harnesses, Lévy bridges and Monsieur Jourdain. Stochastic Process. Appl. **115**(2), 329–338 (2005)
41. Millar, P.W.: Random times and decomposition theorems. In: Probability (Proc. Sympos. Pure Math., Vol. XXXI, Univ. Illinois, Urbana, Ill., 1976), pp. 91–103. Amer. Math. Soc., Providence, R. I. (1977)

42. Millar, P.W.: A path decomposition for Markov processes. Ann. Probability **6**(2), 345–348 (1978)
43. Nguyen-Ngoc, L., Yor, M.: Some martingales associated to reflected Lévy processes. In: Séminaire de Probabilités XXXVIII, *Lecture Notes in Math.*, vol. 1857, pp. 42–69. Springer, Berlin (2005)
44. Nualart, D.: The Malliavin calculus and related topics, second edn. Probability and its Applications (New York). Springer-Verlag, Berlin (2006)
45. Obłój, J.: The Skorokhod embedding problem and its offspring. Probab. Surv. **1**, 321–390 (electronic) (2004)
46. Petiau, G.: La théorie des fonctions de Bessel exposée en vue de ses applications à la physique mathématique. Centre National de la Recherche Scientifique, Paris (1955)
47. Pitman, J., Yor, M.: Itô's excursion theory and its applications. Jpn. J. Math. **2**(1), 83–96 (2007)
48. Rosen, J., Yor, M.: Tanaka formulae and renormalization for triple intersections of Brownian motion in the plane. Ann. Probab. **19**(1), 142–159 (1991)
49. Royer, G., Yor, M.: Représentation intégrale de certaines mesures quasi-invariantes sur $\mathscr{C}(R)$; mesures extrémales et propriété de Markov. Ann. Inst. Fourier (Grenoble) **26**(2), ix, 7–24 (1976)
50. Roynette, B.: Marches aléatoires sur les groupes de Lie. Ecole d'Été de St Flour VII, *Lecture Notes in Math.*, vol. 678. Springer, Berlin (1978)
51. Salminen, P.: On local times of a diffusion. In: Séminaire de Probabilités, XIX, 1983/84, *Lecture Notes in Math.*, vol. 1123, pp. 63–79. Springer, Berlin (1985)
52. Salminen, P.: On last exit decompositions of linear diffusions. Studia Sci. Math. Hungar. **33**(1-3), 251–262 (1997)
53. Sharpe, M.J.: Some transformations of diffusions by time reversal. Ann. Probab. **8**(6), 1157–1162 (1980)
54. Stroock, D.W., Varadhan, S.R.S.: Multidimensional diffusion processes. Classics in Mathematics. Springer-Verlag, Berlin (2006). Reprint of the 1997 edition
55. Watanabe, S.: Bilateral Bessel diffusion processes with drifts and time-inversion. Preprint (1997–1998)
56. Watanabe, S.: Invariants of one-dimensional diffusion processes and applications. J. Korean Math. Soc. **35**(3), 637–658 (1998). International Conference on Probability Theory and its Applications (Taejon, 1998)
57. Werner, W.: Girsanov's transformation for SLE($\kappa, \rho$) processes, intersection exponents and hiding exponents. Ann. Fac. Sci. Toulouse Math. (6) **13**(1), 121–147 (2004)
58. Widder, D.V.: The heat equation. Academic Press [Harcourt Brace Jovanovich Publishers], New York (1975). Pure and Applied Mathematics, Vol. 67
59. Williams, D.: Brownian motion as a harness. University of Swansea. *Unpublished*
60. Williams, D.: Some basic theorems on harnesses. In: Stochastic analysis (a tribute to the memory of Rollo Davidson), pp. 349–363. Wiley, London (1973)
61. Williams, D.: Path decomposition and continuity of local time for one-dimensional diffusions. I. Proc. London Math. Soc. (3) **28**, 738–768 (1974)
62. Yor, M.: Rappels et préliminaires généraux. Temps locaux, Astérisque **52**, 23–36 (1978)
63. Yor, M.: Some remarkable properties of gamma processes. In: Advances in mathematical finance, Appl. Numer. Harmon. Anal., pp. 37–47. Birkhäuser Boston, Boston, MA (2007)

# Index

Azéma-Yor martingale, 30
Azéma supermartingale, 57

Balayage formula, 51
Bessel process, 255
    with a drift, 192
Beta r.v., 62
Biane's transformation, 222
Black-Scholes formula, v, 3
    Generalized, 21
Brownian bridge, 34
Brownian motion (with drift), 1

Call associated to a martingale, 24
Cameron-Martin formula, 2
Change of numéraire, 24
Ciesielski-Taylor formula, 185
Class $\mathscr{D}$, 82

Dambis, Dubins, Schwarz's Theorem, 58
Density of occupation formula, 33
Doob's $h$ transform, 31
Doob's maximal identity, 22
Dupire formula, 246

Enlargement of filtration formula, 57, 70
Exponential martingale, v
Extremal PFH function, 134

Feller's criterion, 205
First passage times, 2
Fokker-Planck formula, 171

Gamma r.v., 90
Gamma subordinator, 62
Generalized Ornstein-Uhlenbeck process, 193
Girsanov density, 34

Hankel function, 104
Harness, 141
Hartman-Watson laws, 168
Hirsch-Song formula, 170, 206, 225

Infinite divisibility, 189
Infinitesimal generator, 37

Knight's representation Theorem, 57

Lévy measure, 61
Lévy process, 60
Last passage time up to a finite horizon, 115
Last passage times, 2
Lipschitz-Hankel formula, 177
Local time-space calculus, 152
Local times of a martingale, 33
Local time of a diffusion, 38

McDonald function, 251
Modified Bessel function, 251

Natural filtration, 1
Non-stopping timeness, 242

Orthogonal local martingales, 55

Past-future (sub)-martingale, 127
Past-future filtration, 115
Perpetuity, 89
PFH-function, 127
Pitman's Theorem, 211
Poisson process, 63
Post $\mathscr{G}_K$-processes, 70
Pre $\mathscr{G}_K$-processes, 70
Predictable compensator, 74
Predictable process, 51
Pseudo-inverse, 162
Put associated to a martingale, 22

Ray-Knight Theorems, 254
Resolvent kernel, 168

Scale function, 37
Scaling, 4
Sized-biased sampling, 240
Skorokhod submartingale, 79
Speed measure, 37
Squared Bessel process, 253
Strict local martingale, 239

Symmetry principle, 6

Time inversion, 11
Time reversal, 233

Williams' time reversal Theorem, 28

Yuri's formula, 117

Zolotarev's identity, 234